Handbook of Computer Simulation in Radio Engineering, Communications, and Radar

For a listing of recent titles in the *Artech House Radar Library*,
turn to the back of this book.

Handbook of Computer Simulation in Radio Engineering, Communications, and Radar

Sergey A. Leonov
Alexander I. Leonov

Artech House
Boston • London
www.artechhouse.com

Library of Congress Cataloging-in-Publication Data
Leonov, S. A. (Sergei Aleksandrovich)
    Handbook of computer simulation in radio engineering, communications, and radar /
Sergey A. Leonov, Alexander I. Leonov.
        p. cm. — (Artech House radar library)
    Includes bibliographical references and index.
    ISBN 1-58053-280-2 (alk. paper)
        1. Radio—Computer simulation. 2. Telecommunication—Computer simulation. 3.
    Radar—Computer simulation. 4. Radar—Mathematics. I. Leonov, A. I. (Aleksandr
    Ivanovich) II. Title. III. Series.

    TK6553 .L43 2001
    621.384'01'13—dc21   2001022595

British Library Cataloguing in Publication Data
Leonov, S. A. (Sergei Aleksandrovich)
    Handbook of computer simulation in radio engineering,
    communications, and radar. — (Artech House radar library)
    1. Radio—Mathematical models
    I. Title II. Leonov, A. I. (Aleksandr Ivanovich)
    621.3'84

    ISBN 1-58053-280-2

Cover design by Igor Valdman

© 2001 ARTECH HOUSE, INC.
685 Canton Street
Norwood, MA 02062

International Standard Book Number: 1-58053-280-2
Library of Congress Catalog Card Number: 2001022595

10 9 8 7 6 5 4 3 2 1

To Galina Leonova

CONTENTS

PART 2   RADAR SYSTEM SIMULATION

APPENDIXES

PREFACE

Computer simulation is the most common and powerful instrument of modern radio system design and analysis in communications, radar, antennas and other related disciplines that are typically referred to as various branches of *radio engineering*. Computer simulation in radio engineering may be defined as running computer-based algorithms that reconstruct mathematical equations and operators describing a system or its performance. Thus two basic steps are involved: to describe a system or its performance figures (detection probability, data transmission quality, measurement error, etc.) by a set of formulas, logic operators, equations to create a relevant *mathematical model*; to program the mathematical model in some computer language that makes it possible to run computer simulation routines. Nowadays *Mathcad* and *Matlab* are the most common simulation software packages used by radio engineers and researchers.

Many excellent texts in various branches of radio engineering are available. Most of these books cover fundamentals, the theory of operation, some aspects of practical applications to specific systems and environmental conditions, and they contain basic formulas and equations in the discipline. However, every engineer and researcher involved in computer simulation knows that they have a long way to go from finding a set of equations and formulas in different sources (often using different notations, reference frames, normalization, background assumptions, etc.) to creating an operational, error-free mathematical model of a system or its characteristics. This handbook intends to fill in this gap and provide a single-source systematic reference to the major mathematical models used in radio engineering (Part 1) and computer simulation algorithms often used to design and analyze radar systems and their performance (Part 2). Simulation algorithms are written in *Mathcad 2000* (an easy-to-learn, easy-to-use versatile programming environment where equations look the way one might see them on a blackboard or in a reference book). That is why provided Mathcad routines can be easily used as is for obtaining immediate simulation results (a software package is available as an optional supplement to the handbook) or as the basic building blocks to create modified custom-made simulation software that better answers the specific needs of the user.

While working on this handbook the authors used, sorted and systematized material from a large number of books, monographs and professional journals

xix

both in English and Russian languages as well as their own works (the authors have more than 25 years of experience in mathematical modeling of state-of-the-art radio systems and were awarded "All-Russian Honorable" titles for their achievements in this field of radio engineering).   Because the theory of mathematical modeling and computer simulation was well developed in the former Soviet Union, the handbook offers many methods and algorithms that have previously only been available in Russian-language sources.

ACKNOWLEDGMENTS

The authors extend their appreciation to Mr. David Barton for his comments, which were of great value in the preparation of this handbook, and to Ms. Svetlana Yakovenko and Mr. Victor Yakovenko for their help in the preparation of the camera-ready copy of the manuscript.

Dr. Sergey A. Leonov
Dr. Alexander I. Leonov
Waterloo, Canada,
May 2001

PART 1

MATHEMATICAL MODELING IN RADIO ENGINEERING

Chapter 1

Deterministic Modeling

1.1 Basic Definitions

Definition # 1. The *mathematical model* is the formal description of an object or a phenomenon by means of functions, logical operators, algebraic, integral, differential equations or other mathematical expressions.

Definition # 2. The *mathematical modeling* is the research of an object or a phenomenon using mathematical models that reconstruct the most important features of the original.

Definition # 3. *Computer simulation* is an application of mathematical modeling technique to the research of an object or a phenomenon by running computer-based algorithms.

Definition # 4. A *radio (frequency) system* is one that performs its operational tasks by radiating and/or receiving radio-frequency (RF) oscillations (signals) in order to obtain, transmit or exchange information.

The ability to obtain, transmit or exchange *information* is the major distinctive feature of a radio system that differentiates it from an electrical engineering system, which transmits and exchanges energy (power). The major types of radio systems are:

1) *Communication systems* that transmit information from a sender to a receiver;

2) *Radar systems* that perform the tasks of detection, tracking and discrimination of the objects of interest (the targets);

3) *Radio navigation systems* that determine positions of ships, aircraft, spacecraft and other vehicles and objects by means of exchanging RF signals;

4) *Radio control systems* that perform remote control functions of the objects and technological processes;

5) *Electronic intelligence systems* that collect electromagnetic radiation data for intelligence purposes;

6) *Electronic warfare systems* that are intended to prevent the hostile use of the electromagnetic spectrum while retaining friendly use of it.

Two basic approaches are typically used for radio systems modeling and simulation. One is to create a complete model of the system and data streams entering and/or exiting it. The level of modeling may be different: at top system level, at the level of subsystems, at the level of devices, at the level of circuits and components. Obviously, the accuracy of system description increases from upper to lower levels but the complexity of the model, computer resources required to run it and the cost also increase. On one hand, a complete model of the system is the most flexible and accurate one. It gives the possibility to estimate the status and performance of the system under the conditions of variation of input data streams and other important factors. On the other hand, complexity, cost and time required to develop such a model may be excessively high, especially for sophisticated multifunctional systems. Monte Carlo simulation is often used to model radio systems.

The second approach is to develop a specific model of the particular system characteristic of interest (e.g., throughput capability, probability of detection, measurement error, etc.) as the function of the parameters of the system and environment that affect this particular characteristic. This approach requires less development time and cost, but it is applicable only to the estimation of the specific characteristics and the model is limited by assumptions introduced (e.g., a uniform flow of input data, Gaussian noise, etc.). In general, the main trade-off that has to be considered is a trade between accuracy of the model and its complexity resulting in development time and cost.

1.2 Mathematical Description of Signals

The majority of computer simulation tasks involve mathematical descriptions of how the radio systems, devices and circuits transform signals and interference that pass through them. There are two basic ways to describe signals and interference depending on its nature and a source of origin: as a deterministic function of time (deterministic process) or a random function of time (a random process). The latter way is covered in Chapter 2.

Definition. *A signal is a useful portion of an electromagnetic oscillation containing information or a message to be processed by a radio system.*

Typically, two types of signals are distinguishable from the point of view of mathematical description: *low-frequency oscillations* with the spectrum close to zero frequency range and *narrowband oscillations* with the spectrum concentrated around the carrier frequency. The former ones are often described through the decomposition models (Section 1.2.1.6), Fourier series, Fourier integrals, Fourier, Laplace transform models (Sections 1.2.2.1–1.2.2.2), or Z-transforms for digital signals (Section 1.2.2.3). Its envelope and phase typically represent the narrowband signals. They can be described:

A. By means of two orthogonal signals (see Section 1.2.1.3):

$$U(t) = A(t)\cos \Phi(t) = A(t)\cos\left[\omega_0 t - \varphi(t)\right]$$

$$V(t) = A(t)\sin \Phi(t) = A(t)\sin\left[\omega_0 t - \varphi(t)\right]$$

where $A(t)$ is the envelope, $\Phi(t)$ is the complete phase, ω_0 is the carrier frequency and $\varphi(t)$ is the phase. The envelope $A(t)$, the complete phase $\Phi(t)$ and the instantaneous frequency $\omega(t)$ are functions of time t and can be modeled as:

$$A(t) = \left[U(t)^2 + V(t)^2\right]^{1/2}$$

$$\Phi(t) = \omega_0 t - \varphi(t) = \arctan\frac{V(t)}{U(t)}$$

$$\omega(t) = \frac{d\Phi(t)}{dt} = A^{-1}(t)\left[U(t)\frac{dV(t)}{dt} - V(t)\frac{dU(t)}{dt}\right]$$

B. By means of the complex envelope vector projections:

$$a(t) = \mathrm{Re}\,\overline{A}(t) = A(t)\cos\varphi(t)$$

$$b(t) = \mathrm{Im}\,\overline{A}(t) = A(t)\sin\varphi(t)$$

that gives:

$$A(t) = \left[a(t)^2 + b(t)^2\right]^{1/2}$$

$$\varphi(t) = \omega_0 t - \Phi(t) = \arctan\frac{b(t)}{a(t)}$$

$$\omega(t) = \omega_0 - A^{-1}(t)\left[a(t)\frac{db(t)}{dt} - b(t)\frac{da(t)}{dt}\right]$$

C. By means of the narrowband signal derivatives:

$$A(t) = \left[U^2(t) + \left(\frac{1}{\omega_0^2} \right) \dot{U}^2(t) \right]^{1/2}$$

$$\varphi(t) = \omega_0 t + \arctan \left[\dot{U}(t) \Big/ \omega_0 U(t) \right]$$

where

$$U(t) = U = A \cos (\omega_0 t - \varphi)$$

$$\dot{U}(t) = -\omega_0 A \sin (\omega_0 t - \varphi)$$

$$\ddot{U}(t) = -\omega_0^2 A \cos (\omega_0 t - \varphi)$$

and the following equations are valid:

$$A(t)\,\dot{A}(t) = \dot{U}(t) \left[\ddot{U}(t) + \omega_0^2 U(t) \right]$$

$$\dot{\varphi}(t) A^2(t) \omega_0 = U(t) \left[\ddot{U}(t) + \omega_0^2 U(t) \right]$$

D. By means of the Hilbert transform (see Section 1.2.2.4) and complex analytical signals:

$$\overline{U}(t) = U(t) + jV(t)$$

$$\overline{U}^*(t) = U(t) - jV(t)$$

that gives

$$\mathrm{Re}\{\overline{U}(t)\} = U(t) = \frac{1}{2} \left[\overline{U}(t) + \overline{U}^*(t) \right]$$

$$\mathrm{Im}\{\overline{U}(t)\} = V(t) = \frac{1}{2j} \left[\overline{U}(t) - \overline{U}^*(t) \right]$$

or via complex envelope

$$a(t) = \text{Re}\left\{\overline{A}(t)\right\} = \frac{1}{2}\left[\overline{A}(t) + \overline{A}^*(t)\right]$$

$$b(t) = \text{Im}\left\{\overline{A}(t)\right\} = \frac{1}{2j}\left[\overline{A}^*(t) - \overline{A}(t)\right]$$

$$\overline{U}(t) = \overline{A}(t) \cdot e^{j\omega_0 t} \qquad \overline{U}^*(t) = \overline{A}^*(t) \cdot e^{-j\omega_0 t}$$

To transform the complex spectra of a narrowband signal the following models are typically used:

$$S_V(jf) = \begin{cases} -j \cdot S_u(jf), & f > 0 \\ j \cdot S_u(jf), & f < 0 \end{cases}$$

$$S_V(f) = S_u(f), \quad \varphi_V(f) = \begin{cases} \varphi_u(f) + \dfrac{\pi}{2}, & f > 0 \\ \varphi_u(f) - \dfrac{\pi}{2}, & f < 0 \end{cases}$$

E. By means of differential equations (see 3.2.1.2).

1.2.1 Time-Domain Description

1.2.1.1 Basic Definitions

The time-domain model of the signal is a deterministic function of time $s(t)$. The signal can be periodic or nonperiodic. The periodic signal is modeled as:

$$s(t) = s(t + kT)$$

where T is a finite time interval (period), and k is any integer. The signal is called a nonperiodic one if this condition is not met. The simplest periodic signal is a harmonic oscillation:

$$s(t) = A \cos\left(\frac{2\pi}{T} t + \varphi\right) = A \cos\left(\omega t + \varphi\right) \tag{1.1}$$

where A, T, ω, φ are constant amplitude, period, angular frequency and initial phase of the signal. The angular ω and linear frequencies f are linked as:

$$\omega = 2\pi f = 2\pi \frac{1}{T}$$

The harmonic oscillation of infinite duration is called a monochromatic signal. This term is taken from optics and emphasizes that the spectrum of such a signal consists of a single line at the frequency f. In practice, a monochromatic signal does not exist since any real signal is a finite one and its spectrum inevitably "smears". But in practical applications the signals defined by model (1.1) are called the harmonic and monochromatic ones if the finite interval they are defined at is large enough to neglect the effects of the interval ends.

The signals are also often classified as analog, discrete and digital. An analog signal is a signal that continuously fills in the range of possible values in magnitude and time (it may take an infinite number of magnitude and time counts). A discrete signal is a signal that continuously fills in the range of possible values in magnitude while taking only specified numbers of values in time (it may take an infinite number of magnitude and a finite number of time counts, so it is quantified in time which is typically referred to as sampling). A digital signal is a signal that takes only specified numbers of values both in magnitude and time (it may take only finite numbers of magnitude and time counts, so it is quantified both in amplitude, which is typically referred to as quantization, and time, which is typically referred to as sampling). Thus the digital signal is virtually a digital code specified at some discrete moments of time. The first two classes – analog and discrete signals – form the linear space with respect to the operations of summation and multiplication. Strictly speaking, the digital signals do not form the linear space when the number of digits (number of bits in quantization device) is limited. But the linear discrete models are also considered to be applicable to the digital signals when the number of bits is large enough (see Section 1.4.1).

1.2.1.2 Energy and Power of the Signals

The basic energy/power characteristics of the signal are:

instantaneous power: $p(t) = s^2(t)$

(it is not additive: $[s_1(t) + s_2(t)]^2 \neq s_1^2(t) + s_2^2(t)$)

average power: $P_{av} = \dfrac{1}{T} \displaystyle\int_{t_a}^{t_b} s^2(t)dt, \quad T = t_b - t_a$

energy: $E = \displaystyle\int_{t_a}^{t_b} p(t)dt = \int_{t_a}^{t_b} s^2(t)dt$

For the discrete signals taken with increment Δt in N points at the interval $T = N\Delta t$:

$$p(n\Delta t) = s^2(n\Delta t)$$

$$P_{av} = \frac{1}{N} \sum_{n=0}^{N-1} s^2(n\Delta t)$$

$$E = \Delta t \sum_{n=0}^{N-1} s^2(n\Delta t)$$

1.2.1.3 Orthogonality and Coherency

Two signals $s_1(t)$ and $s_2(t)$ are called the orthogonal ones if one of the following conditions is met:

$$E_{12} = \int_{t_a}^{t_b} s_1(t)s_2(t)dt = 0$$

$$P_{12} = \frac{1}{T} \int_{t_a}^{t_b} s_1(t)s_2(t) = 0$$

For orthogonal signals energy and power are the additive functions:

$$E = E_1 + E_2, \quad P = P_1 + P_2$$

The two signals $s_1(t)$ and $s_2(t)$ are called the coherent ones if they are not orthogonal.

1.2.1.4 Basic Discontinuous Time-Domain Functions

The following elementary discontinuous time-domain functions are widely used to model the actual signals. The basic functions are:

A. Signature function *sign (t)*:

$$sign\,(t) = \begin{cases} -1, & t < 0 \\ 0, & t < 0 \\ 1, & t > 0 \end{cases}$$

B. Step function $\sigma(t)$

$$\sigma\,(t - t_0) = \begin{cases} 1, & t > t_0 \\ \frac{1}{2}, & t = t_0 \\ 0 & t > t_0 \end{cases}$$

where t_0 is a delay of the function $\sigma(t)$ with respect to $t = 0$.
The typical model of the signal *s(t)* is:

$$s(t) \cdot \sigma\,(t - t_0) = \begin{cases} s(t), & t \geq t_0 \\ 0 & , t < t_0 \end{cases}$$

which is equivalent to the switching of the signal $s(t)$ on at the moment t_0. The correspondence between $\sigma(t)$ and sign (t):

$$\sigma(t) = \frac{1}{2}\left[1 + sign\,(t)\right]$$

C. Delta function (Dirac function) $\delta(t)$

The δ-function is widely used in the mathematical modeling of signals. It is the impulse of infinitesimally short duration that has infinitely large amplitude at the moment $t = t_0$ and an area equal to unity. By Dirac's definition:

$$\delta(t - t_0) = \begin{cases} \infty, & t = t_0 \\ 0, & t \neq t_0 \end{cases}$$

$$\int_{-\infty}^{\infty} \delta(t - t_0)\, dt = 1$$

In the theory of radio signals these two definitions are typically complemented by the other two definitions:

$$\delta(t) = \delta(-t)$$

which means it is an even function of argument, and

$$\lim_{T \to \infty} \frac{1}{T} \int_{-T/2}^{T/2} \delta^2(t) = 1$$

which means that it has a unity average power at the infinite interval T. With these definitions the following equations are valid:

$$\int_{-\infty}^{0} \delta(t)\, dt = \int_{0}^{\infty} \delta(t)\, dt = \frac{1}{2}$$

$$\int_{-\infty}^{t} \delta(t - t_0)\, dt = \begin{cases} 0, & t < t_0 \\ 1/2, & t = t_0 \\ 1, & t > t_0 \end{cases}$$

The basic properties of delta-function are given in Appendix 2. The correspondence between delta-function and step function is:

$$\delta(t - t_0) = \frac{d\,\sigma(t - t_0)}{dt}$$

Some models of the signals using the discontinuous time-domain function are given in Appendix 3.

1.2.1.5 Harmonic and Complex Signals

The model of harmonic signals is as follows:

$$s(t) = A \cos[\omega_0 t + \varphi(t)] = A \cos \Phi(t)$$

where A, ω_0, $\varphi(t)$ are amplitude, angular frequency and phase. The latter is a function of time in a generic case. In this case the complete phase

$$\Phi(t) = \int \omega(t)dt$$

and the instantaneous frequency:

$$\omega(t) = \frac{d\,\Phi(t)}{dt} = \omega_0 + \frac{d\,\varphi(t)}{dt}$$

The harmonic signal is typically modeled as the real part of the complex signal

$$\overline{s(t)} = A \cdot e^{\,j\varphi(t)} \cdot e^{\,j\omega_0 t} = \overline{A(t)} \cdot e^{\,j\omega_0 t}$$

where $\overline{A(t)}$ is called a complex envelope of the signal $s(t)$. Thus:

$$s(t) = \mathrm{Re}\left\{ \overline{s(t)} \right\} = \mathrm{Re}\left\{ A\cos\left[\omega_0 t + \varphi(t)\right] + j\left[A\sin\left[\omega_0 t + \varphi(t)\right]\right]\right\} \qquad (1.2)$$

Another equivalent model is:

$$s(t) = \frac{1}{2}\left[\overline{s(t)} + \overline{s^*(t)}\right] = A\cos\left[\omega_0 t + \varphi(t)\right], \qquad (1.3)$$

The complex envelope $\overline{A(t)}$ can be expressed through in-phase (I) and quadrature (Q) components:

$$s(t) = A\cos\left[\omega_0 t + \varphi(t)\right] = A \cdot \cos\varphi(t)\cdot\cos\omega_0 t - A\cdot\sin\varphi(t)\cdot\sin\omega_0 t$$
$$= A_I \cdot \cos\omega_0 t - A_Q \sin\omega_0 t$$

$$\overline{A(t)} = A_I - jA_Q = \left|\overline{A(t)}\right| \cdot e^{j\varphi(t)}$$

$$\left|\overline{A(t)}\right| = A = \sqrt{A_I^2 + A_Q^2}$$

$$\varphi(t) = \arctan\left(A_Q / A_I\right)$$

Model (1.2) is typically used for analysis of the harmonic signals passing through linear circuits while model (1.3) is often used to describe the harmonic signals passing through nonlinear circuits.

When a complex model of the signal is used the instantaneous power $p(t)$, average power P_{av} and energy E of the signal are given as:

$$p(t) = \bar{s}(t) \cdot \bar{s}^{*}(t)$$

$$P_{av} = \frac{1}{T} \int_{t_a}^{t_b} \bar{s}(t) \cdot \bar{s}^{*}(t) \, dt \quad , \quad T = t_b - t_a$$

$$E = \int_{t_a}^{t_b} \bar{s}(t) \cdot \bar{s}^{*}(t) \, dt$$

For discrete signals correspondingly:

$$p(n\Delta t) = \bar{s}(n\,\Delta t) \cdot \bar{s}^{*}(n\,\Delta t) = \left| s(n\Delta t) \right|^2$$

$$P_{av} = \frac{1}{N} \sum_{n=0}^{N-1} \bar{s}(n\Delta t) \cdot \bar{s}^{*}(n\Delta t) = \frac{1}{N} \sum_{n=0}^{N-1} \left| \bar{s}(n\Delta t) \right|^2$$

$$E = \Delta t \sum_{n=0}^{N-1} \bar{s}(n\Delta t) \cdot \bar{s}^{*}(n\Delta t) = \Delta t \sum_{n=0}^{N-1} \left| \bar{s}(n\Delta t) \right|^2$$

The definition of orthogonality is the same as for the real signals: two complex signals $\bar{s}_1(t)$ and $\bar{s}_2(t)$ are orthogonal at an interval $T = (t_b - t_a)$ if

$$P_{12} = \frac{1}{T} \int_{t_a}^{t_b} \bar{s}_1(t) \cdot \bar{s}_2^{*}(t) \, dt = \frac{1}{T} \int_{t_a}^{t_b} \bar{s}_1^{*}(t) \cdot \bar{s}_2(t) \, dt = 0$$

1.2.1.6 The Decomposition Models of the Signals

The common way to model time-domain signals $s(t)$ is to represent them as a series:

$$s(t) = \sum_{n=0}^{\infty} C_n \cdot \varphi_n (t) \qquad\qquad (1.4)$$

where C_n are some coefficients and $\varphi_0 (t)$, ... $\varphi_n (t)$ is a system of real functions. Typically $\{\varphi_n (t)\}$ is a system of orthogonal functions at some interval a, b:

$$\int_a^b \varphi_n (t)\, \varphi_m (t) = 0 \qquad n \neq m$$

When the signal is modeled as a series (1.4) two basic approaches may be used. First is to achieve an accurate decomposition using relatively simple orthogonal functions. In this case the most common is to use the orthogonal system of basic trigonometric functions: sines and cosines. The choice of trigonometric functions is accounted for by the fact that a harmonic signal is the only function of time that retains its shape when passing through any linear system with constant parameters (only amplitude and phase vary). The second approach is to create a model with the minimum amount of factors in the series (1.4). To achieve this, orthogonal systems based on different functions (Chebyshev, Laguerre, Hermitian, etc.)   are used. The basic decomposition models are given in Appendix 4.

1.2.1.7 Modulation

Modulation is a fundamental concept in radio engineering since practically all the signals involved in information delivery in communications, radar and other disciplines are modulated ones.

Definition. *Modulation is a process by which some characteristic of the carrier is varied in accordance with a modulating wave.*

The basic model of a harmonic signal (see Section 1.2.1.5) can be written as:

$$s(t) = A(t) \cdot \cos \Phi(t)$$

which leads to the following possible types of modulation:

1) Amplitude modulation when the amplitude of the signal $A(t)$ varies according to some specified law. Typically,

$$A(t) = A_0 + k_A \cdot U_m(t)$$

where A_0 is an average amplitude of the signal $s(t)$, k_A some coefficient and $U_m(t)$ is the modulating signal. If modulating signal

$$U_m(t) = U_0 \cos(\Omega t + \Psi)$$

is also a low-frequency harmonic oscillation, the model takes the form

$$s(t) = [A_0 + k_A \cdot U_0 \cos(\Omega t + \Psi)]\cos \Phi(t)$$

$$= A_0 \left[1 + \frac{k_A U_0}{A_0} \cos(\Omega t + \Psi)\right]\cos \Phi(t)$$

where $M = \dfrac{k_A U_0}{A_0}$ is a modulation coefficient. When modulating signal $U_m(t)$ is a single pulse or a sequence of pulses the modulation is often referred to as an amplitude-pulse or pulse modulation.

2) Complete phase modulation is a type of modulation in which the complete phase $\Phi(t) = \omega_0 t + \varphi(t)$ varies according to some specified law. It is usually distinguished as frequency modulation and phase modulation.

Frequency modulation is a complete phase modulation when the instantaneous frequency

$$\omega(t) = \frac{d\,\Phi(t)}{d\,t} = \omega_0 + \frac{d\,\varphi(t)}{d\,t} = \omega_0 + k_f \cdot U_m(t)$$

varies according to the specified modulating function $U_m(t)$. Three major laws of phase $\varphi(t)$ variation are used:

A. Constant phase law $\varphi(t) = \varphi_0 = \text{const}$. In this case

$$\omega(t) = \omega_0$$

which results in the signal with the constant frequency ω_0 (no frequency modulation).

B. Quadratic law of phase variation in time $\varphi(t) = C \cdot t^2$. In this case

$$\omega(t) = \omega_0 + 2Ct$$

which results in a linear frequency modulation (LFM).

C. Phase variation according to some arbitrary function of time $\varphi(t) = F(t)$. In this case

$$\omega(t) = \omega_0 + \frac{d\,F(t)}{d\,t}$$

which results in a nonlinear frequency modulation (NFM). Often $F(t)$ has the form:

$$F(t) = F_L(t) + F_N(t) = C \cdot t^2 + F_N(t)$$

i.e., consists of the linear $F_L(t)$ and nonlinear $F_N(t)$ factors. The common functions to be used as $F_N(t)$ are error function $erf(t)$, trigonometric functions $\sin(t)$, $\arctan(t)$ and some other nonlinear functions of time.

Phase modulation is a complete phase modulation when the phase $\varphi(t)$ varies according to the specified modulation function $U_m(t)$

$$\varphi(t) = \varphi_0 + k_p \cdot U_m(t)$$

Actually, the phase $\varphi(t)$ is the parameter that varies for both types of modulation and they differ in how the modulating function $U_m(t)$ is defined. For phase modulation the phase has to be directly controlled by modulating signal $U_m(t)$ while for frequency modulation the modulating signal $U_m(t)$ has to be integrated first and then used to control $\varphi(t)$.

1.2.1.8 Discrete and Digital Signals

Since most of the modern signal processing devices employ a digital technique, the analog signal is digitized prior to processing, which leads to the concepts of discrete and digital signals (see Section 1.2.1.1 for a definition). The discrete signal $s_\Delta(t)$ can be modeled as

$$s_\Delta(t) = x(n\,\Delta t) = s(t) \cdot y_\Delta(t)$$

where Δt is the sampling interval, $n = 0, ... N - 1$ is a number of samples, $s(t)$ is an analog signal before sampling and $y_\Delta(t)$ is a clock pulse sequence (sampling function). The most common model for sampling function is:

$$y_\Delta(t) = \delta[(n-m)\Delta t] = \begin{cases} 1, & n = m \\ 0, & n \neq m \end{cases}$$

i.e., $y_\Delta(t)$ is a discrete delta-function. Thus, the final model of a discrete signal is:

$$x(n\Delta t) = \sum_{m=0}^{\infty} s(m\Delta t) \cdot \delta[(n-m)\Delta t]$$

i.e., it is a series of discrete delta functions with weighting coefficients that are the signal samples at $t = m\Delta t$.

The discrete signals in a general case are modeled as a sequence $\{x(n\Delta t)\}$, $n = 0,...N-1$, that can take any values at some interval $x \in [-\infty, \infty]$ while argument n takes only discrete values, $n = 0,1,2,...N-1$. The other notations often used for $x(n\Delta t)$ are $x(n)$, x_n or $\vec{x} = (-1,1,2)$, which means that the finite sequence consisting of three samples is described as follows: $x(0) = -1$, $x(\Delta t) = 1$, $x(2\Delta t) = 2$.

The digital signal $x_d(n\Delta t)$ is modeled as a function $x(n\Delta t)$ that takes a finite amount of discrete values: $h_1, h_2,...h_k$, while the argument n takes discrete values, $n = 0,1,2,...N-1$. The model is

$$x_d(n\Delta t) = F_N[x(n\Delta t)]$$

where F_N is a nonlinear function. The common model for F_N is:

$$x_d(n\Delta t) = ceil\{x(n\Delta t) \cdot 2^K - 0.5\}$$

where ceil (A) is an integer number that is bigger than A and the closest to A, K is the number of bits in a device quantizing the signal level, and factor 0.5 centers the signal between two quantization levels. Thus, for example, if discrete signal $0 \leq x(n\Delta t) \leq 1$, $K = 10$, the digital signal can take $2^K = 1024$ values between 0 and 1.

Both discrete and digital signals can be represented by the real or complex models depending on which initial analog signal model is used.

1.2.2 Frequency-Domain Description

The frequency-domain description of the signals is implemented by means of a spectrum.

Definition. The spectrum of the signal is the amplitude-phase distribution of its components as a function of frequency.

Transform pairs are the fundamental concepts used to link time-domain $s(t)$ and frequency-domain $S(\omega)$ representations of the signal. The most common transforms are the Fourier transform, Laplace transform and Z-transform, the latter one is a basic transform used to model discrete and digital signals. Some other transforms are also used in special applications.

1.2.2.1 Fourier Transform

For analog signal $s(t)$ and its spectrum $S(\omega)$ the Fourier transform pair is:

$$S(\omega) = \int_{-\infty}^{\infty} s(t)\, e^{-j\omega t}\, dt$$

$$s(t) = \frac{1}{2\pi} \int_{-\infty}^{\infty} S(\omega)\, e^{j\omega t}\, d\omega$$

For discrete signals $x(n\,\Delta t) = s(t) \cdot y_\Delta(t)$ the sampling function $y_\Delta(t)$ can be represented via a Fourier series:

$$y_\Delta(t) = \frac{1}{\Delta t} \sum_{m=-\infty}^{\infty} \exp(jm\omega_1 t)\,, \qquad \omega_1 = \frac{2\pi}{\Delta t}$$

that gives the spectrum of the discrete signal as:

$$X(e^{j\omega \Delta t}) = \frac{1}{\Delta t} \sum_{m=-\infty}^{\infty} S\left[j(\omega + m\frac{2\pi}{\Delta t}) \right]$$

Thus the spectrum of a discrete signal is a sequence of spectra $S(\omega)$ of the initial signal $s(t)$ shifted with respect to each other by $\Delta\omega = 2\pi/\Delta t$. In practical applications another representation of the discrete signal and its spectrum is often used that is termed a discrete Fourier transform (DFT):

direct DFT:
$$X(k) = \sum_{n=0}^{N-1} x(n\Delta t) \cdot W_N^{kn} \qquad k = 0,..., N-1;$$

inverse DFT:
$$x(n\Delta t) = \frac{1}{N} \sum_{k=0}^{N-1} X(k) \cdot W_N^{-kn} \qquad n = 0,..., N-1$$

where $x(n\Delta t)$ is a sequence of N time-domain counts, $X(k)$ is a sequence of N frequency-domain counts,

$$W_N = e^{-j \cdot \frac{2\pi}{N}}, \quad j = \sqrt{-1}$$

The basic features of the discrete Fourier transforms are given in Appendix 5. In a matrix form DFT can be described by a compact model:

$$\vec{X} = \vec{W}_N \cdot \vec{x}, \quad \vec{x} = \vec{W}_N^{-1} \cdot \vec{X}$$

where $\vec{X}$ and $\vec{x}$ are N-dimensional vectors

$$\vec{X} = [X(0), X(1),...X(N-1)]^T,$$

$$\vec{x} = [x(0), x(1),...x(N-1)]^T,$$

$\vec{W}_N$ is a matrix of dimension $N \times N$ with the elements

$$d(n,k) = W_N^{nk} = W_N^{nk(\bmod N)}, \quad n, k = 0,..., N-1.$$

$\vec{W}_N^{-1}$ is a matrix reciprocal to $\vec{W}_N$ with the elements

$$d^{-1}(n,k) = \frac{1}{N} W_N^{-nk} = \frac{1}{N} W_N^{-(nk(\bmod N))}$$

The computation of DFT requires a significant number of operations to be performed. The special algorithms are used to reduce the number of operations required that are referred to as fast Fourier transform (FFT). The basic FFT models are given in Appendix 6. The multidimensional (1-dimensional) DFT of 1-dimensional sequences is the following pair (direct DFT and inverse DFT correspondingly):

$$X(k_1, k_2, ...k_l) = \sum_{n_1=0}^{N_1-1} ... \sum_{n_l=0}^{N_l-1} x(n_1 \Delta t, ..., n_l \Delta t) W_{N_1}^{k_1 n_1} ... W_{N_l}^{k_l n_l}$$

$$x(n_1 \Delta t, ..., n_l \Delta t) = \frac{1}{N_1 ... N_l} \sum_{k_1=0}^{N_1-1} ... \sum_{k_l=0}^{N_l-1} X(k_1, ..., k_l) W_{N_1}^{-k_1 n_1} ... W_{N_l}^{-k_l n_l}$$

$$k_1 = 0, ..., N_1 - 1, ..., k_l = 0, ..., N_l - 1$$

$$n_1 = 0, ..., N_1 - 1, ..., n_l = 0, ..., N_l - 1$$

The comparison of operations required for FFT algorithms with different bases is given in Appendix 6.

1.2.2.2 The Laplace Transform

The Laplace transform of the analog signal $s(t)$ is defined as:

$$F(p) = \int_0^\infty s(t) e^{-pt} dt$$

where $p = \sigma + j\omega$ is a complex variable. The basic advantage of going from a real variable ω to a complex one p is that all restrictions imposed by the requirement that the function $s(t)$ has to be completely integrable are eliminated. The discrete Laplace transform is:

$$X(p) = \sum_{n=0}^\infty x(n\Delta t) \cdot e^{-pn\Delta t}$$

1.2.2.3 Z-Transform

The change of variables $z = e^{p\Delta t}$, $p = \dfrac{1}{\Delta t} \ln z$ changes the Laplace transform to a Z-transform. The direct Z-transform is defined by a model:

$$X(z) = Z\{x(n\Delta t)\} = \sum_{n=0}^{\infty} x(n\Delta t) \cdot z^{-n}$$

The main features of a direct Z-transform are:

A. Linearity. If $x_3(n\Delta t) = \bar{n}_1 x_1(n\Delta t) + c_2 x_2(n\Delta t)$, then
   $$X_3(z) = c_1 X_1(z) + c_2 X_2(z)$$

B. Shift theorem. If $x_1(-m\Delta t) = x_1((-m+1)\Delta t) = \ldots x_1(-\Delta t) = 0$, then
   $$X_2(z) = Z^{-m} \cdot X_1(z)$$

C. If $x_2(n\Delta t) = x_1[(n-m)\Delta t]$, then
   $$X_2(z) = x_1(-m\Delta t) + x_1[(-m+1)\Delta t] \cdot z^{-1} + \ldots + z^{-m} \cdot X_1(z)$$

D. If $x_3(n\Delta t) = x_1(n\Delta t) \cdot x_2(n\Delta t)$, then

$$X_3(z) = \frac{1}{2\pi j} \int_C X_1(\vartheta) X_2(z/\vartheta) \vartheta^{-1} d\vartheta$$

An inverse Z-transform is defined as:

$$x(n\Delta t) = Z^{-1}\{X(z)\} = \frac{1}{2\pi j} \int_C X(z) z^{n-1} dz$$

where C is a closed-loop contour in the Z plane.
   There are two basic ways to model the inverse Z-transform:

$$x(n\Delta t) = \sum_{k=1}^{P} \frac{1}{(l_k - 1)!} \lim_{z \to z_k^{(l_k)}} \frac{d^{l_k-1}[(z - z^{(l_k)})^{(l_k)} \cdot F(z)]}{d z^{l_k-1}} \tag{1.5}$$

$$x(n\Delta t) = \frac{1}{n!}\left[\frac{d^n X(z^{-1})}{d z^n}\right]_{z=0} \tag{1.6}$$

where $F(z) = X(z) \cdot z^{n-1}$, $z_1^{(l_1)}$, $z_2^{(l_2)}$, ... $z_P^{(l_p)}$ are $F(z)$ poles not equal to each other ; l_k is a multiple of pole $Z_k^{(l_k)}$, and $0! = 1$, $d^o \varphi(z)/d z^o = \varphi(z)$. Model (1.5) makes it possible to derive analytical dependence of $x(n\Delta t)$ upon

n and calculate it for an arbitrary n while the model (1.6) makes it possible to calculate $x(n\Delta t)$ without finding the poles of the function $F(z)$. The Z-transform images for some common functions $f(t)$ are given in Appendix 7.

1.2.2.4 Other Transform Functions

Other special types of transform are sometimes used to describe the signal $s(t)$ and its image $S(\omega)$.

A. The *Hilbert transform* relates the signal $u(t)$ and its image $\hat{u}(t)$ as:

$$\hat{u}(t) = \frac{1}{\pi} \int_{-\infty}^{\infty} \frac{u(z)}{t-z} \, dz$$

$$u(t) = \frac{1}{\pi} \int_{-\infty}^{\infty} \frac{\hat{u}(z)}{z-t} \, dz$$

which is used to model complex signal $\overline{U}(t)$ as:

$$\overline{U}(t) = u(t) + j\hat{u}(t),$$

and real signal:

$$u(t) = \mathrm{Re}\left\{\overline{U}(t)\right\} = \frac{1}{2}\left[\overline{U}(t) + \overline{U}^{*}(t)\right]$$

B. The *Hartly transform* is typically applied to real discrete waveforms. The convenience of this transform results from the fact that direct and inverse transforms match with accuracy up to the constant multiplier $1/N$:

$$X(k) = \sum_{n=0}^{N-1} x(n\Delta t)\left[\cos\left(\frac{2\pi n k}{N}\right) + \sin\left(\frac{2\pi n k}{N}\right)\right]$$

$$x(n\Delta t) = \frac{1}{N}\sum_{k=0}^{N-1} X(k)\left[\cos\left(\frac{2\pi n k}{N}\right) + \sin\left(\frac{2\pi n k}{N}\right)\right]$$

C. *Digital convolution.* The concept of linear (aperiodic) and circular (periodic) discrete convolutions is widely used in fast Fourier transform (FFT) and Winograd-Fourier transform (WFT) models. The linear convolution of two finite sequences $x(n\Delta t)$ and $h(n\Delta t)$ each   having N_1 and N_2 counts correspondingly is a sequence $y(n\Delta t)$:

$$y(n\Delta t) = \sum_{l=0}^{n} h(l\Delta t) x[(n-l)\Delta t] = \sum_{l=o}^{n} x(l\Delta t) h[(n-l)\Delta t]$$
$$n = 0,...,N_1 + N_2 - 2$$

The sequence $y(n\Delta t)$ is also a finite one and has a length of $N_1 + N_2 - 1$ counts.

The circular convolution of the periodic sequences $x(n\Delta t)$ and $h(n\Delta t)$ having N-count periods each is a sequence $y(n\Delta t)$:

$$y(n\Delta t) = \sum_{l=0}^{N-1} h(l\Delta t) x[(n-l)\Delta t] = \sum_{l=0}^{N-1} x(l\Delta t) h[(n-l)\Delta t]$$

The sequence $y(n\Delta t)$ is also a periodic one with a period equal to N counts. So it can be modeled only for a single period, e.g., for $n = 0,...,N-1$. In a matrix representation the circular convolution can be modeled as:

$$\vec{y} = \vec{H} \cdot \vec{x} = \vec{X} \cdot \vec{h}$$

where

$$\vec{y} = \{y(0), y(\Delta t),... y[(N-1)\Delta t]\}^T$$
$$\vec{h} = \{h(0), h(\Delta t),...h[(N-1)\Delta t]\}^T$$
$$\vec{x} = \{x(0), x(\Delta t),...x[(N-1)\Delta t]\}^T$$

are finite vectors, and $\vec{H}, \vec{x}$ are periodic matrices $N \times N$ of the following type $(\vec{A} = \vec{H} \ or \ \vec{A} = \vec{X})$:

$$\vec{A} = \begin{bmatrix} a(0), & a(\Delta t) & ...a[(N-1)\Delta t] \\ a(\Delta t), & a(2\Delta t) & ...a(0) \\ a(2\Delta t), & a(3\Delta t) & ...a(\Delta t) \\ .... & .... & .... \\ a[(N-1)\Delta t], & a[(N-2)\Delta t] & ... \end{bmatrix}$$

If $x_1(n\Delta t)$ and $h_1(n\Delta t)$ have length $N_1 + N_2 - 1$ counts and

$$x_1(n\Delta t) = \begin{cases} x(n\Delta t), & n = 0,..., N_1 - 1 \\ 0, & n = N_1,..., N_1 + N_2 - 2 \end{cases}$$

$$y_1(n\Delta t) = \begin{cases} y(n\Delta t), & n = 0,..., N_2 - 1 \\ 0, & n = N_2,..., N_1 + N_2 - 2 \end{cases}$$

then a linear convolution of $x(n\Delta t)$ and $y(n\Delta t)$ is equal to an $(N_1 + N_2 - 1)$-point circular convolution of the sequences $x_1(n\Delta t)$ and $h_1(n\Delta t)$:

$$y(n\Delta t) = \sum_{l=0}^{N_1+N_2-2} x_1(l\Delta t) h_1 [(n-l)\Delta t], n = 0,..., N_1 + N_2 - 2$$

and can be calculated by using the $(N_1 + N_2 - 1)$-point DFT. The following operations are done to perform a circular convolution of sequences $x_1(n\Delta t)$ and $h_1(n\Delta t)$ by DFT:

1) $X_1(k) = \sum\limits_{n=0}^{N-1} x_1(n\Delta t) W_N^{nk}, \; k = 0, ..., N-1$

2) $H_1(k) = \sum\limits_{n=0}^{N-1} h_1(n\Delta t) W_N^{nk}, \; k = 0, ..., N-1$

3) $Y(k) = X_1(k) \cdot H_1(k), \; k = 0,..., N-1$

4) $y(n\Delta t) = \dfrac{1}{N} \sum\limits_{k=0}^{N-1} Y(k) W_N^{-nk}, n = 0,..., N-1$

If one sequence is much longer than another one ($N_1 \gg N_2$, or $N_2 \gg N_1$), the long sequence is broken into short sections and partial convolutions are calculated. Then these partial convolutions are used to obtain the final one. Fast convolution methods also can be used [1].

D. *The Winograd-Fourier transform* (WFT) is a special algorithm of DFT that minimizes the number of multiplications required for a given n-point FFT. It is based on representation of the matrix $N = N_1 \times N_2 \times ... N_Y$-point DFT (where all N_l are mutually prime numbers) as a direct product of the matrices of an N-point DFT:

$$\vec{W}_N \Rightarrow \vec{W}_{N_1} con\, \vec{W}_{N_2}\, con...\vec{W}_{N_l}$$

where *con* is a convolution sign. Typically, the Winograd-Fourier transform is more efficient than classical FFT algorithms and while the number of additions is comparable to that of conventional FFT, the number of multiplications is approximately 80% less.

In WFT the computation of an N_1-point DFT is done by calculation of circular convolutions based on the following approach. The sequence $y(n\Delta t)$ that is a circular convolution of sequences $x(n\Delta t)$ and $h(n\Delta t)$, $n = 0, ..., N-1$, is a sequence of polynomial coefficients:

$$Y(z) = X(z)H(z)\,(\mathrm{mod}\,(z^N - 1))$$

where $X(z) = \displaystyle\sum_{l=0}^{N-1} x(l\Delta t)\, z^l$

$$H(z) = \sum_{l=0}^{N-1} h(l\Delta t)\cdot z^l \qquad Y(z) = \sum_{l=0}^{N-1} y(l\Delta t)\, z^l$$

The computation of $Y(z)$ is based on the Chinese remainder theorem. The polynomial $z^N - 1$ can be broken down into k mutually prime polynomials:

$$z^N - 1 = \prod_{l=1}^{k} P_l(z), \qquad (P_1(z),... P_k(z)) = 1$$

Then $Y(z) = \left(\displaystyle\sum_{l=1}^{k} Y_l(z)Q_l(z)S_l(z) \right)\!\left(\mathrm{mod}\,(z^N - 1)\right)$

where $Y_l(z) = X(z)H(z)(\mathrm{mod}\,P_l(z))$,

$S_l(z) = (z^N - 1)/P_l(z)$,

and the polynomials $Q_l(z)$ must meet the equality:

$$Q_l(z)S_l(z) = 1(\text{mod } P_l(z)), l = 1,...,k$$

E. *Coordinate rotation digital computer* (CORDIC) methods are the methods used to rotate the vector $(x, y) = x + jy$ by the angle θ by means of operations of summation and multiplication only, and they can be efficiently used as rotation factors in FFT algorithms [1].

1.2.2.5 Discrete Band-Limited Signals

While modeling discrete signals it is important to see how close the discrete signal $x(n\Delta t)$ reproduces the initial analog signal $s(t)$ that existed before discretization. For band-limited signals the function $s(t)$ is completely defined by the sequence $x(n\Delta t)$ if $\Delta t \le 0.5 f_m$, where f_m is the maximum frequency in the signal spectrum:

$$S(\omega) = \begin{cases} S(\omega), & \omega < 2\pi f_m \\ 0, & \omega \ge 2\pi f_m \end{cases}$$

For a band-limited signal with maximum frequency $\omega_m = 2\pi f_m$, the following model is valid:

$$s(t) = \sum_{n=-\infty}^{\infty} x(n\Delta t) \cdot \varphi_n(t)$$

$$\varphi_n(t) = \frac{\sin \omega_m (t - n\Delta t)}{\omega_m (t - n\Delta t)}$$

The function $\varphi_n(t) = 1$ in points $t = n\Delta t$ and has a spectral density:

$$S_\varphi(\omega) = \begin{cases} \dfrac{1}{2 f_m} \cdot e^{-jn\Delta t \omega} = \Delta t \cdot e^{-jn\Delta t \omega}, & -\omega_m \le \omega \le \omega_m \\ 0, & \omega < -\omega_m, \omega > \omega_m \end{cases}$$

Theoretically, the band-limited signal always has an infinite duration. Vice versa the limited-duration signal has an infinite bandwidth. In practice, the duration of the signal T_s is limited so it is always possible to define maximum frequency f_m beyond which the energy of the remaining spectrum components is negligible with respect to the main portion of the spectrum at frequencies less than f_m. In this case the number of counts $N = T_s/\Delta t = 2 f_m T_s$, and the model takes the form:

$$s(t) = \sum_{n=0}^{N-1} x(n\Delta t) \frac{\sin \omega_m (t - n\Delta t)}{\omega_m (t - n\Delta t)}$$

Analogously, because of interchangeability of the arguments t and ω in the Fourier transform, in the frequency domain the following equations are valid:

$$\Delta \omega = \frac{2\pi}{T_s}, \Delta f = 1/T_s$$

$$S(\omega) = \sum_{n=0}^{N-1} X(n\Delta \omega) \frac{\sin \frac{T_s}{2}(\omega - n\Delta \omega)}{\frac{T_s}{2}(\omega - n\Delta \omega)}$$

where $X(n\Delta \omega)$ is the sequence of spectrum samples at points $\omega = n \cdot \Delta \omega$, $n = 0, \ldots N - 1$.

1.2.2.6 Energy and Power

The energy of the signal $s(t)$ with the spectral density $S(\omega)$ is defined by Parseval's equality:

$$E = \int_{-\infty}^{\infty} s^2(t) \, dt = \frac{1}{2\pi} \int_{-\infty}^{\infty} |\overline{S}(\omega)|^2 \, d\omega$$

that shows equivalency of time-domain and frequency-domain representations of the signals (see Section 1.2.1.2). Here $|S(\omega)|^2$ is energy per unity frequency band (the spectral density). If the spectrum is a function which is symmetrical around zero, then

$$E = \frac{1}{2\pi} \int_{-\infty}^{\infty} |\overline{S}(\omega)|^2 \, d\omega = \frac{1}{\pi} \int_{0}^{\infty} |\overline{S}(\omega)|^2 \, d\omega$$

1.3 Mathematical Description of Noise

Definition. Noise is the unwanted disturbances superimposed upon a useful signal that tend to obscure its information content.

The primary sources of noise in modern radio engineering systems are external noise, intrinsic noise, active devices noise, and digital circuits noise.

1.3.1 External Noise

The external noise is originated by human activity (man-made noise), Earth atmosphere (atmospheric noise) and outer space (galactic or cosmic noise, in which solar noise sometimes is separately distinguished). The external noise is typically modeled via its contribution to the equivalent noise temperature of an antenna (see Section 11.4).

1.3.2 Intrinsic Noise

The intrinsic noise is originated by the passive components of the electronic circuits and typically is classified as *thermal noise, shot noise, contact noise* and *popcorn noise (burst noise)*.

1.3.2.1 Thermal Noise

Thermal noise comes from thermal agitation of electrons within a resistance (it is often also termed Johnston noise). The basic model for thermal noise is:

$$V = \sqrt{4kTBR}$$

where V is rms noise voltage [Volts], $k = 1.38 \cdot 10^{-23}$ joules/°K is Boltzmann's constant, T[°K] is an absolute temperature, B[Hz] is noise bandwidth, R is resistance [Ohm]. The available noise power

$$P_a = \frac{V^2}{4R} = kTB$$

Thermal noise is white noise with a Gaussian distribution.

1.3.2.2 Shot Noise

Shot noise comes from a current flowing across a potential barrier in active semiconductor or tube devices, and is originated by the fluctuation of current around an average value resulting from the random emission of electrons (or holes). The basic model is:

$$I = \sqrt{2qI_{dc}B}$$

where I is rms noise current [A], $q = 1.6 \cdot 10^{-19}$ is an electron charge [coulombs], I_{dc} is an average direct current [A], B is a noise bandwidth [Hz]. Shot noise is white noise with a Gaussian distribution.

1.3.2.3 Contact Noise

Contact noise is caused by fluctuating conductivity due to imperfect contact between two materials, and occurs anywhere where two conductors are joined together (e.g. in a switch). In vacuum tubes it is also called a flicker noise. The basic model is:

$$I = C_0 \cdot I_{dc} \sqrt{B} / \sqrt{f}$$

where I is rms noise current [A], C_0 is a constant depending on the type of material and its geometry, I_{dc} is an average value of direct current [A], f is frequency [Hz], and B is a bandwidth centered around f [Hz]. The magnitude of contact noise has a Gaussian distribution.

1.3.2.4 Popcorn Noise

Popcorn noise (also known as a burst noise) is noise that is originated by a manufacturing defect in the junction of a semiconductor device (usually a metallic impurity). It causes a discrete change in level, typically 2–100 times the thermal noise, with the power density following $1/f^n$ characteristics where f is frequency and n is typically equal to 2.

1.3.3 Active Devices Noise

There are three basic methods to describe the internal noise sources in active devices (transistors, integrated circuits, operational amplifiers, etc.): by noise factor (noise figure), noise voltage and current, or noise temperature. The basic models for these approaches are given in Appendix 8.

1.3.4 Digital Circuits Noise

There are two basic types of noise introduced by digital circuits: digital logic noise and the noise originated by analog-to-digital conversion (ADC). The primary source of noise introduced by digital logic gates is their high switching speed combined with inductance and conductors that interconnect them. The noise voltage V generated when current I changes through inductance L is:

$$V = L \cdot \frac{dI}{dt}$$

where dI/dt is the rate of current change. Although the current itself is rather small, the voltage generated may be high because of the high rate of current change.

ADC noise (quantization noise) depends on the number of ADC bits N and has a root-mean-square value:

$$\sigma = \frac{2}{(2^N - 1)\sqrt{12}}$$

1.4 Mathematical Description of Signal Processing

1.4.1 Basic Models of Signal Processing Filters

Definition. *The signal processing filter is a two-port device that transforms the input signal $u_1(t)$ into desired output signal $u_2(t)$.*

The mathematical model of a linear analog filter is:

$$u_2(t) = -\sum_{i=1}^{M-1} a_i \cdot \frac{d^i u_2(t)}{dt^i} + \sum_{l=0}^{N-1} b_l \cdot \frac{d^l u_1(t)}{dt^l}$$

where a_i, b_l are constant or time-dependent coefficients.

The mathematical model of a linear discrete filter is:

$$y(n\Delta t) = -\sum_{i=1}^{M-1} a_i \cdot y[(n-i)\Delta t] + \sum_{l=0}^{N-1} b_l \cdot x[(n-l)\Delta t] \qquad (1.7)$$

where $x(n\Delta t)$, $y(n\Delta t)$ are the nth counts of the input and output signals, a_i, b_l are the constants or the counts of the functions depending on n only. If a_i, b_l are the constants, the filter is called a filter with constant parameters, and if not, the filter is called a filter with variable parameters. If at least one of the coefficients $a_i \neq 0$, the filter is called a recursive (a feedback) filter, and if all

$a_i = 0$ the filter is called a nonrecursive filter (a filter without a feedback). For nonrecursive filter the model is:

$$y(n\Delta t) = \sum_{l=0}^{N-1} b_l \cdot x[(n-l)\Delta t] \qquad (1.8)$$

The digital filter is a filter implementing algorithm (1.7). The input and output signals are the digital codes. Strictly speaking, digital filters are nonlinear devices but the number of bits in the signals passing through the filters is typically large. In this case the operation of digital filtering can be described by model (1.7) without considerable error. The mathematical description of the filters is typically performed by two basic characteristics: filter impulse response $h(t)$ (the time-domain description) and filter transfer function $H(j\omega)$ (the frequency-domain description).

1.4.2 The Impulse Response

The impulse response $h(t)$ is the time response of the filter to an impulse input (δ-function). The mathematical models of the linear filter output signal for analog $u_2(t)$ and discrete $y(n\Delta t)$ cases are as follows:

$$u_2(t) = \int_{-\infty}^{\infty} u_1(t-x) \cdot h(x)dx$$

$$y(n\Delta t) = \sum_{k=-\infty}^{\infty} x[(n-k)\Delta t] \cdot h(k\Delta t)$$

1.4.3 The Transfer Function

The transfer function $H(j\omega)$ describes filter response in the frequency domain. It is a complex quantity:

$$H(j\omega) = H(\omega) \cdot e^{j\phi(\omega)}$$

where $H(\omega)$, $\phi(\omega)$ are the amplitude-frequency and phase-frequency responses correspondingly. The impulse response $h(t)$ and the transfer function $H(j\omega)$ are the Fourier transform pairs. Output signal $u_2(t)$ and its complex spectrum $S_2(j\omega)$ are related to the input signal $u_1(t)$ and its spectrum $S_1(j\omega)$ as:

$$u_2(t) = u_1(t) \ conv \ h(t)$$

$$S_2(j\omega) = S_1(j\omega) \cdot H(j\omega)$$

For a digital filter with sampling increment Δt :

$$H_d(j\omega) = \sum_{n=-\infty}^{\infty} H\left[j\left(\omega - n \cdot \frac{2\pi}{\Delta t}\right)\right]$$

or in terms of the Z-transform:

$$H_d(z) = \frac{U_2(z)}{U_1(z)}$$

where $U_2(z), U_1(z)$ are Z-transforms of the output $u_2(t)$ and input $u_1(t)$ signals under zero initial conditions.

Thus for a recursive filter:

$$H_d(z) = \frac{\displaystyle\sum_{l=0}^{N-1} b_l \cdot z^{-l}}{1 + \displaystyle\sum_{i=1}^{M-1} a_i z^{-i}} \tag{1.9}$$

For a nonrecursive filter:

$$H_d(z) = \sum_{l=0}^{N-1} b_l \cdot z^{-l} \tag{1.10}$$

The complex, amplitude and phase responses for recursive and nonrecursive filters based on models (1.9) and (1.10) are given in Appendix 9. For digital signal processing filters the impulse response $h(n\Delta t)$ and transfer function $H_d(z)$ are related by direct and inverse Z-transforms:

$$h(n\Delta t) = Z^{-1}\{H_d(z)\}, \quad H_d(z) = Z\{h(n\Delta t)\}$$

which gives:

$$h(n\Delta t) = \frac{\Delta t}{2\pi} \int_{-\pi/\Delta t}^{\pi/\Delta t} H(e^{j\omega\Delta t}) \cdot e^{jn\omega\Delta t} \, d\omega$$

$$H_d(e^{j\omega\Delta t}) = \sum_{n=0}^{\infty} h(n\Delta t) \cdot e^{-jn\omega\Delta t}$$

Transfer functions for cascaded filters F_1 and F_2 with $H_1(z)$ and $H_2(z)$ are given in Appendix 10.

For digital complex sequences $\bar{x}(n\Delta t)$ and $\bar{y}(n\Delta t)$ with spectra $\bar{X}(e^{j\omega\Delta t})$ and $\bar{Y}(e^{j\omega\Delta t})$

$$\sum_{n=0}^{\infty} \bar{x}(n\Delta t)\bar{y}^*(n\Delta t) = \frac{\Delta t}{\pi} \int_{0}^{\pi/\Delta t} \left| \bar{X}(e^{j\omega\Delta t}) \right|^2 d\omega$$

If $\bar{x}(n\Delta t) = \bar{y}(n\Delta t)$, Parseval's theorem states that:

$$\sum_{n=0}^{\infty} \left| \bar{x}(n\Delta t) \right|^2 = \frac{\Delta t}{\pi} \int_{0}^{\pi/\Delta t} \left| \bar{X}(e^{j\omega\Delta t}) \right|^2 d\omega$$

Thus for any signal processing filter with real coefficients:

$$\sum_{n=0}^{\infty} h^2(n\Delta t) = \frac{\Delta t}{\pi} \int_{0}^{\pi/\Delta t} \left| \bar{H}_d(e^{j\omega\Delta t}) \right|^2 d\omega$$

Chapter 2

Stochastic Modeling

Statistical Synthesis of Random Functions

2.1 Random Events and Variables

2.1.1 Statistical Description

2.1.1.1 A Random Event

Definition. *A random event is any occurrence that can happen or not happen as the result of a test.*

The event can be totally deterministic (*D*), impossible (*I*) or random (*A, B, C…*). Probabilities of the events *D, I,* and *A* are:

$$P(D) = 1; \quad P(I) = 0; \quad 0 \leq P(A) \leq 1$$

2.1.1.2 Equivalent Events

If event *B* occurs any time when event *A* occurs, it is denoted $A \subset B$. If $A \subset B$ and at the same time $B \subset A$, the events *A* and *B* are called the equivalent events:

$$A = B \quad P(A) = P(B)$$

2.1.1.3 Sum and Product of Random Events

The sum of events $A_1, A_2, A_3, \ldots A_n$ is the event A that occurs when at least one event from this series happens:

$$A = \sum_n A_n = A_1 \bigcup A_2 \bigcup A_3 \ldots \bigcup A_n = \bigcup_n A_n$$

35

The product of events $A_1, A_2, A_3, \ldots A_n$ is the event A that occurs when all events from this series happen simultaneously:

$$A = \prod_n A_n = A_1 \bigcap A_2 \bigcap A_3 \cdots \bigcap A_n = \bigcap_n A_n$$

The following rules are applicable to the operations of addition and multiplication of the random events:

$$A + A = A \quad A + D = D \quad A + I = A \quad A + B = B + A$$

$$(A + B) + C = A + (B + C)$$

$$AA = A \quad AD = D \quad AI = A \quad AB = BA \quad (AB)C = A(BC)$$

$$(A + B)C = AC + BC \quad AB + C = (A + C)(B + C)$$

2.1.1.4 Disjoint Events

If mutual occurrence of events A and B is impossible, the events are called disjoint or mutually exclusive ones. In this case:

$$A \cdot B = I$$

2.1.1.5 Complete Group of Events

Events A, B, C... form the complete group of events if as the result of the test at least one of them definitely occurs:

$$A + B + C + \ldots = D$$

2.1.1.6 Complement Events

Two events are called complement events if they are mutually exclusive and form the complete group: ($\overline{A}$ is a complement of A):

$$P(\overline{A}) = 1 - P(A)$$

The following rules are applicable to complement events:

$$\overline{\overline{A}} = A \quad \overline{D} = I \quad \overline{I} = D \quad A + \overline{A} = D \quad A\overline{A} = I$$

$$\overline{A+B} = \overline{A} \cdot \overline{B} \quad \overline{AB} = \overline{A} + \overline{B} \quad A+B = A + \overline{A} \cdot B$$

$$A+B = \overline{\overline{A} \cdot \overline{B}} \quad AB = \overline{\overline{A} + \overline{B}}$$

2.1.1.7 Conditional Probability

Probability that event A occurs under the condition that event B has already occurred is denoted $P(A|B)$ and is called a conditional probability of the event A:

$$P(A|B) = P(A \cdot B)/P(B)$$

2.1.1.8 Independent Events

If event A statistically does not depend on event B, events are called independent:

$$P(A \cdot B) = P(A) \cdot P(B); \quad P(A|B) = P(A); \quad P(B|A) = P(B)$$

2.1.1.9 Probability of the Sum of the Events

The probability of the sum of two events A and B:

$$P(A+B) = P(A) + P(B) - P(AB)$$

If the events are disjoint:

$$P(A+B) = P(A) + P(B)$$

In the general case:

$$P\left(\sum_{n=1}^{N} A_n\right) = \sum_{n=1}^{N} P(A_n) - \sum_{n=1}^{N-1} \sum_{j=n+1}^{N} P(A_n A_j) +$$
$$+ \sum_{n=1}^{N-2} \sum_{j=n+1}^{N-1} \sum_{i=j+1}^{N} P(A_n A_j A_i) - \dots + (-1)^{N-1} P\left(\prod_{n=1}^{n} A_n\right)$$

The sum of disjoint events that form a complete group is equal to unity:

$$\sum_{n=1}^{N} P(A_n) = 1$$

The sum of two complement events is equal to unity:

$$P(A) + P(\overline{A}) = 1$$

2.1.1.10 The Probability of the Product of Events

The *probability of the product* of two events A and B is

$$P(AB) = P(A) \cdot P(B|A) = P(B) \cdot P(A|B)$$

For independent events:

$$P(AB) = P(A) \cdot P(B)$$

In the general case:

$$P(A_1 A_2 \dots A_n) = P(A_1) P(A_2|A_1) P(A_3|A_1 A_2) \dots P(A_n|A_1 A_2 \dots A_{n-1})$$

For N independent events:

$$P(\prod_{n=1}^{N} A_n) = \prod_{n=1}^{N} P(A_n)$$

2.1.1.11 An Event Conditional on the Set of Other Events

If event B occurs as the result of an event A occurrence, and event B is a set of N events $B_1, B_2, \dots B_N$, then

$$P(A) = \sum_{i=1}^{N} P(B_i) \cdot P(A|B_i)$$

2.1.1.12 Bayes Rule

$$P(B_i/A) = \frac{P(B_i) \cdot P(A|B_i)}{\sum_{j=1}^{N} P(B_j) \cdot P(A|B_j)} = \frac{P(B_i) P(A|B_i)}{P(A)}$$

2.1.1.13 A Random Variable

Definition. *A random variable is a variable that can take a random value in each trial.*

The basic types of random variables are discrete random variables and continuous random variables. The discrete random variable can take only a particular finite or countable infinite set of values. For continuous random variables the set of values cannot be defined and it fills in some interval continuously.

2.1.1.14 Probability Distribution Law

The complete statistical description of a one-dimensional random variable can be given by probability distribution law. For a discrete random variable X this law defines the linkage between possible values x_i the variable X can take and its probabilities $p_i = p(x_i)$. The general description both for discrete and continuous variables is based on the concept of probability distribution function $F_1(x)$. This function defines the probability P of the fact that a random variable X will be less than some value x:

$$F_1(x) = P(X < x)$$

The probability distribution function (PDF) has the following properties:

1) $F_1(-\infty) = 0$ 2. $F_1(+\infty) = 1$

2) $F_1(x_2) \geq F_1(x_1)$ *if* $x_2 > x_1$

3) $P(x_1 \leq X < x_2) = F(x_2) - F(x_1)$

The PDF of a discrete random variable is a stepped function with the steps in points $x_1, x_2, ...,$ and the PDF of a continuous random variable is a continuous function (Figure 2.1).

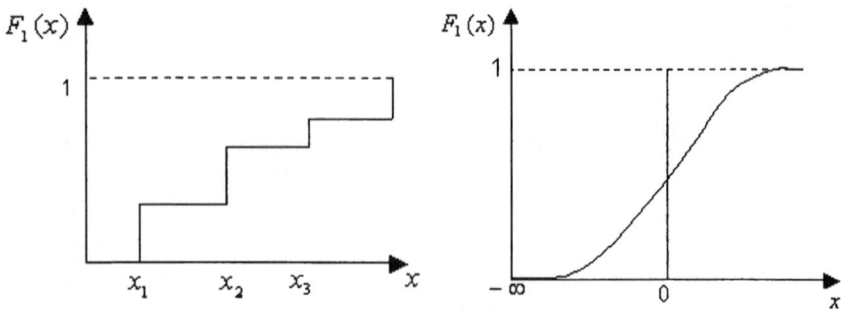

Figure 2.1 Probability distribution functions of discrete and continuous random variables.

2.1.1.15 Probability Density Function

If a probability distribution function has a derivative over the entire range of the possible values of random variables, the concept of a probability density function (pdf) is introduced:

$$f_1(x) = d\,F_1(x)/d\,x$$

A probability density function has the following basic properties:

1) $f_1(x) \geq 0$

2) $P(x_1 \leq X < x_2) = \int_{x_1}^{x_2} f_1'(x)\,d\,x = F_1(x_2) - F_1(x_1)$

3) $\int_{-\infty}^{\infty} f_1(x)\,d\,x = 1$

For a discrete random variable:

$$f_1(x) = \sum_{i=1}^{N} p_i \cdot \delta(x - x_i)$$

where x_i are the possible values of random variables (N values), p_i is the probability of the possible values, and $\delta(u - u_0)$ is delta-function:

$$\delta(u - u_0) = \begin{cases} \infty, & if\ u = u_0 \\ 0, & if\ u \neq u_0 \end{cases}$$

$$\int_{U_0 - \alpha}^{U_0 + \alpha} \delta(u - u_0)\, d u = 1 \ \text{ for any } \ \alpha > 0$$

$$\int_{U_0 - \alpha}^{U_0 + \alpha} f(u) \cdot \delta(u - u_0)\, d u = f(u_0)$$

The common probability distribution functions (PDF) and probability density functions (pdf) for discrete and continuous random variables are given in Appendixes 11 and 12, respectively.

2.1.1.16 Gaussian Probability Distribution Function

Gaussian (normal) PDF is one of the most important distribution laws used in practical applications (see Appendix 12). In this case the probability that a random variable will fall within the interval (a, b) is:

$$P(a \leq X < b) = \Phi(\frac{b - m}{\sigma}) - \Phi(\frac{a - m}{\sigma})$$

where $\Phi(z)$ is the error function integral (see Appendix 1 for definition of all special functions used in the handbook).

Typically, different integral forms for Gaussian pdfs are used in different texts. The common formulas to transform the integral forms from one to another are given in Appendix 13.

2.1.1.17 Statistical Parameters

In practice it is difficult or often impossible to find distribution functions for many classes of random variables of interest. In this case the description of a random variable can be done at the level of some numerical values (nonrandom numbers), called statistical parameters. The most important statistical parameters are:

A. Mathematical expectation:

$$m_x = M(x) = \sum_{n=1}^{N} x_n \cdot p_n \quad \text{(discrete variable)}$$

$$m_x = M(x) = \int_{-\infty}^{\infty} x f_1(x) dx \quad \text{(continuous variable)}$$

B. Variance:

$$D_x = \sigma_x^2 = M\left[(x - m_x)^2\right] = \sum_{n=1}^{N} (x_n - m_x)^2 \cdot p_n \quad \text{(discrete variable)}$$

$$D_x = \sigma_x^2 = \int_{-\infty}^{\infty} (x - m_x)^2 f_1(x) dx \quad \text{(continuous variable)}$$

C. The median is the value of a random variable when

$$P(X < M_l) = P(X > M_l) = 0.5$$

Thus $F_1(M_l) = 0.5$ and

$$\int_{-\infty}^{M_l} f_1(x) dx = \int_{M_l}^{\infty} f_1(x) dx$$

D. The mode is the value of a random variable when for discrete variable $P(X = M)$, and for continuous variable $f_1(M)$ reaches a maximum. The distribution is called unimodal if a single mode exists, and it is called multimodal if there are several modes in the distribution function.

E. The fractile is the value $x = x_P$ when $F_1(x_P) = P$.

2.1.1.18 Moments of Distribution

Moments are the generalized numerical values used for description of a random variable. The moment of order k of a random variable X with respect to an arbitrary point a is a mathematical expectation:

$$m_k(a) = M\left\{(X - a)^k\right\}$$

If $a = 0$, the moment m_k is called the initial one, and if $a = m_x$, the moment M_k is called a central one. The basic formulas to calculate the moments of a random variable are given in Appendix 14.

Typically, the following moments are used to describe a random variable:

A. m_1 = mathematical expectation (m)

B. M_2 = variance (σ^2)

C. M_3 leading to asymmetry coefficient:

$$\gamma_1 = M_3 \Big/ \sqrt{M_2^3} = M_3 \big/ \sigma^3$$

D. M_4 leading to the coefficient of excess:

$$\gamma_2 = \frac{M_4}{M_2^2} - 3 = \frac{M_4}{\sigma^4} - 3$$

The latter describes the deviation of a distribution function from a Gaussian one for which $\gamma_2 = 0$.

The basic moments for the common distribution laws are given in Appendix 15 (discrete variables) and Appendix 16 (continuous variables).

2.1.1.19 Functional Transform of Random Variables

A common task in simulation of the signals passing through radio circuits is to determine pdf $f_1(y)$ of the continuous random variable Y (output signal) when pdf $f_1(x)$ of the random variable X (input signal) and deterministic dependence $Y = \varphi(X)$ are known.

If the reciprocal function $X = h(Y)$ is single valued:

$$f_1(y) = f_1(x) \cdot \left| \frac{dx}{dy} \right| = f_1[h(y)] \cdot \left| \frac{dh(y)}{dy} \right|$$

If the reciprocal function is ambiguous:

$$f_1(y) = \sum_{n=1}^{N} f_1[h_n(y)] \cdot \left| \frac{dh_n(y)}{dy} \right|$$

where n is the number of ambiguities. The functional transform of the discrete random variable does not change the distribution function while the numerical values of the distribution are changed.

The moments of the random variable $Y = \varphi(X)$ can be determined in two ways: 1) by applying the pdf $f_1(y)$; and 2) by averaging dependence $\varphi(X)$. Applying these approaches to mean value and variance results in:

$$m_y = \int_{-\infty}^{\infty} y f_1(y) dy = \int_{-\infty}^{\infty} \varphi(x) f_1(x) dx$$

$$\sigma_y^2 = \int_{-\infty}^{\infty} (y - m_y)^2 f_1(y) dy = \int_{-\infty}^{\infty} [\varphi(x) - m_y]^2 f_1(x) dx$$

For a discrete random variable:

$$m_y = M[\varphi(X)] = \sum_n \varphi(x_n) \cdot f_n$$

$$\sigma_y^2 = D[\varphi(X)] = \sum_n [\varphi(x_n) - m_y]^2 \cdot f_n$$

The probability distribution functions for some common functional transforms of random variables are given in Appendix 17.

2.1.1.20 The Characteristic Function

There is a special type of functional transform that is widely used:

$$Y = e^{j\vartheta x}$$

The characteristic function $\theta_1(j\vartheta)$ is defined as a mathematical expectation of a random variable $e^{j\vartheta x}$

$$\theta_1(j\vartheta) = M\left(e^{j\vartheta x}\right) = \int_{-\infty}^{\infty} e^{j\vartheta x} f_1(x) dx$$

where ϑ is a real variable, $j = \sqrt{-1}$.

$$\theta_1(j\vartheta) = \sum_{n=1}^{N} p_i \cdot e^{j\vartheta x_n}$$

Then:

$$f_1(x) = \frac{1}{2\pi} \int_{-\infty}^{\infty} \theta_1(j\vartheta) e^{-j\vartheta x} d\vartheta$$

The moments can be defined as:

$$m_k = \frac{1}{j^k} \frac{d^k \theta_1(j\vartheta)}{d\vartheta^k} \bigg|_{\vartheta = 0}$$

The characteristic functions for common distributions are given in Appendixes 11 and 12.

2.1.1.21 A Multidimensional Random Variable

A multidimensional random variable is a system (set) of two or more one-dimensional random variables: $X_1, X_2, ... X_n$. It is described by the distribution function of nth order :

$$P\{X_1 < x_1, X_2 < x_2, ... X_n < x_n\} = F_n(x_1, x_2, ... x_n)$$

and corresponding probability density function:

$$f_n(x_1, x_2, ... x_n) = \frac{d^n F_n(x_1, x_2, ... x_n)}{d(x_1, x_2, ... x_n)}$$

2.1.1.22 Two-Dimensional Distribution Functions

Two-dimensional probability distribution functions $F_2(x_1, x_2)$ and corresponding density functions $f_2(x_1, x_2)$ are widely used in stochastic simulation of random variables. They have the following properties, hereinafter $(F_2(x, y) = F_2(x_1, x_2), f_2(x, y) = f_2(x_1, x_2))$:

1) $F_2(x_2, y) \ge F_2(x_1, y)$ if $x_2 > x_1$

   $F_2(x, y_2) \ge F_2(x, y_1)$ if $y_2 > y_1$

2) $F_2(x, -\infty) = F_2(-\infty, y) = F_2(-\infty, -\infty) = 0$

3) $F_2(\infty, \infty) = 1$

4) $F_2(x, \infty) = F_1(x) \quad F_2(\infty, y) = F_1(y)$

5) $F_2(x, y) = \int\limits_{-\infty}^{x} \int\limits_{-\infty}^{y} f_2(u, \vartheta)\, du\, d\vartheta$

6) $P(x_1 \le X \le x_2,\ y_1 \le Y \le y_2) =$

   $F_2(x_2, y_2) - F_2(x_1, y_2) - F_2(x_2, y_1) + F_2(x_1, y_1)$

7) $f_2(x, y) \ge 0$

8) $\int\limits_{-\infty}^{\infty} \int\limits_{-\infty}^{\infty} f_2(x, y)\, dx\, dy = 1$

9) $\theta_2(j\vartheta_1, j\vartheta_2) = M\left[e^{j(\vartheta_1 x + \vartheta_2 y)} \right] = \int\limits_{-\infty}^{\infty} \int\limits_{-\infty}^{\infty} e^{j(\vartheta_1 x + \vartheta_2 y)} \cdot f_2(x, y)\, dx\, dy$

The one-dimensional distribution and density functions are expressed through two-dimensional ones as follows:

$$F_1(x) = F_2(x, \infty) = \int\limits_{-\infty}^{x} \int\limits_{-\infty}^{\infty} f_2(x, y)\, dx\, dy$$

$$f_1(x) = \frac{dF_1(x)}{dx} = \int\limits_{-\infty}^{\infty} f_2(x, y)\, dy$$

$$F_1(y) = F_2(\infty, y) = \int\limits_{-\infty}^{y} \int\limits_{-\infty}^{\infty} f_2(x, y) \, dx \, dy$$

$$f_1(y) = \frac{dF_1(y)}{dy} = \int\limits_{-\infty}^{\infty} f_2(x, y) \, dx$$

2.1.1.23 The Conditional Probability Density Function

The conditional pdfs are defined as:

$$f_1(x|y) = \frac{f_2(x, y)}{f_1(y)} = \frac{f_2(x, y)}{\int\limits_{-\infty}^{\infty} f_2(x, y) \, dx}$$

$$f_1(y|x) = \frac{f_2(x, y)}{f_1(x)} = \frac{f_2(x, y)}{\int\limits_{-\infty}^{\infty} f_2(x, y) \, dy}$$

If X and Y are independent random variables:

$$f_2(x, y) = f_1(x) \cdot f_2(y)$$

2.1.1.24 The Moments: Mathematical Expectation, Variance, Correlation

The moments are defined as:

1) the initial moment of order $k_1 + k_2$

$$m_{k_1, k_2} = M\left[X^{k_1} Y^{k_2}\right] = \sum_i \sum_j x_i^{k_1} y_j^{k_2} p_{ij} \qquad \text{(discrete variable)}$$

$$m_{k_1, k_2} = \int\limits_{-\infty}^{\infty} \int\limits_{-\infty}^{\infty} x^{k_1} \cdot y^{k_2} \cdot f_2(x, y) \, dx \, dy \qquad \text{(continuous variable)}$$

2) the central moment of order $k_1 + k_2$

$$M_{k_1,k_2} = M\left[X_0^{k_1}Y_0^{k_2}\right] = \sum_i \sum_j (x_i - m_x)^{k_1}(y_j - m_y)^{k_2} P_{ij}$$

$$M_{k_1,k_2} = \int\limits_{-\infty}^{\infty}\int\limits_{-\infty}^{\infty}(x - m_x)^{k_1}(y - m_y)^{k_2}f(x,y)dxdy$$

In practice, widely used moments are:

$$m_{10} = M\left(X^1Y^0\right) = m_x \qquad m_{01} = M\left(X^0Y^1\right) = m_y$$

i.e., mathematical expectations of random variables X and Y;

$$M_{20} = M\left[(X - m_x)^2(Y - m_y)^0\right] = D_x = \sigma_x^2$$

$$M_{02} = M\left[(X - m_x)^0(Y - m_y)^2\right] = D_y = \sigma_y^2$$

i.e., the variances of random variables X and Y, and

$$M_{11} = M\left[(X - m_x)\cdot(Y - m_y)\right] = K_{xy}$$

i.e., correlation moment of random variables X and Y, that can be written as:

$$K_{xy} = \sum_i \sum_j (x_i - m_x)(y_j - m_y)P_{ij} \qquad \text{(discrete variable)}$$

$$K_{xy} = \int\limits_{-\infty}^{\infty}\int\limits_{-\infty}^{\infty}(x - m_x)(y - m_y)f_2(x,y)dxdy \qquad \text{(continuous variable)}$$

The dimensionless value $R_{xy} = \dfrac{K_{xy}}{\sigma_x\sigma_y}$ is called a correlation coefficient for variables X and Y.

2.1.1.25 The Conditional Moments

The conditional moments of the random variable X with respect to the random variable Y are defined as follows:

1) conditional mathematical expectation:

$$m\left(X|y\right)= \int\limits_{-\infty}^{\infty} x\, f_1\left(x|y\right)dx = \frac{\int\limits_{-\infty}^{\infty} xf_2\left(x,y\right)dx}{\int\limits_{-\infty}^{\infty} f_2\left(x,y\right)dx}$$

2) conditional variance:

$$D\left(X|y\right)= \int\limits_{-\infty}^{\infty}\left[x-M\left(X|y\right)\right]^2 f_1\left(x|y\right)dx$$

2.1.1.26 The Numerical Parameters

Basic numerical parameters of a set of n random variables $X_1, X_2,...X_n$ are mathematical expectations (means) $m\left(X_k\right)= m_k$, variances $D\left(X_k\right)= D_k$, $k =1,2,...n$, and correlation moments $K\left(X_i\, X_j\right)= K_{ij}$, or corresponding correlation coefficients $R_{ij} = K_{ij}/\left(\sigma_i\sigma_j\right)$. For the sake of convenience the correlation moments and coefficients are typically written in a matrix form:

$$\vec{K}_{XY} = \begin{vmatrix} K_{11} & K_{12}...K_{1n} \\ K_{21} & K_{22}...K_{2n} \\ .... \\ K_{n1} & K_{n2}...K_{nn} \end{vmatrix}; \quad \vec{R}_{XY} = \begin{vmatrix} 1 & R_{12}...R_{1n} \\ R_{21} & 1...R_{2n} \\ .... \\ R_{n1} & R_{n2}...1 \end{vmatrix}$$

2.1.1.27 Functional Transform of Two Random Variables

If random variables X_1, X_2 have to be transformed into random variables Y_1, Y_2, and functions

$$Y_1 = g_1\left(X_1, X_2\right) \quad Y_2 = g_2\left(X_1, X_2\right)$$

are known (g_1, g_2 are deterministic functions), probability density functions are linked as:

$$f_2(y_1, y_2) = \frac{d^2 F_2(y_1, y_2)}{dy_1 \, dy_2}$$

$$f_2(y_1, y_2) = f_2[h_1(y_1, y_2), h_2(y_1, y_2)] \cdot \left| \frac{d(x_1, x_2)}{d(y_1, y_2)} \right|$$

if the reciprocal functions $X_1 = h_1(Y_1, Y_2)$, $X_2 = h_2(Y_1, Y_2)$ are single-valued. If ambiguity originates several branches h_i, the sum for each branch has to be taken in the previous formula for $f_2(y_1, y_2)$.

The common simulation task is to find *pdf* for two random variables X_1 and X_2 when the following is known:

1) mutual *pdf* $f_2(x_1, x_2)$
2) function $Y = Y_1 = g_1(X_1, X_2)$
3) function $Y_2 = g_2(X_1, X_2) = X_2$ (or X_1)

If the reciprocal function $X_1 = h_1(Y_1, Y_2) = h(Y_1, Y_2)$ is a single-valued one, then:

$$f_1(y) = \int_{-\infty}^{\infty} f_2(y_1, y_2) \, dy_2 = \int_{-\infty}^{\infty} f_2[h(y, y_2), y_2] \cdot \left| \frac{dh(y, y_2)}{dy} \right| dy_2$$

The probability distribution functions of sum, difference, product and quotient of two random variables are given in Appendix 18.

2.1.1.28 The Moments of Transformed Variables

The moments of the variables Z_1, Z_2 that are the functions of two random variables X, Y

$$Z_1 = g_1(X, Y) \quad Z_2 = g_2(X, Y)$$

can be found directly from mutual *pdf* $f_2(x, y)$:

$$m_{Z_k} = M(Z_k) = \sum_i \sum_j g_k(x_i, y_j) \cdot p_{ij} \qquad \text{(discrete variable)}$$

$$m_{Z_k} = M\left(Z_k\right) = \int\limits_{-\infty}^{\infty} \int\limits_{-\infty}^{\infty} g_k\left(x, y\right) f_2\left(x, y\right) dx dy \qquad \text{(continuous variable)}$$

$$D\left(Z_k\right) = \sum_i \sum_j \left[g_k\left(x_i, y_j\right) - m_{Z_k}\right]^2 \cdot p_{ij} \qquad \text{(discrete variable)}$$

$$D\left(Z_k\right) = \int\limits_{-\infty}^{\infty} \int\limits_{-\infty}^{\infty} \left[g_k\left(x, y\right) - m_{Z_k}\right]^2 \cdot f_2\left(x, y\right) dx dy \qquad \text{(continuous variable)}$$

$$K_{Z_1 Z_2} = \sum_i \sum_j \left[g_1\left(x_i, y_j\right) - m_{Z_1}\right]\left[g_2\left(x_i, y_j\right) - m_{Z_2}\right] \cdot p_{ij} \qquad \text{(discrete variable)}$$

$$K_{Z_1 Z_2} = \int\limits_{-\infty}^{\infty} \int\limits_{-\infty}^{\infty} \left[g_1\left(x, y\right) - m_{Z_1}\right]\left[g_2\left(x, y\right) - m_{Z_2}\right] \cdot f_2\left(x, y\right) dx dy$$

$$\text{(continuous variable)}$$

The following formulas that are often used in practical applications can be derived based on these expressions:

1) $M\left(C\right) = C \quad M\left(C \cdot X\right) = C \cdot M\left(X\right)$

where C is a nonrandom parameter.

2) $M\left(X \pm Y\right) = M\left(X\right) \pm M\left(Y\right)$

3) $M\left(\sum\limits_{i=1}^{n} a_i X_i + b\right) = \sum\limits_{i=1}^{n} a_i \cdot M\left(x_i\right) + b$

where a_i, b are nonrandom coefficients.

4) $M\left(XY\right) = M\left(X\right) \cdot M\left(Y\right) + K_{xy}$

If X and Y are uncorrelated $\left(K_{xy} = 0\right)$, then $M\left(XY\right) = M\left(X\right) \cdot M\left(Y\right)$

5) $D\left(C\right) = 0$, where C is a nonrandom parameter

6) $D\left(C \cdot X\right) = C^2 \cdot D\left(X\right) \quad \sigma\left(CX\right) = |C| \cdot \sigma\left(X\right)$

7) $D(X \pm Y) = D(X) + D(Y) \pm 2 \cdot K_{xy}$

If X and Y are uncorrelated $(K_{xy} = 0)$, then $D(X \pm Y) = D(X) + D(Y)$

8) $D\left(\sum\limits_{i=1}^{n} a_i \cdot X_i + b\right) = \sum\limits_{i=1}^{n} a_i^2 \cdot D(X_i) + 2\sum\limits_{i<j} a_i a_j K_{ij}$

If $X_1, X_2,...X_n$ are uncorrelated $D\left(\sum\limits_{i=1}^{n} a_i X_i + b\right) = \sum\limits_{i=1}^{n} a_i^2 \cdot D(X_i)$

9) If X and Y are independent, $D(XY) = D(X)D(Y) + m_x^2 \, D(Y) + m_y^2 \, D(X)$

2.1.1.29 The Multidimensional Gaussian Distribution

The multidimensional Gaussian pdf of order n is:

$$f_1(x_1,...x_n) = \frac{1}{\sigma_1 \sigma_2 ... \sigma_n \sqrt{(2\pi)^n \, D}} \times \exp\left\{-\frac{1}{2D}\sum_{i=1}^{n}\sum_{k=1}^{n} D_{ik} \frac{x_i - a_i}{\sigma_i} \cdot \frac{x_k - a_k}{\sigma_k}\right\}$$

$$D = \begin{vmatrix} 1 & R_{12}...R_{1n} \\ R_{21} & 1...R_{2n} \\ \cdots\cdots\cdots\cdots \\ R_{n1} & R_{n2}...1 \end{vmatrix}, \quad R_{ik} = R_{ki}$$

For the two-dimensional case ($n=2$):

$$f_2(x_1, x_2) = \frac{1}{2\pi \, \sigma_1 \sigma_2 \sqrt{1 - R^2}}$$

$$\times \exp\left\{-\frac{1}{2(1-R^2)}\left[\frac{(x_1 - a_1)^2}{\sigma_1^2} - 2R\frac{(x_1 - a_1)(x_2 - a_2)}{\sigma_1 \sigma_2} + \frac{(x_2 - a_2)^2}{\sigma_2^2}\right]\right\}$$

2.1.1.30 The Modulus and Phase of a Random Vector

There is a special type of functional transform that is of major interest in radio engineering applications:

$$Z = \sqrt{X^2 + Y^2} \quad \Phi = \arctan\left(\frac{Y}{X}\right)$$

The reciprocal transformation is:

$$X = Z \cos \Phi \quad Y = Z \sin \Phi$$

This transform is used to describe the pdf of the modulus Z and phase Φ of a random vector with random Cartesian coordinates X, Y. The one-dimensional probability density functions are:

$$f_1(z) = z \int_0^{2\pi} f_2(z \cos \varphi, z \sin \varphi) d\varphi, \quad z > 0$$

$$f_1(\varphi) = \int_0^{\infty} z \cdot f_2(z \cos \varphi, z \sin \varphi) dz, \quad 0 \le \varphi \le 2\pi$$

If the coordinates X and Y are independent Gaussian random variables with parameters (a, σ) and (b, σ), the modulus of the random vector follows a Ricean pdf:

$$f_1(z) = \frac{z}{\sigma^2} e^{-\frac{z^2 + \alpha^2}{2\sigma^2}} \cdot I_0\left(\frac{\alpha z}{\sigma^2}\right) \quad z > 0$$

$$\alpha = \sqrt{a^2 + b^2}$$

$I_0(u)$ is a Bessel function of zero order:

$$I_0(u) = \frac{1}{2\pi} \int_{-\varphi_0}^{2\pi - \varphi_0} e^{-ju\cos t} dt$$

The phase is uniformly distributed within the interval $(-\pi, \pi)$:

$$f_1(\varphi) = \frac{1}{2\pi} \quad |\varphi| \le \pi$$

For $a=b=0$ a Ricean pdf $f_1(z)$ reduces to a Rayleigh pdf:

$$f_1(z) = \frac{z}{\sigma^2} \cdot e^{-z^2/2\sigma^2}$$

2.1.2 Simulation Algorithms

2.1.2.1 Common Simulation Algorithms

In practice, the majority of computer simulation tasks dealing with stochastic modeling involve the generation of random variables with specified probability distribution laws.

Definition. *Random variable simulation is the process of obtaining sample values* $z_1,...z_n$ *of a random variable* $Z = \xi$.

The values $z_1,...z_n$ are statistically independent and all have the same probability distribution function equivalent to that of the variable Z.

Uniform distribution is the basic distribution used to generate random variables with different probability distribution functions. All modern programming packages have system-supplied library routines to generate the random variable $\xi = \{z_u\}$ uniformly distributed within the range (0,1). The algorithms to simulate random variables with other distribution laws are typically based on the functional transformation of z_u. Algorithms to simulate random variables with the most common distribution laws are cited in Appendix 19. Some common simulation approaches are considered in subsequent paragraphs. Many additional algorithms are available in [6–12].

2.1.2.2 Simulation of Discrete Random Variables

A. The method of reciprocal function. If $\xi = \{x_u\}$, then a random variable η can be simulated based on the transformation $\eta = F_\eta^{-1}(\xi)$, where F_η^{-1} is the function reciprocal to the probability distribution function F_η of the random variable η:

$$F_\eta(y) = F_1(y) = \sum_{j=1}^{m} p_j$$

where $y_1 \le y_2 \le ...y_j \le ...$ are possible values of the random variable η with probabilities $p_1, p_2,...p_j$. The simulation algorithm in this case follows the rule:

if $\xi < p_1$, then $\eta = y_1$, else

if $\xi < p_1 + p_2$, then $\eta = y_2$, else

....

if $\xi < \sum_{j=1}^{m} p_j$, then $\eta = y_m$, else

....

B. The method of direct simulation. In some cases the discrete variable can be simulated based on the direct simulation of the sequence of the random events $A_i = (\xi_i \leq p)$, $i = 1, 2, ...n$. If we use the expression from the previous section N times and calculate the number of the events A_i that have occurred, ·the samples of the random variable η with binomial pdf may be obtained.

2.1.2.3 Simulation of Continuous Random Variables

The algorithms used to simulate a continuous variable are usually based on some functional transforms of the random variables with standard distribution laws. The following notations are used hereinafter. The random variables are denoted with letters of the Greek alphabet: ξ, ζ, η, etc., and corresponding samples that can occur as the result of a statistic test (nonrandom values) are denoted with letters of the Roman alphabet: x, y, z, etc., with corresponding indices (u for uniform pdf, N for Gaussian (normal) pdf, etc.). The typical distribution laws and its parameters are denoted as follows: $UN(a, b)$ stands for uniform pdf at interval (a, b), and $N(a, b)$ stands for normal pdf with mean a and variance b. Typically, $a = 0$, $b = 1$. The fact that the random variable has the specified distribution is denoted as $\xi \in UN(a,b)$, or $\xi \in N(a,b)$. Typically, the following notations are used:

1) for a uniform pdf: $\gamma = \{x_u\}$, $\gamma \in UN(0,1)$

2) for a Gaussian pdf: $\varepsilon = \{x_N\}$, $\varepsilon \in N(0,1)$

Other probability density functions are noted specifically if used, e.g., $\xi = \{z_\Gamma\}$, $\xi \in \Gamma(\alpha, \beta)$, where Γ is the gamma pdf.

The algorithms to simulate continuous variables with the specified distribution functions are usually based on some functional transform of uniformly or normally distributed variables γ, ε:

$$\xi = \text{function}(\gamma), \text{ or } \xi = \text{function}(\varepsilon)$$

that can be written also as:

$$y = \text{function}(x_u), \text{ or } y = \text{function}(x_N)$$

Basic expressions for these transforms are given in Appendix 19.

A. *Variables with uniform pdf.* The algorithms to simulate a continuous variable $\gamma \in UN(0,1)$ are the fundamental building blocks in theory of computer simulation since the majority of the simulation algorithms are derived from them. Actually, the computer-generated sequence of numbers $\gamma = \{x_{u_k}\}$ $k = 1, 2,...n$, is a pseudo-random one, and sometimes this fact is blamed for all results in doubt in stochastic modeling. Basic computer algorithms to generate a random variable $\gamma \in UN(0,1)$ are well described in the literature [9], and they are the part of standard library routines in all modern programming packages.

B. *Variables with a Gaussian pdf.* The random variable $\varepsilon_N = \{x_n\}$ with pdf

$$f_1(x) = \frac{1}{\sigma\sqrt{2\pi}} e^{-\frac{(x-a)^2}{2\sigma^2}}, \quad -\infty < x < \infty$$

and distribution function

$$P\{\varepsilon_N \leq x\} = \Phi\left(\frac{x-a}{\sigma}\right)$$

$$\Phi(y) = \int_{-\infty}^{y} \frac{1}{\sqrt{2\pi}} e^{-y^2/2} dy$$

can be obtained:

$$\varepsilon_N = a + b \cdot \varepsilon$$

where $\varepsilon \in N(0,1)$. Basic algorithms that can be used to simulate the random variable ε via transformation of a variable with uniform pdf are given in Appendix 20.

C. *The method of reciprocal function.* If $f_1(y)$ and $F_1(y)$ are probability density and probability distribution functions of a continuous random variable y, and $F_1^{-1}(y)$ is the function reciprocal to $F_1(y)$, then the random variable

$$y = F_1^{-1}(x_u) \tag{2.1}$$

has the specified distribution law $f_1(y)$ if the random variable $\xi = \{x_u\}$ is uniformly distributed on the interval $(0,1)$:

$$\xi = \{x_u\} \in UN(0,1)$$

Equation (2.1) involves the reciprocal function F_1^{-1} that often cannot be expressed in a closed-loop analytical form. In this case the function F_1^{-1} has to be approximated by another function that can be described analytically. The simulation algorithms for some variables with probability distribution functions that allow closed-loop forms for reciprocal functions are given in Appendix 21.

   D. *Neumann's method.* This method is applied to simulation of random variables with truncated distribution functions (the variables that do not exceed some interval a, b), or to the variables that have distribution functions which can be approximated by the truncated ones. The principle of the method is as follows:

   1) a pair of random numbers $x_1, x_2 \in UN(0,1)$ is generated;
   2) the transformed pair is formed as:

$$y_1 = a + (b-a) \cdot x_1 \quad y_2 = f_{max} \cdot x_2$$

where (a, b) is an interval of possible values of the random variable y with the specified pdf $f_1(y)$; f_{max} is maximum of the function $f_1(y)$;

   3) if the following inequality is met:

$$y_2 \le f_1(y_1)$$

then the number y_2 is taken as a sample value from this pair. The pairs failing to meet the above-mentioned inequality are disregarded.

   E. *The method of superposition.* If the probability density function is:

$$f_1(y) = \sum_{i=1}^{n} p_i \cdot f_i(y), \quad p_i > 0, \quad \sum_{i=1}^{n} p_i = 1$$

then simulation can be done in two phases. First, the discrete random variable with the distribution

$$\begin{pmatrix} 1, & 2, & \dots & n \\ p_1, & p_2, & \dots & p_n \end{pmatrix}$$

i.e., having values $1, 2, ... n$ with the probabilities p_k, $k = \overline{1, n}$, is simulated.

Second, after the value k is determined, the value with pdf $f_k(y)$ is simulated. The obtained value is taken as the sample y of the random variable ξ.

This simulation algorithm is convenient when combination of different probability distribution functions is investigated. It is also applicable to the distribution expressed as:

$$f(y) = \int_{-\infty}^{\infty} f(y|\lambda) \cdot \varphi(\lambda) d\lambda$$

where $\varphi(\lambda)$ is the pdf of some random variable η, and $f(y|\lambda)$ is the *pdf* for ξ under the condition that $\eta = \lambda$. Examples for distribution laws that are combinations of two Cauchy distributions with different shifts $b = \pm\beta$, and two Gaussian distributions with equal variances α^2 and means $\pm\beta$, $p_1 = p_2 = 0.5$, are given in Appendix 21.

F. *Simulation by means of gamma distribution.* The random variables with desirable distribution laws can be simulated based on standard distributions that differ from uniform or Gaussian. One of them is gamma distribution. Some typical simulation algorithms are given in Appendix 22.

The random variable ξ is described by gamma distribution function $(\xi \in \Gamma(\alpha, \beta))$ when:

$$f_1(x) = \frac{\beta^\alpha}{\Gamma(\alpha)} \cdot x^{\alpha-1} \cdot e^{-\beta x}, \quad x > 0$$
$$0, \quad x \leq 0$$

where $\alpha, \beta > 0$

$\Gamma(\alpha) = \int_0^\infty t^\alpha e^{-t} dt$ is a gamma function.

When $\beta = 1/2$, $\alpha = n/2$ the pdf $f_1(x)$ is reduced to:

$$x_n^2 = \sum_{k=1}^n \varepsilon_k^2, \quad \varepsilon_k \in N(0,1)$$

The corresponding random variable

$$x_n^2 = \sum_{k=1}^{n} \varepsilon_k^2, \quad \varepsilon_k \in N(0,1)$$

The corresponding random variable

$$\xi = \bar{\xi}/\beta, \quad \bar{\xi} \in \Gamma(\alpha,1)$$

and pdf for $\bar{\xi}$ when $x > 0$

$$f_1(x) = \frac{x^{\alpha-1}}{\Gamma(\alpha)} e^{-x}$$

For integer $\alpha = n$:

$$\bar{\xi} = -\ln\left(\prod_{k=1}^{n} \gamma_k\right), \quad \gamma_k \in UN(0,1) \tag{2.2}$$

For nonlinear transform $\eta = \xi^\mu, \quad \mu \neq 0, \quad \xi \in \Gamma(\alpha, \beta)$:

$$f_1(x) = \frac{\beta^\alpha}{|\mu| \Gamma(\alpha)} \cdot x^{\frac{\alpha}{\mu}-1} \cdot e^{-\beta \cdot x^{1/\mu}}$$

When $\mu = \dfrac{1}{2}$:

$$f_1(x) = \frac{2\beta^\alpha}{\Gamma(\alpha)} \cdot x^{2\alpha-1} \cdot e^{-\beta x^2}$$

which corresponds to the Rayleigh pdf

$$f_1(x) = \frac{x}{\alpha^2} \exp\left(-\frac{x^2}{2\alpha^2}\right), \quad x > 0, \alpha > 0$$

The random variable $\xi = \alpha\sqrt{2\bar{\xi}}, \quad \bar{\xi} \in \Gamma(1,1)$ when $n = 1$ in (2.2) leads to the standard simulation algorithm:

$$\xi = \alpha\sqrt{-2\ln\gamma}$$

or in terms of nonrandom samples:

$$y = \alpha\sqrt{-2\ln x_u}$$

Fisher probability distribution function (F-distribution) and several other similar distributions can be also derived by means of gamma-distribution. If $\xi_1 \in \Gamma(\alpha_1, \beta_1)$ and $\xi_2 \in \Gamma(\alpha_2, \beta_2)$ are independent random variables, the pdf of the variable $\xi = \xi_1/\xi_2$ is equal to:

$$f_1(x) = \begin{cases} \dfrac{\Gamma(\alpha_1 + \alpha_2)}{\Gamma(\alpha_1)\,\Gamma(\alpha_2)} \cdot \beta^{\alpha_1} \dfrac{x^{\alpha_1 - 1}}{(1 + \beta x)^{\alpha_1 + \alpha_2}}, & x \geq 0 \\ 0 & x < 0 \end{cases}$$

which is a general form for the F-distribution $(\beta = \beta_1/\beta_2)$.

2.2 Random Processes

2.2.1 Statistical Description

2.2.1.1 Basic Definitions

Definition. A random process is a random function of a single argument – time.
A random process is considered to be determined within the interval $(0, T)$ if for any n and for any moments of time $t_1, t_2, ... t_n$ the n-dimensional pdf $f_n(x_1, x_2, ... x_n; t_1, t_2, ... t_n)$ or n-dimensional characteristic function $\theta_n(j\vartheta_1, ... j\vartheta_n; t_1, ... t_n)$ is defined.

2.2.1.2 Probability Distribution Function

The pdf of a random process has to meet the following requirements:

1) $f_n(x_1, ... x_n; t_1, ... t_n) \geq 0$

2) $\displaystyle\int_{-\infty}^{\infty} ... \int_{-\infty}^{\infty} f_n(x_1, ... x_n; t_1, ... t_n) = 1$

3) $f_n(x_1,...x_n;t_1,...t_n)$ has to be symmetrical with respect to its arguments $x_1,...x_n$, i.e., it is not supposed to change when position of arguments change;

4) for any $m < n$

$$f_m(x_1,...x_m;t_1,...t_m) = \int\limits_{-\infty}^{\infty}...\int\limits_{-\infty}^{\infty} f_n(\xi_1,...,\xi_m,\xi_{m+1},...\xi_n)d\,\xi_{m+1}...d\,\xi_n$$

2.2.1.3 The Moments

The most important statistical parameters of a random process are the following moments:

1) initial moment of the first order

$$m_1(t) = M\,[\xi(t)] = \int\limits_{-\infty}^{\infty} x\,f_1(x,t)dx = m_\xi(t), \text{ i.e. mathematical expectation (mean)}$$

2) two-dimensional initial moment of the second order:

$$m_{1,1}(t_1,t_2) = M\,[\xi(t_1)\cdot\xi(t_2)] = \int\limits_{-\infty}^{\infty}\int\limits_{-\infty}^{\infty} x_1\,x_2\,f_2(x_1,x_2;t_1,t_2)dx_1\,dx_2 = B_\xi(t_1,t_2)$$

i.e., covariation function;

3) two-dimensional central moment of the second order:

$$M_{1,1}(t_1,t_2) = M\,\{[\xi(t_1) - m_\xi(t_1)][\xi(t_2) - m_\xi(t_2)]\}$$

$$= \int\limits_{-\infty}^{\infty}\int\limits_{-\infty}^{\infty}[x_1 - m_\xi(t_1)]\times[x_2 - m_\xi(t_2)]f_2(x_1,x_2;t_1,t_2)dx_1\,dx_2 = K_\xi(t_1,t_2)$$

i.e., correlation function;

If $t_1 = t_2 = t$, $K_\xi(t,t) = D_\xi(t)$ is the variance, and $\sqrt{D_\xi(t)} = \sigma_\xi(t)$ is the root-mean-square (rms) deviation.

2.2.1.4 A Complex Random Process

For a complex random process

$$\xi(t) = \eta(t) + j \cdot \xi(t)$$

where $\eta(t)$ and $\xi(t)$ are real random processes; the major moments are:

$$m_\xi(t) = M\left[\xi(t)\right] = m_\eta(t) + jm_\xi(t)$$

$$B_\xi(t_1, t_2) = M\left[\xi(t_1) \cdot \xi^*(t_2)\right] = K_\xi(t_1, t_2) + m_\xi(t_1) \cdot m_\xi^*(t_2)$$

$$K_\xi(t_1, t_2) = M\left\{\left[\xi(t_1) - m_\xi(t_1)\right]\left[\xi^*(t_2) - m_\xi(t_2)\right]\right\} = B_\xi(t_1, t_2) - m_\xi(t_1) \cdot m_\xi^*(t_2)$$

2.2.1.5 A Stationary Random Process

Definition. *A stationary random process is a random process having statistical parameters that are invariant with respect to time shift.*
For a stationary random process:

$$m_\xi(t) = M\left[\xi(t)\right] = \int_{-\infty}^{\infty} x\, f_1(x)\, dx = m_\xi$$

$$B(t_1, t_2) = M\left[\xi(t_1)\xi(t_2)\right]$$

$$= \int_{-\infty}^{\infty}\int_{-\infty}^{\infty} x_1 x_2 f_2(x_1, x_2)\, dx_1\, dx_2 = K(t_2 - t_1) + m_\xi^2 = B_\xi(\tau) \quad \tau = t_2 - t_1$$

$$K(t_1, t_2) = M\left\{\left[\xi(t_1) - m_\xi\right]\left[\xi(t_2) - m_\xi\right]\right\}$$

$$= \int_{-\infty}^{\infty}\int_{-\infty}^{\infty}(x_1 - m_\xi)(x_2 - m_\xi) f_2(x_1, x_2)\, dx_1\, dx_2 = B_\xi(\tau) - m_\xi^2 = K_\xi(\tau)$$

The correlation function (sometimes also termed the autocorrelation function) of the stationary random processes $\xi(t)$ has the following properties:

1) $K_\xi(\tau) = K_\xi(-\tau)$

2) $\left|K_\xi(\tau)\right| \le K_\xi(0) = D_\xi$

3) $\lim\limits_{\tau \to \infty} K_\xi(\tau) = 0$

The normalized correlation function is

$$R_\xi(\tau) = \frac{K_\xi(\tau)}{K_\xi(0)} = \frac{K_\xi(\tau)}{D_\xi} \le 1$$

or in other terms

$$K_\xi(\tau) = D_\xi \cdot R_\xi(\tau).$$

2.2.1.6 The Mutual Correlation Function

The mutual correlation function describes the statistical linkage between two random processes $\xi(t)$ and $\eta(t)$:

$$K_{\xi\eta}(t_1, t_2) = M\left\{ \left[\xi(t_1) - m_\xi\right]\left[\eta(t_2) - m_\eta\right]\right\}$$

$$K_{\eta\xi}(t_1, t_2) = M\left\{ \left[\eta(t_1) - m_\eta\right]\left[\xi(t_2) - m_\xi\right]\right\}$$

If mutual correlation function depends on the difference $\tau = t_2 - t_1$ only, the processes $\xi(t)$ and $\eta(t)$ are termed stationary-linked ones, and in this case:

$$K_{\xi\eta}(t_1, t_2) = K_{\xi\eta}(\tau) = K_{\eta\xi}(-\tau)$$

$$R_{\xi\eta}(\tau) = \frac{K_{\xi\eta}(\tau)}{\sqrt{D_\xi D_\eta}} = \frac{K_{\xi\eta}(\tau)}{\sigma_\xi \sigma_\eta}$$

2.2.1.7 Correlation Interval

Typically, in radio engineering applications the normalized correlation functions are monotonous damping or oscillating-damping functions of τ. The degree of correlation in this case is described by the correlation interval τ_c:

$$\tau_c = \frac{1}{2} \int\limits_{-\infty}^{\infty} |R(\tau)| \, d\tau = \int\limits_{0}^{\infty} |R(\tau)| \, d\tau$$

Sometimes τ_c is defined simply as the time shift for which the normalized correlation function drops below some specified value:

$$|R(\tau)| \le R_0, \quad \text{e.g.,} \quad \frac{|R(\tau_c)|}{R(0)} \le 0.1$$

2.2.1.8 The Spectral Density

The spectral density (power spectrum) $S_\xi(\omega)$ of the stationary random process $\xi(t)$ and its covariation function $B_\xi(\tau)$ are linked by a Fourier transform pair:

$$S_\xi(\omega) = \int\limits_{-\infty}^{\infty} B_\xi(\tau) e^{-j\omega\tau} \, d\tau$$

$$B_\xi(\tau) = \frac{1}{2\pi} \int\limits_{-\infty}^{\infty} S_\xi(\omega) e^{j\omega\tau} \, d\omega$$

Correspondingly, for the centered random process $\xi^0(t) = \xi(t) - m_\xi$

$$S_{\xi^0}(\omega) = \int\limits_{-\infty}^{\infty} K_\xi(\tau) e^{-j\omega\tau} \, d\tau = S_\xi(\omega) - 2\pi m_\xi^2 \delta(\omega)$$

$$K_\xi(\tau) = \frac{1}{2\pi} \int\limits_{-\infty}^{\infty} S_{\xi^0}(\omega) e^{j\omega\tau} \, d\omega$$

Because $B(\tau)$ and $K(\tau)$ are the even functions of argument τ:

$$S_\xi(\omega) = 2 \int\limits_{0}^{\infty} B_\xi(\tau) \cos \omega\tau \, d\tau$$

$$S_{\xi^0}(\omega) = 2 \int\limits_{0}^{\infty} K_\xi(\tau) \cos \omega\tau \, d\tau$$

$$B_\xi(\tau) = \frac{1}{\pi} \int_0^\infty S_\xi(\omega) \cos \omega\tau\, d\omega$$

$$K_\xi(\tau) = \frac{1}{\pi} \int_0^\infty S_{\xi^0}(\omega) \cos \omega\tau\, d\omega$$

Because $S_{\xi^0}(\omega)$ and $S_\xi(\omega)$ differ only in the constant factor $2\pi m_\xi^2 \cdot \delta(\omega)$, hereinafter we will use only the $S(\omega) = S_\xi(\omega)$ representation $\left(S_{\xi^0}(\omega) = S_\xi(\omega) \text{ if } m_\xi = 0\right)$. The main features of the spectral density of the stationary random process are as follows:

1) $S(\omega) \geq 0$ for any ω ;

2) for real random processes

$$S(\omega) = S(-\omega)$$

Sometimes instead of the double-sided spectrum the one-sided "physical" spectral density is used:

$$S(f) = S(\omega) + S(-\omega) = \begin{cases} 2S(\omega) & f \geq 0 \\ 0 & f < 0 \end{cases}$$

which gives the expressions:

$$S(f) = 4\int_0^\infty B(\tau) \cos 2\pi f\tau\, d\tau$$

$$B(\tau) = \int_0^\infty S(f) \cos 2\pi f\tau\, df$$

Note: when the random processes are simulated based on their spectral density, it is very important to make sure that the proper normalization is used since different reference sources may provide different definitions of the spectral density as in the examples above.

The most common numerical parameters of the spectrum are:

1) effective spectrum width: $\Delta\omega_e = \dfrac{1}{S(0)} \displaystyle\int_{-\infty}^{\infty} S(\omega)d\omega$

2) mean frequency: $m_{1\omega} = M\left[\omega\right] = \dfrac{2}{\sigma^2} \displaystyle\int_{0}^{\infty} \omega S(\omega)d\omega$

3) mean-square frequency: $m_{2\omega} = M\left[\omega^2\right] = \dfrac{2}{\sigma^2} \displaystyle\int_{0}^{\infty} \omega^2 S(\omega)d\omega$

4) root-mean-square width:

$$\sigma_\omega = \sqrt{M\left[\omega^2\right] - M^2\left[\omega\right]} = \sqrt{\dfrac{2}{\sigma^2} \int_{0}^{\infty} (\omega - m_\omega)^2 S(\omega)d\omega}$$

The correspondence between some typical correlation functions and its spectral densities is given in Appendix 23.

2.2.1.9 A Derivative of a Random Process

The random process $\xi(t)$ has a derivative $\dot{\xi}(t) = d\xi(t)/dt$ if

$$\lim_{T \to 0} M\left\{\left[\frac{\xi(t+T) - \xi(t)}{T} - \dot{\xi}(t)\right]^2\right\} = 0$$

The mathematical expectation $m_{\dot{\xi}}(t)$, covariation $B_{\dot{\xi}}(t_1, t_2)$, and correlation function $K_{\dot{\xi}}(t_1, t_2)$ are defined as:

$$m_{\dot{\xi}}(t) = M\left[\dot{\xi}(t)\right] = \frac{dm_{\xi}(t)}{dt}$$

$$B_{\dot{\xi}}(t_1, t_2) = M\left[\dot{\xi}(t_1) \cdot \dot{\xi}(t_2)\right] = \frac{\partial^2 B_{\xi}(t_1, t_2)}{\partial t_1 \partial t_2}$$

$$K_{\dot{\xi}}(t_1, t_2) = M\left[\dot{\xi}^0(t_1) \cdot \dot{\xi}^0(t_2)\right] = \frac{\partial^2 K_{\xi}(t_1, t_2)}{\partial t_1 \partial t_2}$$

For the stationary random process:

$$m_{\xi}(t)=0$$

$$B_{\dot{\xi}}(\tau)=K_{\dot{\xi}}(\tau)=-\frac{d^2 B_{\xi}(\tau)}{d\tau^2}=-\frac{d^2 K_{\xi}(\tau)}{d\tau^2}$$

$$S_{\dot{\xi}}(\omega)=\omega^2 \cdot S_{\xi^0}(\omega)$$

The mutual covariation $B_{\xi\dot{\xi}}(t_1,t_2)$ and correlation $K_{\xi\dot{\xi}}(t_1,t_2)$ functions are:

$$B_{\xi\dot{\xi}}(t_1,t_2)=M\left[\xi(t_1)\cdot\dot{\xi}(t_2)\right]=\frac{dB_{\xi}(t_1,t_2)}{dt_2}$$

$$K_{\xi\dot{\xi}}(t_1,t_2)=M\left[\xi^0(t_1)\cdot\dot{\xi}^0(t_2)\right]=\frac{dK_{\xi}(t_1,t_2)}{dt_2}$$

For the stationary random process:

$$B_{\xi\dot{\xi}}(\tau)=K_{\xi\dot{\xi}}(\tau)=-B_{\dot{\xi}\xi}(\tau)=-K_{\dot{\xi}\xi}(\tau)=\frac{dB_{\xi}(\tau)}{d\tau}=\frac{dK_{\xi}(\tau)}{d\tau}$$

$$S_{\xi\dot{\xi}}(\omega)=j\omega S_{\xi^0}(\omega)$$

Because $dK_{\xi}(0)/d\tau=0$, the stationary process $\xi(t)$ and its derivative $\dot{\xi}(t)$ are always uncorrelated in any coincident moments of time $(\tau=0)$.

The general expressions for derivatives of nth order $\xi^{(n)}(t)$ are given in Appendix 24.

2.2.1.10 An Integral of a Random Process

A random function $\xi(t)$ can be integrated, i.e., the random function

$$\eta(t)=\int_0^t \xi(u)\,du$$

exists, if its correlation function can be integrated:

$$K_\eta(t_1,t_2) = \int_0^{t_1} \int_0^{t_2} K_\xi(u,\vartheta)\,du\,d\vartheta$$

For the stationary random process $\xi(t)$:

$$K_\eta(t_1,t_2) = \int_0^{t_1} \int_0^{t_2} K_\xi(\vartheta-u)\,du\,d\vartheta$$

$$D_\eta(t) = \int_0^t (t-\tau)[K_\xi(\tau) + K_\xi(-\tau)]\,d\tau$$

These expressions show that in the general case $\eta(t)$ is not necessarily a stationary process even when $\xi(t)$ is a stationary one. For the real stationary random process $\xi(t)$:

$$D_\eta(t) = 2\int_0^t (t-\tau)K_\xi(\tau)\,d\tau$$

For any random function $\xi(t)$:

$$m_\eta(t) = \int_0^t m_\xi(u)\,du$$

2.2.1.11 A Gaussian Random Process

A Gaussian random process is the most common process widely used in simulation of radio systems and circuits. The probability density function and characteristic function for this process are given in Appendix 25. There Δ is the determinant of nth order:

$$\Delta = \left\| K_\xi(t_\mu, t_\gamma) \right\|$$

where $\Delta_{\mu\gamma}$ is a cofactor of an element $K_\xi(t_\mu, t_\gamma)$ in this determinant, and $m_\xi(t_\mu)$ is a mathematical expectation of the random variable $\xi_\mu = \xi(t_\mu)$. For a stationary process m is mathematical expectation, $\sigma^2 = D = R(0)$ is a variance of the process $\xi(t)$, $D = \|R(\tau_{\mu\gamma})\|$ is a determinant of the order n composed from the correlation coefficients

$$R(\tau_{\mu\gamma}) = R(|t_\mu - t_\gamma|) = \frac{K_\xi(|t_\mu - t_\gamma|)}{\sigma^2}$$

2.2.1.12 A Markovian Random Process

The fundamental role in random process simulation belongs to a Markovian random process. There are four basic types of Markovian processes depending on whether the random variable $\xi(t)$ and its argument t are discrete or continuous: the discrete process and discrete time (Markovian chain), the continuous process and discrete time (Markovian sequence), discrete process and continuous time (discrete Markovian process) and continuous process and continuous time (continuous Markovian process).

Definition. *The random process is called a Markovian one if for any n moments of time $t_1 < t_2 < ...t_n$ at the interval (0, T) the conditional distribution function of $\xi(t_n)$ depends only on $\xi(t_{n-1})$ under the conditions of fixed $\xi(t_1), \xi(t_2)...\xi(t_{n-1})$, i.e.,*

$$P\left\{\xi(t_n) \le \xi_n \middle| \xi(t_1) = \xi_1, ...\xi(t_{n-1}) = \xi_{n-1} \right\} = P\left\{\xi(t_n) \le \xi_n \middle| \xi(t_{n-1}) = \xi_{n-1} \right\}$$

In other words: the status of the Markovian random process in the future (t_{n+1}) depends only on its status in the present (t_n) and does not depend on the status in the past (t_{n-1}). This assumption makes it possible to reduce considerably the number of computations in simulation algorithms. That is why this class of random processes is widely used in computer simulation of radio engineering systems.

In practical applications, the process is considered to be the Markovian one if the three following conditions are met:

1) The process is a stationary one;
2) The process is a Gaussian one;
3) Spectral density of the process is a fractional-rational function of frequency:

$$S(\omega) = \frac{\left|P_m(j\omega)\right|^2}{\left|Q_n(j\omega)\right|^2}, \quad m < n$$

where P_m, Q_n are some polynomials:

$$P_m(u) = \beta_0 u^m + \beta_1 u^{m-1} + ... + \beta_m$$

$$Q_n(u) = \alpha_0 u^n + \alpha_1 u^{n-1} + ... + \alpha_n$$

and α_i, β_i are some real coefficients.

The random process $\xi(t)$ that is the stationary Gaussian random process with the correlation function

$$K_\xi(t) = \sigma^2 \cdot e^{-\alpha|\tau|}$$

is always a Markovian one.

2.2.1.13 A Random Flow

Another important class of random processes is a point random process (a random flow)

Definition. *A point random process is a sequence of random events (requests, failures, etc.) with the random moments* $t_1, t_2, t_3 ...$ *of its occurrence.*

There are two basic ways to describe a point random process. The first one considers a random sequence of points in time. The second one considers the integer random process $N(t)$ that is the number of events (points) at the half-interval $(0, t)$. The sequence of points can be described by n-dimensional pdf $f_n(\tau_1, ... \tau_n)$ of intervals between the points:

$$\tau_i = t_i - t_{i-1} > 0, \quad i = 1, 2, 3, ...$$

or distribution function $F_n(t_1, ..., t_n)$ of the coordinates of the points:

$$F_n(t_1, ... t_n) = f_n(t_1, t_2 - t_1, ..., t_n - t_{n-1})$$

$$f_n(\tau_1, \tau_2, ..., \tau_n) = F_n(\tau_1, \tau_1 + \tau_2, ..., \tau_1 + \tau_2 + ... \tau_n)$$

For the process $N(t)$, if t_i is a coordinate of the ith point that occurred $t_1 < t_2 < ..t_i < ...,$ then $N(t) = 0$ only if $\tau_i > t$, and $N(t) < n$ if $\tau_1 + \tau_2 + ...\tau_n > t$:

$$P\{N(t) = 0\} = P\{\tau_1 > t\}$$

$$P\{N(t) < n\} = P\{\tau_1 + \tau_2 + ...\tau_n > t\}   n = 1,2,3,...$$

2.2.1.14 Poisson Flow

Poisson random flow is the common flow that is typically used in simulation tasks. The Poisson flow $N(t)$, $0 < t < \infty$, has three basic features:

1) The flow is an ordinary one, i.e., probability that more than one event occurs at any small interval Δt is of a higher order being infinitesimal than Δt :

$$P\{N(t + \Delta t) - N(t) = 1\} = P\{N(\Delta t) = 1\} = \gamma \cdot \Delta t + o(\Delta t)$$

$$P\{N(t + \Delta t) - N(t) > 1\} = P\{N(\Delta t) > 1\} = o(\Delta t)$$

where γ is a positive value with dimension reciprocal to that of time and $o(t)$ are factors of the series at least an order less than Δt . Thus,

$$P\{N(t + \Delta t) - N(t) = 0\} = P\{N(\Delta t) = 0\} = 1 - \gamma \Delta t + o(\Delta t)$$

2) The flow is a stationary one, i.e., its probabilistic characteristics do not change when all points are shifted along the time axis by an arbitrary value Δ .
3) The flow increments are independent at nonoverlapping time intervals (the absence of aftereffects).

For these conditions the probability $P_k(t)$ that k points will occur at the half-interval $(0, t]$ is covered by Poisson law:

$$P_k(t) = (\gamma t)^k \cdot \frac{e^{-\gamma t}}{k!}, \quad \begin{matrix} k = 0,1,2,... \\ t \geq 0 \end{matrix}$$

The probability that there are no points ($k=0$) at some half-interval τ is

$$P_0(\tau) = e^{-\gamma \tau}$$

and that there is only one point ($k=1$) is

$$P_1(\tau) = \gamma \tau e^{-\gamma\tau}$$

The mathematical expectation m and variance D are equal to each other:

$$m = D = \gamma t$$

Because $\gamma = m/t$, it can be considered to be the average number of points at the unity interval and is termed a flow intensity. The probability distribution function is:

$$f_{tk}(t) = \gamma e^{-\gamma t} \cdot (\gamma t)^{k-1} \cdot \frac{1}{(k-1)!}$$

$$m_{tk} = \frac{k}{\gamma} \quad D_{tk} = \frac{k}{\gamma^2}$$

where t_k is time of kth point occurrence.

The sequence $\tau_i > i = 1,2,3\ldots$ of the time intervals between adjacent points of Poisson flow is a sequence of independent random variables with exponential pdf:

$$f(\tau) = \gamma \cdot e^{-\gamma\tau}, \quad \tau > 0$$

$$m_\tau = 1/\gamma, \quad D_\tau = 1/\gamma^2$$

2.2.1.15 A Random Pulsed Process

Definition. *A random pulsed process* $\xi(t)$ *is a sequence of pulses (in the general case of different shapes) following each other at some time interval.*

If pulse shape is known, the different parameters of the pulse can be random: amplitude A_γ, width τ_γ, time of occurrence t_γ, and so forth. The condition that the pulses do not overlap can be defined as:

$$t_\gamma + \tau_{\gamma+1} \leq t_{\gamma+1}, \quad \gamma = 0,1,2,\ldots$$

Obviously, if a sequence of nonoverlapping pulses passes through a radio system containing inertial circuits, the pulses may overlap at the system output. If pulses do not overlap, the major task typically is to determine the probability distribution function for the parameters of a random pulse: amplitude, pulse width, etc., and to determine corresponding spectral density. For overlapping pulses the resultant random process is formed by superposition of overlapped pulses and in this case stochastic analysis and simulation are much more complicated tasks.

The spectral density of the stationary sequence of the mutually independent nonoverlapping pulses is:

$$S(\omega) = \lim_{T \to \infty} \frac{1}{T} M \left\{ \left| F_T(j\omega) \right|^2 \right\}$$

$$F_T(\omega) = \int_{-\frac{T}{2}}^{T/2} \xi(t) e^{-j\omega t} dt$$

where $F_T(\omega)$ is the spectrum of the truncated random pulsed process $\xi(t)$ at the interval $(-T/2, T/2)$.

Let us assume that truncated function $\xi(t)$ consists of n pulses. If t_γ is the moment of time when the γ th pulse begins, then $-T/2 \le t_0 < t_1 < t_2 < ... < t_{n-1} < t_n \le T/2$. Duration of interval depends on n, $T = T_n$. To denote duration of the interval between two adjacent pulses as $\vartheta_\gamma = t_{\gamma+1} - t_\gamma$, then $T_n = \sum_{\gamma=1}^{n} \vartheta_\gamma$.

The single pulse in the sequence can be denoted $A_\gamma \cdot s(t - t_\gamma, \tau_\gamma)$, where

$$s(t - t_\gamma, \tau_\gamma) = \begin{cases} s_0(t, \tau_\gamma) & t_\gamma \le t \le t_\gamma + \tau_\gamma \\ 0 & t < t_\gamma, \, t > t_\gamma + \tau_\gamma \end{cases}$$

Maximum of the function is $s_0(t, \tau_\gamma) = 1$. Since $s_0(t, \tau_\gamma)$ is a deterministic function, the random nature of the single pulse is defined by the fact that its amplitude A_γ, duration τ_γ and moment of occurrence t_γ are random variables. The spectral density of the single pulse can be found via its spectrum function:

$$F_T(\omega) = \sum_{\gamma=1}^{n} A_\gamma F_1(\omega, \tau_\gamma) e^{-j\omega t_\gamma}$$

where $F_1(\omega, \tau_\gamma) = \int_{-\infty}^{\infty} s_0(t, \tau_\gamma) e^{-j\omega t} dt$

The spectrum functions for some common pulse shapes $s_0(t)$ are given in Appendix 26.

The spectral density of the function $F_1(\omega, \tau_\gamma)$ is:

$$S(\omega) = \frac{1}{m_\vartheta} \left[\begin{array}{c} M\left\{A^2 |F_1(\omega,\tau)|^2\right\} + 2M\left\{AF_1(\omega,\tau)\right\} \times \\ \times M\left\{AF_1^*(\omega,\tau)\right\} \mathrm{Re}\, \dfrac{\theta_\vartheta(\omega)}{1-\theta_\vartheta(\omega)} \end{array} \right] + 2\pi m_\xi^2 \delta(\omega)$$

where $m_\vartheta = M\{\vartheta\} = \int\limits_0^\infty \vartheta f(\vartheta)d\vartheta$ is the mathematical expectation of the interval between pulses;

$m_\xi = M\{\xi(t)\} = \dfrac{1}{m_\vartheta} M\{AF_1(0,\tau)\}$ is the mathematical expectation of the random pulsed process $\xi(t)$; and

$\theta_\vartheta(\omega) = M\left\{e^{j\omega\vartheta}\right\} = \int\limits_0^\infty e^{j\omega\vartheta} \cdot f(\vartheta)d\vartheta$ is a characteristic function of the intervals between pulses.

The spectral densities based on these expressions for some common pulsed random processes used in communications are given in Appendix 27. The probability distribution functions for common amplitude distributions are given in Appendix 28. The spectral densities for the amplitude-pulse modulation case based on the amplitude distribution cited in Appendix 28 and the last expression in Appendix 27 are given in Appendix 29.

Pulse-time modulation is widely used in communication theory. Typically, four basic types of pulse-time modulation are used:

1) *pulse-position modulation* (PPM): the pulses have constant amplitude and pulse width, but their position changes from period to period based on transmitted message content;

2) *single-sided pulse-width modulation* (SSPWM): all pulses start at the time with constant offset ϑ_0, and duration varies within some interval $(0, \tau_m) < \vartheta_0$;

3) *double-sided pulse-width modulation 1* (DSPWM1): the interval between the middle points of two adjacent pulses is the same $(\vartheta_0 = \text{const})$, and pulse durations have random variation;

4) *double-sided pulse-width modulation 2* (DSPWM2): pulse duration and position of the pulse edge varies, but each pulse does not go beyond its repetition interval.

The expressions for spectral densities of the stationary sequences of pulses for the cases of PPM, SSPWM, and DSPWM 1 for different amplitude probability distribution functions are given in Appendix 30.

2.2.2 Simulation Algorithms

2.2.2.1 Basic Approach

To simulate a random process $\xi(t)$ means to obtain the samples of this process $\xi(t) = \{x(t)\}$. Typically, the task of simulation is as follows. Using the specified statistic parameters of the process $\xi(t)$: mean $m_\xi(t)$, variance $D_\xi(t)$, correlation function $K_\xi(t_1, t_2)$ (or correspondingly the spectral density $S_\xi(\omega)$) a researcher needs to come up with the numerical algorithm that makes it possible to obtain the required number of samples $x_k(t)$, $k = 0,1,2,...$, for the process $\xi(t)$.

The main assumption typically is that the process $\xi(t)$ is a stationary Gaussian random process. In this case the model defined by mathematical expectation m_ξ and correlation function $K_\xi(\tau)$ is completely sufficient to describe the process $\xi(t)$. Typically, at the first stage it is assumed that $m_{\xi_0} = 0$, $D_{\xi_0} = 1$, and the process $\xi_0(t)$ with these parameters is simulated.

Then the transform algorithm

$$\xi(t) = m_\xi + \sigma_\xi \cdot \xi_0(t)$$

makes it possible to obtain the process $\xi(t)$ with required statistical parameters.

All methods and algorithms of random process simulation can be divided into two groups: rigorous methods (the basic method being that of recurrent algorithms) and approximate methods (the basic methods being that of formation filter and sliding summation). In the former case the specified correlation function $K_\xi(\tau)$ is reproduced without a methodological error, while approximate methods introduce some error that leads to inequality between specified $K_\xi(\tau)$ and simulated $\hat{K}_\xi(\tau)$. The most reliable way to estimate this error is to perform statistical analysis of obtained samples $\xi(t) = \{x(t)\}$ (see Section 2.4).

2.2.2.2 The Recurrent Methods

The most common recurrent method is based on the algorithm:

$$\xi_k = \sum_{j=0}^{l} a_j \cdot \varepsilon_{k-j} - \sum_{j=1}^{m} b_j \cdot \xi_{k-j}, \quad k = m, m+1,\ldots,$$

Here l, m, a_j, b_j are parameters defined by the correlation function $K_\xi(\tau)$.

Typically, in radio engineering applications the correlation function is approximated as:

$$K_\xi(\tau) = \sum_{j=1}^{n} \left[A_j(\tau) \cdot \cos \beta_j \tau + B_j \cdot \sin |\beta_j| \tau \right] \cdot e^{-\alpha_j |\tau|}$$

which results in the fractional-rational spectral densities. This method is very efficient with regard to the simulation of the processes with typical correlation functions $K_i(\tau)$ (see Appendix 23), but for correlation functions other than those evaluation of the coefficients a_j, b_j may become complicated. The simulation algorithms for a stationary Gaussian random process with typical correlation functions $K_i(\tau)$, $i = \overline{1,4}$, are given in Appendix 31. For the processes with typical correlation functions, the recurrent method provides simple and efficient simulation algorithms that are free from methodological errors inherent to approximate simulation methods.

2.2.2.3 The Method of Formation Filter

The formation filter is a filter that transforms the random process $\eta(t)$ of white noise type into the random process $\xi(t)$ with the specified statistical parameters.

Definition. *White noise is the stationary random process with constant spectral density* $S(\omega) = S_0$ *and correlation function*

$$K_\eta(\tau) = 2\pi S_0 \cdot \delta(\tau)$$

where $\delta(\tau)$ is a delta function.

Typically, the process $\eta(t)$ is considered to be the stationary Gaussian one, $S_0 = 1/2\pi$, $M\{\eta(t)\} = 0$, and the spectral density of $\xi(t)$ is the fractional-rational function of frequency. Then the frequency response $\Phi(p)$ of the formation filter is:

$$\Phi(p) = \frac{1}{\sqrt{S_0}} \cdot \frac{F_m(p)}{H_n(p)}$$

where $F_m(p), H_n(p)$ are the polynomials of m and n order correspondingly, $m < n$.

In this case the process $\xi(t)$ is simulated as the first component of an n-dimensional Markovian process:

$$\vec{x}(t) = (x_1(t), ..., x_n(t)), \quad \xi(t) = x_1(t)$$

that satisfies the equation:

$$\frac{d\vec{x}}{dt} = \vec{A} \cdot \vec{x} + \vec{B} \cdot \eta(t)$$

Usually,

$$\sqrt{2\pi} F_m(p) = \beta_0 p^m + \beta_1 \cdot p^{m-1} + ... \beta_m$$

$$H_n(p) = p^n + \alpha_1 \cdot p^{n-1} + ... \alpha_n,$$

where α_i, β_i are some real coefficients. Thus:

$$\vec{A} = \begin{pmatrix} 0 & 1 & 0 & . & 0 \\ 0 & 0 & 1 & . & 0 \\ ...... & .... & .... & . & ..... \\ 0 & 0 & 0 & . & 1 \\ -\alpha_n & -\alpha_{n-1} & -\alpha_{n-2} & . & -\alpha_1 \end{pmatrix}; \quad \vec{B} = \begin{pmatrix} 0 \\ 0 \\ ... \\ b_{n-m} \\ ... \\ b_{n-1} \\ b_n \end{pmatrix}$$

where $b_{n-m} = \beta_0$,

$$b_k = \beta_{k-(n-m)} - \sum_{j=1}^{k-(n-m)} \alpha_j \cdot b_{k-j}, \quad k = n-m+1, ..., n$$

Since white noise with infinite variance is an abstract process that cannot be simulated in practice, some preliminary restrictions have to be imposed to simulate the feasible formation filter. The following approach can be recommended.

1) Discrete white noise $\varepsilon_k \in N(0,1)$ is simulated, $k = 0,1,2,..., \varepsilon_k$ are uncorrelated;

2) The stepped process $\varepsilon_h(t)$ with an increment h derived from ε_k is considered:

$$\varepsilon_h(t) = \varepsilon_k, \quad t \in [kh, (k+1)h]$$

The spectral density of the process $\varepsilon_h(t)$:

$$S_h(\omega) = \frac{h}{2\pi} \left[\frac{\sin(\omega h/2)}{\omega h/2} \right]^2$$

and relative error to simulate white noise by the process $\varepsilon_h(t)$ is:

$$\frac{|S_h(0) - S_h(\omega_0)|}{S_h(0)} \leq \delta$$

where ω_0 is the frequency at the end of the frequency interval and δ is the specified simulation error.

The equation to choose increment h is:

$$h \leq \frac{2\sqrt{3\delta}}{\omega_0}, \quad \sup \{S_\xi(\omega)/S_\xi(0)\} < \Delta \quad \omega \in [0, \omega_0]$$

where Δ is the specified error to reproduce the spectral density.

3) The differential equation to simulate $\xi(t)$ is as follows:

$$\frac{d\vec{x}}{dt} = \vec{A} \cdot \vec{x} + \frac{1}{\sqrt{h}} \cdot \vec{B} \cdot \varepsilon_h(t)$$

Frequency responses of the formation filters and corresponding equations of the filters based on the above formula for Gaussian stationary random processes with typical correlation functions are given in Appendix 32.

2.2.2.4 The Method of Sliding Summation

In this method the discrete values of the simulated process are formed as the sliding sum:

$$\xi_k = \sum_{j=-M}^{M} a_j \cdot \varepsilon_{k-j}, \quad \varepsilon_k \in N(0,1)$$

with weighting coefficients a_j. The most common way to determine a_j is based on the convolution integral:

$$\xi(t) = \int_{-\infty}^{\infty} g(\tau)\eta(t-\tau)d\tau$$

where $\eta(t)$ is normalized white noise and

$$g(\tau) = \sqrt{\frac{2}{\pi}} \int_{0}^{\infty} \sqrt{S_\xi(\omega)} \cdot e^{j\omega t} d\omega$$

is a weighting function of the formation filter.

The discretization of this integral gives:

$$a_j = \sqrt{\Delta t} \cdot g(i\Delta t), \quad i = 0, \pm 1, \pm 2, \ldots \pm M$$

and $g(i\Delta t)$ is typically evaluated by using numerical methods.

The infinite upper limit in the integral is changed to the finite one. The simulated sequence has a discrete correlation function:

$$K_\xi[l] = \sum_{j=-M}^{M-l} a_j \cdot a_{j+l} = \sum_{j=-M}^{M-l} g(j\Delta t)g[(j+l)\Delta t]\Delta t$$

and the continuous correlation function is:

$$K_\xi(\tau) = \int_{-\infty}^{\infty} g(t) \cdot g(t+\tau)dt$$

When $\Delta t \to 0$, $l \cdot \Delta t = \tau = \text{const}$, $K_\xi[l] \to K_\xi(\tau)$.

2.2.2.5 The Method of Canonical and Noncanonical Representations

This method is based on representation of the model of the random process $\xi(t)$ as the deterministic function of some random variables. In such a representation the main objective is to achieve equality of simulated mathematical expectations and correlation functions to the specified values.

Any stationary (in a broad sense) random process can be represented as:

$$\xi(t) = m_\xi + \int_{-\infty}^{\infty} e^{j\omega t} d\Phi_\xi(\omega)$$

where $d\Phi_\xi(\omega)$ is the random function with uncorrelated increments for which

$$M\{d\,\Phi_\xi(\omega)\} = 0,$$
$$M\{d\,\Phi_\xi(\omega) \cdot d\Phi_\xi^*(\omega_1)\} = S_\xi(\omega)\delta(\omega - \omega_1)d\,\omega\,d\,\omega_1$$

A more general representation that can be applied both to stationary and nonstationary processes is termed a canonical decomposition:

$$\xi(t) = m_\xi(t) + \sum_{k=1}^{\infty} \vartheta_k \cdot \varphi_k(t) \tag{2.3}$$

where ϑ_k are uncorrelated variables and $\varphi_k(t)$ are deterministic functions derived from the correlation function $K_\xi(\tau)$.

The process $\xi(t)$ can be decomposed onto the system of orthogonal functions with the weight $\mu(t)$ at the interval $[0, T]$ by means of Fourier series:

$$\xi(t) = m_\xi(t) + \sum_{j=1}^{\infty} \vartheta_j \cdot \psi_j(t) \tag{2.4}$$

$$\int_0^T \psi_j(t)\psi_l(t)\mu(t)dt = \begin{cases} c_j^2, & j = l \\ 0, & j \neq l \end{cases}$$

$$\frac{1}{c_j^2}\int_0^T [\xi(t) - m_\xi(t)]\psi_j(t)\mu(t)dt = \vartheta_j$$

Correlation function $K_\xi(t, \tau)$ satisfies the integral equation:

$$\int_0^T K_\xi(t, \tau) \psi(\tau) d\tau = \lambda \cdot \psi(t)$$

Representation of the random process with models (2.3) or (2.4) introduces a methodological error that depends on the number of factors used in the series.

In noncanonical representation the model of the random process is specified as a nonlinear function of some random variables. The most common algorithms for the noncanonical model are:

$$\zeta(t, \vec{\Omega}) = m_\xi + \sigma_\xi (\lambda_1 \cos \omega t + \lambda_2 \sin \omega t)$$

$$\zeta(t, \vec{\Omega}) = m_\xi + \sigma_\xi [\sin \omega(t + t_0) + \lambda \cos \omega(t + t_0)]$$

$$\zeta(t, \vec{\Omega}) = m_\xi + \sqrt{S_\xi(\omega) / f(\omega)} (\sin \omega t + \lambda \cos \omega t)$$

$$\zeta(t, \vec{\Omega}) = m_\xi + \lambda \sin(\omega t + \varphi)$$

$$\zeta(t, \vec{\Omega}) = m_\xi + \sigma_\xi \sqrt{\frac{2}{N}} \sum_{j=1}^N \cos(\omega_j t + \varphi_j)$$

where $\vec{\Omega}$ is the vector of the model parameters (e.g., for the first model $\vec{\Omega} = (\lambda_1, \lambda_2, \omega)$). These models result in non-Gaussian multidimensional distributions and do not meet the ergodicity requirements, i.e., the estimate of the correlation function based on the sample $\zeta(t, \vec{\Omega})$ with the length T does not converge to the true correlation function when $T \to \infty$.

2.2.2.6 Simulation of a Vector Random Process

Any n-dimensional Gaussian stationary random process $\vec{\xi}(t) = (\xi_1(t), ..., \xi_n(t))$ is completely described by its mathematical expectation $\vec{m}_\zeta = (M\{\xi_1(t)\}, ..., M\{\xi_n(t)\})$ and matrix correlation function:

$$\vec{K}_\xi(\tau) = M\{[\vec{\xi}(t) - \vec{m}_\xi][\vec{\xi}(t + \tau) - \vec{m}_\xi]\} = \| K_{ij}(\tau) \|$$

or equivalent matrix spectral density:

$$\vec{S}_\xi(\omega) = \left\| S_{ij}(\omega) \right\| = \frac{1}{2\pi} \int\limits_{-\infty}^{\infty} \vec{K}_\xi(\tau) e^{-j\omega\tau} \, d\tau$$

Most of the simulation methods are based on a technique that reduces the process $\vec{\xi}(t)$ to m-dimensional white noise $\eta(t) = (\eta_1(t),...,\eta_m(t))$. This transformation is done by means of a convolution integral:

$$\vec{\xi}(t) = \int\limits_{-\infty}^{\infty} \vec{g}(\tau) \cdot \vec{\eta}(t-\tau) \, d\tau + \vec{m}_\xi$$

where $\vec{g}(\tau) = \left\| g_{ij}(\tau) \right\|$ is the matrix weighting function of the formation filter.

The random process $\eta(t)$ has zero mean and correlation function $\vec{K}_\eta(\tau) = \vec{I}_m \cdot \delta(\tau)$, where I_m is a unity matrix and $\delta(\tau)$ is the delta function.

2.2.2.7 Simulation of a Gaussian Markovian Random Process

The common case is when $n = 3$ (three-dimensional process) and the process can be considered to be a stationary, Gaussian and Markovian one:

$$\vec{\xi}(t) = (\xi_1(t), \xi_2(t), \xi_3(t))$$

It is completely defined by its mathematical expectation

$$\vec{m}_\xi = (m_{\xi_1}, m_{\xi_2}, m_{\xi_3})$$

and correlation matrix:

$$\vec{K}_\xi(\tau) = \begin{vmatrix} D_1 \cdot R_{11}(\tau) & \sqrt{D_1 D_2} \cdot R_{12}(\tau) & \sqrt{D_1 D_3} \cdot R_{13}(\tau) \\ \sqrt{D_2 D_1} \cdot R_{21}(\tau) & D_2 \cdot R_{22}(\tau) & \sqrt{D_2 D_3} \cdot R_{23}(\tau) \\ \sqrt{D_3 D_1} \cdot R_{31}(\tau) & \sqrt{D_3 D_2} R_{32}(\tau) & D_3 \cdot R_{33}(\tau) \end{vmatrix}$$

where $D_n = \sigma_n^2$ is the variance of $\xi_n(t), n = \overline{1,3}$; $R_{ij}(\tau) = \rho_{ij} \cdot f_{ij}(\tau)$ is the normalized cross-correlation function of the processes $\xi_i(t)$ and $\xi_j(t)$, $i, j = \overline{1,3}$, having the properties:

$$R_{ij}(\tau) = R_{ji}(\tau), R_{ij}(\tau) = R_{ij}(-\tau),$$

$$\rho_{ij} = \begin{cases} 1, & \text{if } i = j \\ 0 \leq \rho_{ij} < 1, & \text{if } i \neq j \end{cases},$$

$f_{ij}(\tau)$ is the deterministic function describing the dependence of correlation function $R_{ij}(\tau)$ upon parameter τ.

The algorithm to simulate the process $\vec{\xi}(t_{k+1})$ in the moment $k+1$ based on its value $\vec{\xi}(t_k)$ in the preceding moment k is as follows:

A. The conditional mathematical expectation is defined:

$$\vec{m}_{\xi}^{(c)} = M\left\{\vec{\xi}(t_{k+1}) \middle| \vec{\xi}(t_k) = \vec{x}(t_k)\right\} = \vec{m}_{\xi}(t_{k+1}) + \vec{K}_{\xi}(t_{k+1}, t_k) \cdot \left[K_{\xi}(t_k, t_k)\right]^{-1}$$
$$\times \left(\vec{x}(t_k) - \vec{m}_{\xi}(t_k)\right)$$

B. The conditional correlation matrix is defined:

$$\vec{K}_{\xi}^{(c)} = \left\|K_{ij}^{(c)}\right\| = K_{\xi}\left[t_{k+1} \middle/ \vec{\xi}(t_k) = \vec{x}(t_k)\right] = \vec{K}_{\xi}(t_{k+1}, t_{k+1}) - \vec{K}_{\xi}(t_{k+1}, t_k)$$
$$\times \left[\vec{K}_{\xi}(t_k, t_k)\right]^{-1} \cdot \vec{K}_{\xi}(t_k, t_{k+1})$$

C. The vector-column of the random process is defined as:

$$\vec{\xi}(t_{k+1}) = \vec{A} \cdot \vec{\varepsilon} + \vec{m}_{\xi}^{(c)}$$

where

$$\vec{A} = \begin{vmatrix} A_{11} & 0 & 0 \\ A_{12} & A_{22} & 0 \\ A_{31} & A_{32} & A_{33} \end{vmatrix}$$

is the conversion matrix and $\vec{\varepsilon}$ is the vector-column of the independent Gaussian random variables with zero means and unity variances. The elements of the conversion matrix are calculated based on the recurrent formula:

$$A_{ij} = \frac{K_{ij}^{(c)} - \sum\limits_{l=1}^{j-1} A_{il} \cdot A_{jl}}{\sqrt{K_{jj}^{(c)} - \sum\limits_{l=1}^{j-1} A_{jl}^2}}$$

$$\sum_{l=1}^{0} A_{il} \cdot A_{jl} = 0 \quad 1 \le j \le i \le 3$$

Let us consider $\bar{m}_\xi = 0$. Then $\bar{m}_\xi^{(c)}(t_{k+1}) = \left\| m_{ij}(t_{k+1}) \right\| = \left\| m'_{ij}(\tau) \right\| \cdot (x(t_k))^T$,

where

$$m'_{ij}(\tau) = \sqrt{\frac{D_i}{D_j}} \cdot f^{-1}(\rho) \cdot \sum_{l=1}^{3} a_{il}^{(j)} \cdot f_{il}(\tau)$$

$$f(\rho) = 1 + 2\rho_{12}\,\rho_{13}\,\rho_{23} - \rho_{12}^2 - \rho_{13}^2 - \rho_{23}^2$$

The coefficients $a_{il}^{(j)}$ are given in Table 2.1, where

$$a_1(\rho) = 1 - \rho_{23}^2;\ a_2(\rho) = 1 - \rho_{13}^2;\ a_3(\rho) = 1 - \rho_{12}^2;\ a_4(\rho) = \rho_{13}\,\rho_{23} - \rho_{12}$$

$$a_5(\rho) = \rho_{12}\,\rho_{23} - \rho_{13};\ a_6(\rho) = \rho_{12}\,\rho_{13} - \rho_{23}$$

$$K_{ij}^{(c)}(\tau) = \begin{cases} D_i \left\{ 1 - f^{-1}(\rho) \sum\limits_{l=1}^{3} \left[b_{il} \cdot f_{il}^2(\tau) + 2c_{il} \cdot F_{il}(\tau) \right] \right\} & i = j \\ \sqrt{D_i D_j} \left[\rho_{ij} - f^{-1}(\rho) \cdot f_{ij}^{(c)}(\tau) \right] & i \ne j \end{cases}$$

The parameters $b_{il}, c_{il}, f_{if}^{(c)}(\tau)$ are determined by matrix equations:

$$\bar{B}_{il} = \| b_{il} \| = \bar{p}_b \cdot \bar{a}_b(\rho)$$

$$\bar{C}_{il} = \| c_{il} \| = \bar{p}_c \cdot \bar{a}_c(\rho)$$

$$f_{ij}^{(c)}(\tau) = f_{ji}^{(c)}(\tau) = \bar{G}^{(ij)}(\tau) \cdot \bar{a}(\rho)$$

Table 2.1

The Coefficients $a_{il}^{(j)}$

i	j	l		
		1	2	3
	1	$a_1(\rho)$	$\rho_{12}a_4(\rho)$	$\rho_{13}a_5(\rho)$
1	2	$a_4(\rho)$	$\rho_{12}a_2(\rho)$	$\rho_{13}a_6(\rho)$
	3	$a_5(\rho)$	$\rho_{12}a_6(\rho)$	$\rho_{13}a_3(\rho)$
	1	$\rho_{12}a_1(\rho)$	$a_4(\rho)$	$\rho_{23}a_5(\rho)$
2	2	$\rho_{12}a_4(\rho)$	$a_2(\rho)$	$\rho_{23}a_6(\rho)$
	3	$\rho_{12}a_5(\rho)$	$a_6(\rho)$	$\rho_{23}a_3(\rho)$
	1	$\rho_{13}a_1(\rho)$	$\rho_{23}a_4(\rho)$	$a_5(\rho)$
3	2	$\rho_{13}a_4(\rho)$	$\rho_{23}a_2(\rho)$	$a_6(\rho)$
	3	$\rho_{13}a_5(\rho)$	$\rho_{23}a_6(\rho)$	$a_3(\rho)$

where $\vec{a}_B(\rho)$, $\vec{a}_c(\rho)$ are diagonal matrices (3×3):

$$\vec{a}_B(\rho) = \| a_n(\rho) \|, \quad n = \overline{1,3}$$
$$\vec{a}_c(\rho) = \| a_n(\rho) \|, \quad n = \overline{4,6}$$

$\vec{\rho}_B, \vec{\rho}_c$ are matrices of correlation coefficients:

$$\vec{\rho}_B = \begin{vmatrix} 1 & \rho_{12}^2 & \rho_{13}^2 \\ \rho_{12}^2 & 1 & \rho_{23}^2 \\ \rho_{13}^2 & \rho_{23}^2 & 1 \end{vmatrix}; \quad \vec{\rho}_c = \begin{vmatrix} \rho_{12} & \rho_{13} & \rho_{12}\rho_{13} \\ \rho_{12} & \rho_{12}\rho_{23} & \rho_{23} \\ \rho_{13}\rho_{23} & \rho_{13} & \rho_{23} \end{vmatrix}$$

$\vec{G}^{(ij)}(\tau) = \left(g_n^{(ij)}(\tau) \right)$, $n = \overline{1,6}$ is the matrix-row with elements given in Table 2.2;

Table 2.2

Elements of the Matrix $\vec{G}^{(ij)}(\tau)$

ij	$g_1(\tau)$	$g_2(\tau)$	$g_3(\tau)$	$g_4(\tau)$	$g_5(\tau)$	$g_6(\tau)$
12, 21	$\rho_{12}F_{11}(\tau)$	$\rho_{12}F_{21}(\tau)$	$\rho_{13}\rho_{23}\times F_{31}(\tau)$	$\rho_{12}^2 f_{12}^2(\tau)+f_{11}(\tau)\times f_{22}(\tau)$	$\rho_{12}\rho_{13}F_{13}(\tau)+\rho_{23}f_{11}(\tau)\times f_{23}(\tau)$	$\rho_{12}\rho_{23}F_{22}(\tau)+f_{13}(\tau)\rho_{13}\times f_{22}(\tau)$
13, 31	$\rho_{13}F_{12}(\tau)$	$\rho_{23}F_{22}(\tau)\times\rho_{12}$	$\rho_{13}F_{32}(\tau)$	$\rho_{12}\rho_{13}F_{13}(\tau)+\rho_{23}f_{11}(\tau)\times f_{23}(\tau)$	$\rho_{13}^2 f_{13}^2(\tau)+f_{11}(\tau)\times f_{33}(\tau)$	$\rho_{13}\rho_{23}F_{31}(\tau)+\rho_{12}f_{12}(\tau)\times f_{33}(\tau)$
23, 32	$\rho_{12}\rho_{13}\times F_{13}(\tau)$	$\rho_{23}F_{23}(\tau)$	$\rho_{23}F_{33}(\tau)$	$\rho_{12}\rho_{23}F_{22}(\tau)+\rho_{13}f_{13}(\tau)\times f_{22}(\tau)$	$\rho_{13}\rho_{23}F_{31}(\tau)+\rho_{12}f_{21}(\tau)\times f_{33}(\tau)$	$\rho_{23}^2 f_{23}^2(\tau)+f_{22}(\tau)\times f_{33}(\tau)$

$\vec{a}(\rho)=(a_n(\rho))^T$, $n=\overline{1.6}$ is the matrix-column, the functions $F_{il}(\tau)$ are the elements of the matrix:

$$\vec{F}(\tau)=\|F_{il}(\tau)\|=\begin{vmatrix} f_{11}(\tau)f_{12}(\tau) & f_{11}(\tau)f_{13}(\tau) & f_{12}(\tau)f_{13}(\tau) \\ f_{21}(\tau)f_{22}(\tau) & f_{21}(\tau)f_{23}(\tau) & f_{22}(\tau)f_{23}(\tau) \\ f_{31}(\tau)f_{32}(\tau) & f_{31}(\tau)f_{33}(\tau) & f_{32}(\tau)f_{33}(\tau) \end{vmatrix}$$

This algorithm makes it possible to simulate the three-dimensional stationary Gaussian Markovian process with any arbitrary correlation functions $f_{ij}(\tau)$ varying for different combinations of i and j.

If function $f_{ij}(\tau)$ is identical for all combinations i, j ($f_{ij}(\tau)=f(\tau)$ for any i, j), then the simulation algorithm can be simplified considerably. In this case correlation function $\vec{K}_\xi(\tau)$ takes the form:

$$\vec{K}_{\xi}(\tau) = \begin{vmatrix} D_1 & \sqrt{D_1\, D_2}\cdot\rho_{12} & \sqrt{D_1\, D_3}\ \rho_{13} \\ \sqrt{D_2\, D_1}\ \rho_{12} & D_2 & \sqrt{D_2\, D_3}\ \rho_{23} \\ \sqrt{D_3\, D_1}\ \rho_{13} & \sqrt{D_3\, D_2}\ \rho_{23} & D_3 \end{vmatrix}\cdot f(\tau)$$

The algorithm to simulate a process $\xi(t_{k+1})$ with a correlation function $\vec{K}_{\xi}(\tau)$ in any moment of time is given in Appendix 33. The two-dimensional process can be simulated using first two components $\xi_1(t), \xi_2(t)$, and a scalar stationary Gaussian Markovian process $\xi_1(t)$ can be simulated using the first component only.

Correlation function of the process $\xi_1(t)$ is often approximated as:

$$K_1(\tau) = D\cdot f(\tau) = D\cdot e^{-\alpha|\tau|} = \sigma^2\cdot e^{-\alpha|\tau|}$$

which gives the following simple algorithm:

$$\xi(t_{k+1}) = \xi(t_k)\cdot e^{-\alpha|\tau|} + \sigma\sqrt{1 - e^{-2\alpha|\tau|}}\cdot\varepsilon(t_{k+1})$$

The stationary Gaussian process with correlation function $K_1(\tau)$ is always a Markovian one.

2.2.2.8 Simulation of a Stationary Non-Gaussian Process

Typically, simulation of a non-Gaussian random process is a much more complicated and nontrivial task. The best developed approach is the method of nonlinear transformation that enables us to simulate a scalar non-Gaussian process $\xi(t)$ with the specified one-dimensional pdf $f_1(x)$ and autocorrelation function $K_{\xi}(\tau)$. Generally speaking, in opposition to the model of a Gaussian process, the model of a non-Gaussian process is not completely defined by $f_1(x)$ and $K_{\xi}(\tau)$ only. There might be several random processes with different multidimensional pdfs but identical parameters $f_1(x)$ and $K_{\xi}(\tau)$. Typically, only one process which is more convenient for simulation will be chosen from the set of processes providing the specified one-dimensional probability density function and correlation function.

Basic principles of the method of nonlinear transformation are as follows. A stationary Gaussian process $\varepsilon(t)$ is chosen to be the initial one. There is always the nonlinear transformation

$$y = \varphi(x_n)$$

which transforms the normal pdf $N(m_n, D_n)$ associated with $\varepsilon(t) = \{x_n(t)\}$ into specified pdf $f_1(y) = f_1(x)$. If the Gaussian process has correlation function $K_\varepsilon(\tau)$, then some dependence φ_K exists:

$$K_\xi(\tau) = \varphi_K \left[K_\varepsilon(\tau) \right]$$

defined by dependence $y = \varphi(x_n)$.

To obtain a required correlation function one should choose an initial correlation function as:

$$K_\varepsilon(\tau) = \varphi_K^{-1} \left[K_\xi(\tau) \right]$$

where φ_K^{-1} is a function reciprocal to φ_K. The basic steps to implement this method are:

A. To determine the appropriate transformation dependence $y = \varphi(x_n)$ based on the specified pdf $f_1(y)$.

B. To determine the dependence

$$K_\xi(\tau) = \varphi_K \left[K_\varepsilon(\tau) \right]$$

based on the function $\varphi(x_n)$ that satisfies the equation:

$$K_\xi(\tau) = \frac{1}{2\pi D_n \sqrt{1 - \left[\dfrac{K_\varepsilon(\tau)}{D_n} \right]^2}} \int_{-\infty}^{\infty} \int_{-\infty}^{\infty} \varphi(x_{n1})(x_{n2}) \cdot \exp(Q) \, dx_{n1} \, dx_{n2} - m_n^2$$

where

$$Q = \cfrac{1}{2\left\{1 - \left[\dfrac{K_\varepsilon(\tau)}{D_n}\right]^2\right\}}$$

$$\times \frac{(x_{n1} - m_n)^2 + 2\dfrac{K_\varepsilon(\tau)}{D_n}(x_{n1} - m_n)(x_{n2} - m_n) + (x_{n2} - m_n)^2}{D_n}$$

C. To use numerical methods to solve this equation with respect to parameter $K_\varepsilon(\tau)$ and to determine the correlation function of initial normal process $\varepsilon(t)$.

D. To simulate the desired non-Gaussian process by simulating the Gaussian process $\varepsilon(t) = \{x_n(t)\}$ with the estimated correlation function $K_\varepsilon(\tau)$ and to get the samples of the non-Gaussian process through the transformation:

$$y(t_k) = \varphi[x_n(t_k)]$$

2.2.2.9 Simulation of a Nonstationary Process

In general simulation of a nonstationary random process can be done based on a discrete model of the linear nonstationary system. The linear stationary system (formation filter) when input is a stationary random process is described by the equation:

$$\frac{d\bar{x}}{dt} = \bar{A} \cdot \bar{x} + \bar{B} \cdot \bar{\eta}(t) \tag{2.5}$$

or in terms of discrete representation:

$$\bar{x}_{k+1} = \bar{A}_\Delta \cdot \bar{x}_k + \bar{B}_\Delta \cdot \bar{\varepsilon}_{k+1}$$

where

$$\bar{A}_\Delta = \left\| a_{ij}^\Delta \right\| = \exp\left(\bar{A} \cdot \Delta t\right)$$

$$\bar{B}_\Delta = \left\| b_{ij}^\Delta \right\| \text{ is a matrix } (n \times r)$$

Here r is a rank of correlation matrix

$$\bar{K}_\xi = \bar{B}_\Delta \cdot \bar{B}_\Delta^T$$

$\vec{\varepsilon}_{k+1}$ is the vector of independent Gaussian random variables, $\vec{\varepsilon}_{k+1} \in N(0, \bar{K}_\xi)$.

Analogously, for a nonstationary system the discrete model is:

$$\bar{x}_{k+1} = \bar{\Phi}(k+1,k) \cdot \bar{x}_k + \bar{D}_{k+1} \cdot \vec{\varepsilon}_{k+1}$$

where $\bar{\Phi}(k+1,k)$ is a matrix impulse response of the formation filter, and $\bar{D}_{k+1}$ is defined from equation:

$$\bar{K}_{\xi_{k+1}} = \bar{D}_{k+1} \cdot \bar{D}_{k+1}^T$$

which is a discrete representation of the equation:

$$\frac{d\bar{x}}{dt} = \bar{A}(t) \cdot \bar{x} + \bar{B}(t) \cdot \eta(t) \tag{2.6}$$

As opposed to equation (2.5), matrices $A(t)$, $B(t)$ in equation (2.6) are functions of time. If $A(t) = A$, $B(t) = B$, and $\Delta t = t_{k+1} - t_k$, both equations converge. In the generic case, a complicated preliminary analysis to evaluate matrices $\bar{\Phi}(k+1,k)$ and $\bar{D}_{k+1}$ is required. The simulation algorithms for the stationary and nonstationary linear systems of the first order are given in Appendix 34. Specific assumptions about coefficients a, b and $a(t)$, $b(t)$ are made. In the case of nonstationary processes the function $\Phi(\tau,\tau)$ is found from this equation:

$$\dot{\Phi}(t,\tau) = a(t)\Phi(t,\tau), \quad \Phi(\tau,\tau) = 1$$

which gives:

$$\Phi(t,\tau) = \exp\left(\int_\tau^t a(t)dt\right) = \left(\frac{1+\vartheta t}{1+\vartheta \tau}\right)^{\alpha/\vartheta},$$

and

$$K_{\xi_{k+1}} = \int_{t_k}^{t_{k+1}} \exp\left(2\int_\tau^{t_{k+1}} a(t)dt\right) \cdot b^2(\tau)d\tau = b^2 \int_{t_k}^{t_{k+1}} \left(\frac{1+\vartheta t_{k+1}}{1+\vartheta \tau}\right)^{2\alpha/\vartheta} d\tau$$

Another approach to simulate a nonstationary process is based on the linear transformation of the stationary process $\eta(t)$:

$$\xi(t) = f(t) \cdot \eta(t) + g(t)$$

where $f(t), g(t)$ are nonrandom functions of time, $\eta(t)$ is the stationary random process with zero mean, correlation function $K_\eta(t_2 - t_1)$, and variance $K_\eta(0) = \sigma_\eta^2 = 1$. The mathematical expectation and correlation function of the process $\xi(t)$ are:

$$m_\xi(t) = g(t)$$

$$K_\xi(t_1, t_2) = f(t_1) f(t_2) \cdot K_\eta(t_2 - t_1)$$

Thus, the nonstationary process $\xi(t)$ with specified mean $g(t)$ and correlation function $K_\xi(t_1, t_2)$ is simulated as:

$$\xi(t_k) = f(t_k) \cdot \eta(t_k) + g(t_k)$$

where $f(t_k) = \sqrt{K_\xi(t_k, t_k)} = \sigma_\xi(t_k)$.

2.3 Random Fields

2.3.1 Statistical Description

2.3.1.1 Basic Definitions

Definition. *M-dimensional random field is the random function* $\xi(\vec{x})$ *of m variables:* $x_1, x_2, ... x_m$.

A one-dimensional random field $\xi(\vec{x})$ is called a scalar random field, and an n-dimensional random field $\vec{\xi}(\vec{x}) = (\xi_1(\vec{x}), ... \xi_n(\vec{x}))$ is called a vector random field. The parameter $\vec{x}$ is called an argument of the random field. In radio engineering applications the typical case is the random field $\xi(x, y, z, t)$, which is represented as a function of four arguments: x, y, z are three-dimensional Cartesian coordinates of some point in space and t is time (e.g., in case of phased arrays simulation). When a scalar random field is the function of a single argument, time t, it is reduced to a random process.

2.3.1.2 The Statistical Characteristics

The main characteristics to describe a vector random field $\vec{\xi}(\vec{x}) = (\xi_1(\vec{x})...\xi_n(\vec{x}))$ are:

A. The multidimensional probability density function is the distribution of $n \times k$-dimensional vector $\eta = (\vec{\xi}(\vec{x}_1)...\vec{\xi}(\vec{x}_k))$. If the pdf does not depend on the variation of an argument $\Delta \vec{x}$:

$$f_k(\vec{y}_1,...\vec{y}_k;\vec{x}_1 + \Delta \vec{x},...\vec{x}_k + \Delta \vec{x}) = f_k(\vec{y}_1,...\vec{y}_k;\vec{x}_1,...\vec{x}_k)$$

the random field is called an isotropic one.

B. The moments are:

$$m(\vec{x}) = M\left[\vec{\xi}(\vec{x})\right] = (m_1(\vec{x})...m_n(\vec{x})) \text{ is a mathematical expectation,}$$

where $m_i(\vec{x}) = M[\xi_i(\vec{x})]$;

$$K_0(\vec{x}_1,\vec{x}_2) = M\left[\xi(\vec{x}_1) \cdot \xi^T(\vec{x}_2)\right] \text{ is a covariation function;}$$

$$K(\vec{x},\vec{x}_2) = M\left\{[\xi(\vec{x}_1) - m(\vec{x}_1)][\xi(\vec{x}_2) - m(\vec{x}_2)]^T\right\} \text{ is a correlation function.}$$

The vector random field is called an isotropic one in a broad sense if

$$m(\vec{x}) = m = const, \quad K(\vec{x} + \vec{y},\vec{y}) = K(\vec{x})$$

i.e., the mean is constant and correlation function depends only upon the difference of arguments. For an isotropic field:

$$K(\vec{x}) = K^T(-\vec{x}) \quad K_{ij}(\vec{x}) = K_{ji}(-\vec{x})$$

$$K_{ij}(\vec{x}) \le \sqrt{K_{ii}(0)K_{jj}(0)}$$

C. The spectral density and the correlation function are linked as:

$$S(\vec{u}) = \frac{1}{(2\pi)^m} \int_{R^m} K(\vec{x})e^{-j\vec{u}^T\vec{x}} d\vec{x}$$

$$K(\vec{x}) = \int\limits_{R^m} S(\vec{u}) e^{j\vec{u}^T \vec{x}} d\vec{u}$$

For each fixed $\vec{u}$:

$$S(\vec{u}) = S^*(\vec{u}) = \overline{S}^T(\vec{u})$$

$$S_{kl}(\vec{u}) = \overline{S}_{lk}(\vec{u})$$

where $\vec{u}$ denotes the conjugation operation with respect to u. This leads to the conditions:

$$S(\vec{u}) = \overline{S}(-\vec{u})$$

$$\vec{z}^T S(\vec{u}) \cdot \overline{\vec{z}} = \sum_{k,l=1}^{n} z_k S_{kl}(\vec{u}) \overline{z}_l \geq 0$$

for any complex vectors $\vec{z} = (z_1,...z_n)$.

The major statistical characteristics to describe a scalar random field are:

A. N-dimensional probability density function is composed from n sections of the field in the points $\vec{x}_1, \vec{x}_2 ... \vec{x}_n$, i.e., the pdf of the random vector $(\xi(x_1)...\xi(x_n))$ is:

$$f_n(y_1,...y_n; \vec{x}_1,...\vec{x}_n)$$

The scalar random field is called an isotropic one if it is invariant to argument shift, i.e., for any shift of argument $\Delta \vec{x}$ and any n:

$$f_n(y_1,...y_n; \vec{x}_1 + \Delta \vec{x}_1,...\vec{x}_n + \Delta \vec{x}_n) = f_n(y_1,...y_n; x_1,...x_n)$$

B. One-dimensional probability density function is the probability distribution of $\xi(\vec{x})$ when $\vec{x}$ is fixed. The one-dimensional pdf of an isotropic field $f_1(y; \vec{x})$ does not depend on argument x.

C. The moments are:

$m(\vec{x}) = M[\xi(\vec{x})]$ is a mathematical expectation;

$\sigma^2[\xi(\vec{x})] = M[\xi(\vec{x}) - m(\vec{x})]^2$ is a variance;

$K_0(\vec{x}_1 \vec{x}_2) = M[\xi(\vec{x}_1) \cdot \xi(\vec{x}_2)]$ is a covariation function;

$K(\vec{x}_1, \vec{x}_2) = M\{[\xi(\vec{x}_1) - m(\vec{x}_1)] \cdot [\xi(\vec{x}_2) - m(\vec{x}_2)]\}$ is a correlation function.

In practical simulation tasks the random field typically is defined by its correlation function. For a random field that is isotropic in a broad sense:

$$m(\vec{x}) = m = \text{const}$$

$$K(\vec{x}_1, \vec{x}_2) = K(\vec{x}), \quad \vec{x} = \vec{x}_1 - \vec{x}_2$$

The random field is called Gaussian (normal) if all multidimensional probability distribution functions are described by Gaussian distribution law.

D. The spectral density of the random isotropic field is:

$$S(\vec{u}) = \frac{1}{(2\pi)^m} \int_{R^m} K(\vec{x}) e^{-j\vec{u}^T \vec{x}} d\vec{x}$$

where $\vec{u}^T \vec{x} = \sum_{i=1}^{m} u_i x_i$, R^m is the range of existence for argument $\vec{x}$.

For two-dimensional isotropic field $\xi(\vec{x})$, $\vec{x} = (x, y)$ the spectral density and correlation function are a Hankel transform pair:

$$S(u) = \frac{1}{2\pi} \int_0^{\infty} J_0(|u| \cdot \rho) \cdot K(\rho) \cdot \rho \cdot d\rho$$

$$K(\rho) = 2\pi \int_0^{\infty} S(u) \cdot J_0(\rho u) du$$

where $\rho = \sqrt{x^2 + y^2}$;

$$J_0(z) = \frac{1}{2\pi} \int_0^{\infty} e^{-iz \cos t} dt \quad \text{is a Bessel function of zero order.}$$

2.3.2 Simulation Algorithms

2.3.2.1 The Method of Sliding Summation

The following algorithm can be used to simulate the two-dimensional Gaussian field $\xi(\vec{x})$, $\vec{x} \in R^2$ with the specified correlation function $K(\vec{x})$ or spectral density $S(\vec{u})$ by the discrete values of the field in a point $\vec{x} = (m\Delta_1, n\Delta_2)$, where Δ_1, Δ_2 are corresponding increments:

$$\xi_{m,n} = \sum_i \sum_j a_{ij} \cdot \varepsilon_{m-i,n-j}$$

where $m, n = 0,1,....$; ε_{kl} is a sequence of independent standard Gaussian random variables, $\varepsilon_{kl} \in N(0,1)$; and a_{ij} are weighting coefficients.

Calculation of weighting coefficients is based on the specified statistical description of the field. If white-noise approximation is applicable, the coefficients can be calculated based on weighting function $g(\vec{x})$ via the convolution:

$$\xi(\vec{x}) = \int\int\limits_{-\infty}^{\infty} g(\vec{y})\eta(\vec{x} - \vec{y})d\vec{y}$$

where $\eta(\vec{y})$ is two-dimensional white noise: $M[\eta(\vec{y})] = 0$, $K[\eta(\vec{y})] = \delta(\vec{x})$.
In this case: $a_{ij} = g(i\Delta_1, j\Delta_2)\sqrt{\Delta_1\Delta_2}$

$$g(x) = \frac{1}{2\pi} \int\int\limits_{-\infty}^{\infty} \sqrt{S_\xi(\vec{u})} e^{j\vec{u}^T\vec{x}} d\vec{u}$$

Another approach can be implemented based on calculations of the spectra $S(\vec{u})$ of the discrete field ξ_{mn}:

$$S(\vec{u}) = \frac{1}{(2\pi)^2} \sum_{n_1=-\infty}^{\infty} \sum_{n_2=-\infty}^{\infty} K(n_1\Delta_1, n_2\Delta_2) e^{-j\vec{n}^T\vec{u}}, \quad \vec{n} = (n_1, n_2)$$

and

$$a_{ij} = \frac{1}{2\pi} \int\limits_{-\pi}^{\pi}\int\limits_{-\pi}^{\pi} \sqrt{S(\vec{u})}\, e^{jk^T\vec{u}}\, d\vec{u}, \quad \vec{k} = (i,\, j)$$

2.3.2.2 The Parametric Model

The parametric model of a scalar random field $\zeta(\vec{x},\vec{\Omega})$ is a description that is adequate to the simulated field at the level of the first two moments (parameters): mathematical expectation (mean) and correlation function. The model depends on an argument $\vec{x} \in R^m$ and a finite random parameter $\vec{\Omega}$. It is desirable to describe the model $\zeta(\vec{x},\vec{\Omega})$ by means of the most simple analytic algorithms and reduce the dimension of the parameter $\vec{\Omega}$ to the least possible value.

The essence of the parametric approach can be demonstrated on the following model with two parameters $\vec{\Omega} = (z,\vec{v})$:

$$\zeta(\vec{x},\vec{\Omega}) = m_\xi + \sigma_\xi \cdot \sqrt{2} \cdot z \cdot \sin\left[v^T(\vec{x}_0 + \vec{x}) + \frac{\pi}{4}\right]$$

where $\vec{v} \in R^m$ is an m-dimensional random vector with the specified spectral density $S_\xi(\vec{u})$; z is the random variable with probability density function $f_z(\lambda)$ and it is independent from $\vec{v}$, $M[z] = 0$, $M[z^2] = 1$, $\vec{x}_0$ is the deterministic vector of initial parameters, $\vec{x}$ is an argument of the field.

For this model:

$$M\left[\zeta(\vec{x},\vec{\Omega})\right] = m_\xi$$

$$K_\zeta(\vec{y} + \vec{x},\, \vec{y}) = \sigma_\xi^2 M[z^2] K_\chi(\vec{x} + \vec{y},\, \vec{y})$$

where K_χ is the correlation function of the field:

$$\chi(\vec{x}) = \sqrt{2}\, \sin\left[\vec{v}^T(\vec{x}_0 + \vec{x}) + \frac{\pi}{4}\right]$$

It can be shown that the moments of the random field $\zeta(\vec{x},\vec{\Omega})$ for $|\vec{x}_0| \to \infty$ converge to the corresponding moments of the random variable:

$$\xi_\infty = \sqrt{2}\sigma_\xi z \sin\varphi + m_\xi$$

where φ is the random variable, $\varphi \in N(0, 2\pi)$. Thus, at the level of the parametric model simulation of the field $\zeta(\vec{x}, \vec{\Omega})$ becomes equivalent to the simulation of an m-dimensional vector with pdf $f_{\xi_-} = (\lambda)$. In this case

$$f_{\xi_-}(\lambda) = \frac{2}{\pi} \int_{\lambda/2}^{\infty} \frac{f_z(\lambda)}{\sqrt{2y^2 - \lambda^2}} dy, \quad \lambda \geq 0$$

2.3.2.3 Simulation of the Random Spatial Frequency

When a parametric model is applied, simulation of the random field virtually is synthesis of an m-dimensional random vector $\vec{v}$ samples with probability distribution function equal to the spectral density $S(\vec{u})$ of the simulated field. The typical cases are:

The isotropic fields meet the following conditions:

$$R_\xi(\vec{x}) = \rho_0(|\vec{x}|), \quad S_\xi(\vec{u}) = S_0(|\vec{u}|)$$

and the distribution of the vector $\vec{v}$ is also an isotropic function. In this case simulation of vector $\vec{v}$ is equal to simulation of its modulus $|\vec{v}| = \vartheta$; that is, the random variable with probability density function $\varphi_0(u)$ and simulation of the unity vector $\vec{e}$ that defines the isotropic direction in the frequency domain.

For two-dimensional and three-dimensional fields:

$$\varphi_0(u) = \begin{cases} 2\pi u \, S_0(u), & m = 2 \\ 4\pi u^2 \, S_0(u), & m = 3, \quad u \in R^1 \end{cases}$$

Finally, the random vector $\vec{v}$ is simulated as:

$$\vec{v} = \vartheta \cdot \vec{e}$$

The algorithms to simulate ϑ and $\vec{e}$ for two-dimensional and three-dimensional fields with typical correlation functions are given in Appendix 35.

2.3.2.4 Simulation of Gaussian Random Fields

The easiest way to simulate a Gaussian random field is to use the parametric model. The algorithm cited in Section 2.3.2.2 can be used to obtain:

$$\eta_N(\vec{x}) = \sum_{j=1}^{N} A_j \cdot \sin\left[\vec{v}_j^T(\vec{x}_0 + \vec{x}) + \frac{\pi}{4}\right]$$

where $A_j = \sqrt{2}z_j \cdot N^{-1/2}$ are random amplitudes and $\vec{v}_j$ are random spatial frequencies. This representation gives an accurate simulation of the first two moments (mean and correlation function) of the Gaussian distribution for any N. The higher N is, the higher the accuracy of simulation of the Gaussian moments of higher orders. When $N \to \infty$ the multidimensional Gaussian probability distribution function is synthesized with zero error.

Statistical Analysis of Random Functions

2.4 Estimation of Random Function Parameters

2.4.1 Analysis of a Stationary Random Process

The most common process used in simulation of radio circuits and devices is the ergodic stationary random process $\xi(t)$. For such a process all parameters can be estimated by analyzing a single available sample $x(t)$. The four main parameters of an ergodic stationary process that a researcher is typically interested in are:

A. Mathematical expectation (mean) m_ξ ;

B. Variance $D_\xi = \sigma_\xi^2$;

C. Autocorrelation function (ACF) $K_\xi(\tau) = \sigma_\xi^2 R_\xi(\tau)$;

D. Spectral density (power spectrum) $S_\xi(\omega)$.

When the samples are taken with equal increments in time Δt, convenient expressions to estimate these parameters of a stationary ergodic process $\xi(t)$ are given in Appendix 36 ($\hat{u}$ denotes an estimate of u). N is a number of discrete points in a sample $x(t)$ with the length T, where $T = N \cdot \Delta t$.

Accuracy and reliability of estimates depend much on correct choice of the parameters N and Δt. These parameters should be defined before simulation starts. When mathematical expectation m_ξ is known, variances of the estimates are:

$$D_{\hat{m}_\xi} = \frac{D_\xi}{N}\left[1 + \frac{2}{N}\sum_{k=1}^{N-1}\left(1 - \frac{k}{N}\right)R_\xi(k\Delta t)\right]$$

$$D_{\hat{D}_\xi} = \frac{D_\xi^2}{N}\left\{1+4\left[\sum_{k=1}^{N-1}\left(1-\frac{k}{N}\right)\left(R_\xi\left(k\Delta t\right)\right)^2 + D_\xi^2\right]\right\}$$

$$D_{\hat{K}_\xi} = \frac{D_\xi^2}{N-r}\left\{\begin{array}{l}1+R_\xi^2\left(r\Delta t\right)+2\cdot\sum_{k=1}^{N-1-r}\left(1-\frac{k}{N-r}\right)\times\\ \times\left[R_\xi^2\left(k\Delta t\right)+R_\xi\left[(k+r)\Delta t\right]\cdot R_\xi\left[(k-r)\Delta t\right]\right]\end{array}\right\}$$

When m_ξ is unknown, the expressions to estimate variances of the estimates become very bulky. That is why, without considerable loss in accuracy, expressions cited above are usually used to get the estimates for the process with unknown m_ξ too if a sample $x(t)$ is long enough. For the majority of the random processes $\xi(t)$ considered in radio engineering applications the following asymptotic equations are valid:

$$D_{\hat{m}_\xi} = \frac{C_m}{T}+o\left(\frac{1}{T}\right) \quad D_{\hat{D}_\xi} = \frac{C_D}{T}+o\left(\frac{1}{T}\right) \quad D_{\hat{k}_\xi} = \frac{C_K}{T}+o\left(\frac{1}{T}\right)$$

where C_m, C_D, C_K are constants and $o(1/T)$ is the infinitesimal value of the order higher than T. These equations show behavior of the estimate accuracy when the length of sample T grows. When $T > 20\tau_c$, where τ_c is a correlation interval, the factor $o(T)$ can be omitted (the error introduced does not exceed several percent).

The concept of the effective number of samples is often used. The effective number of samples is defined as a number N_{ef} of independent values of the process $\xi(t)$ that gives the same variance for the estimate as the entire sample $x(t)$ would give. For $\hat{m}_\xi$ and $\hat{K}_\xi$:

$$N_{ef}\left(\hat{m}_\xi\right) = \frac{D_\xi}{D_{\hat{m}_\xi}}$$

$$N_{ef}\left(\hat{K}_\xi\right) = \frac{\left(D_\xi\right)^2\cdot\left[1+\left(R_\xi\left(\tau\right)\right)^2\right]}{D_{\hat{K}_\xi}}$$

Typically, statistical parameters of the simulated process are unknown *a priori* (i.e., before analysis starts). That is why it is expedient to divide the procedure into two phases. At the first stage the tentative values of the parameters N, Δt are determined. At the second stage these values are estimated more accurately by using the test samples. At the first stage the following approach can be recommended:

A. First, length T of the sample $x(t)$ has to be chosen. Because the correlation function of the process $\xi(t)$ is unknown before the analysis is performed, the autocorrelation function initially can be approximated by the simplest function:

$$R_\xi(\tau) = e^{-\alpha|\tau|}$$

For $|R_\xi(\tau)| \le 0.05$ (less than 5%) the correlation interval is $\tau_c = 3/\alpha$ and the expressions to estimate T are:

$$D_{\hat{m}_\xi} = \frac{2D_\xi}{\alpha T}; \quad D_{\hat{D}_\xi} = \frac{2D_\xi^2}{\alpha T}$$

which gives:

$$\hat{T} = 2 \bigg/ \left(\alpha \cdot \frac{D_{\hat{m}_\xi}}{D_\xi} \right) \quad \text{for the mean value;}$$

$$\hat{T} = 2 \bigg/ \left(\frac{D_{\hat{D}_\xi}}{D_\xi^2} \right) \quad \text{for the variance.}$$

These expressions can be used to find the required length of the sample $\hat{T}$ based on accuracy of mean or variance estimate ($E = \left(\sigma_{\hat{m}_\xi} / \sigma_\xi \right)$ or $E = \left(\sigma_{\hat{D}_\xi} / D_\xi \right)$):

$$\left(\hat{T}/\tau_c \right) = \frac{2}{3} \cdot \frac{1}{E^2}$$

The dependence of the sample length in fractions of correlation time T/τ_c upon required accuracy E in percent is given in Figure 2.2. Typically, $E = 5\%$ and the length equal to 267 correlation intervals is a reasonable choice.

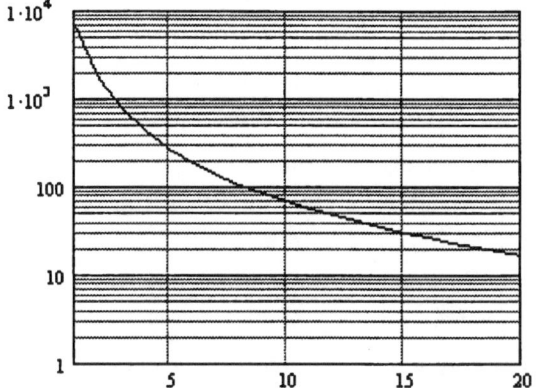

Figure 2.2 Sample length relative to correlation time $\hat{T}/\tau_c$ vs. required accuracy of the estimate (percent).

B. When the estimate of length $\hat{T}$ is defined, the estimate of an increment $\Delta\hat{t}$ has to be selected. After this is done the required number of samples can be defined as:

$$\hat{N} = \frac{\hat{T}}{\Delta\hat{t}}$$

The increment Δt depends on correlation interval τ_c too. The estimates $\Delta\hat{t}_1$ for mean, and the estimates $\Delta\hat{t}_2$ for variance and autocorrelation function can be found by solving the equations:

$$\frac{4}{9}R_\xi\left(\frac{\Delta\hat{t}_1}{2}\right) - \frac{5}{18}R_\xi\left(\Delta\hat{t}_1\right) - \frac{1}{6} = 0$$

$$\frac{4}{9}\left[R_\xi\left(\frac{\Delta\hat{t}_2}{2}\right)\right]^2 - \frac{5}{18}\left[R_\xi\left(\Delta\hat{t}_2\right)\right]^2 - \frac{1}{6} = 0$$

For the process with autocorrelation function $R_\xi(\tau) = e^{-\alpha|\tau|}$ the solution is:

$$\Delta\hat{t}_1 = 1.02/\alpha \; ; \; \Delta\hat{t}_2 = 0.51/\alpha$$

C. The estimate of the spectral density $\hat{S}_\xi(\omega)$ is based on the estimate of the correlation function $K_\xi(\tau)$. The main source of error in this case is the estimation error for an autocorrelation function transformed to the spectrum estimate via model $\hat{S}_\xi(\omega) =$ function $\left(\hat{K}_\xi(\tau)\right)$ in Appendix 36. Analysis of these errors shows that the reliable estimate of the spectrum can be obtained even when the number of samples N_S used to estimate $S_\xi(\omega)$ is much less than the number of samples N_K used to estimate the autocorrelation function. Typically, $N_S \leq 0.1 N_K$. On the other hand, when the number of samples is limited, it shortens the interval $T_S = N_S \cdot \Delta t$ and "smears" the estimate of the function $\hat{S}_\xi(\omega)$. That is why too few samples N_S should not be used. The increment Δt_S is selected to avoid cutoff effects at the higher frequencies:

$$\hat{\Delta t}_S < \frac{\pi}{\beta \cdot \omega_b}$$

where ω_b is a boundary frequency starting from which $S_\xi(\omega_b) = 0$, and $\beta = (2 \div 5)$ is an empirical coefficient.

2.4.2 Analysis of a Nonstationary Random Process

In the general case a single sample is not sufficient to estimate statistical parameters of a nonstationary random process. M samples have to be obtained at the interval T that corresponds to the model $\xi(t) = \{x_m(t)\}$ $m = \overline{0, M-1}$. These give M independent samples $x_m(n \cdot \Delta t)$, $n = \overline{0, N-1}$; $m = \overline{0, M-1}$ of the discrete random function $\xi(n\Delta t)$. Each sample is taken in N points with an increment Δt at the interval T.

Thus, for each moment of time $n \cdot \Delta t = k$ $(k = 0,1,2,...)$ M samples of the random process $x_m(k)$, $m = \overline{0, M-1}$, exist. Algorithms to estimate the first four moments of a nonstationary random process $\xi(t)$ with independent samples $x(k)$ are given in Appendix 37. These algorithms are valid for any probability distribution function that provides finite moments. For some common distributions the expressions can be simplified:

    A. Gaussian pdf:

$$f_1(x) = \frac{1}{\sqrt{2\pi}\sigma} e^{-\frac{(x-a)^2}{2\sigma^2}}$$

$$m_\xi = a, \quad D_\xi = \sigma^2$$

$$\sigma_{\hat{D}_\xi} \big/ D_\xi = \sqrt{2/(M-1)}; \quad \sigma_{\hat{\sigma}_\xi} \big/ \sigma_\xi \approx 1/\sqrt{2M}$$

B. Exponential pdf:

$$f_1(x) = \frac{1}{\lambda} e^{-x/\lambda}$$

$$m_\xi = \lambda \quad \sigma_{\hat{m}_\xi} = \lambda/\sqrt{M-1} \quad \sigma_{\hat{m}_\xi} \big/ m_\xi = 1/\sqrt{M-1}$$

C. Rayleigh pdf:

$$f_1(x) = \frac{\pi x}{(2m)^2} \cdot e^{-\frac{\pi}{4}\cdot\left(\frac{x}{m}\right)^2}$$

$$m_\xi = m \quad \sigma_{\hat{m}_\xi} = m \cdot \frac{\sqrt{4-\pi}}{\sqrt{\pi(M-1)}} \quad \sigma_{\hat{m}_\xi} \big/ m_\xi = \frac{\sqrt{4-\pi}}{\sqrt{\pi(M-1)}}$$

2.4.3 Analysis of the Distribution Functions

The basic approach used to estimate the distribution function of the random variable ξ with independent samples $x_1,...x_N$ is to form the empirical distribution function:

$$\hat{F}_N(x) = \frac{N_x}{N} = \frac{1}{N}\sum_{n=1}^{N}\theta(x - x_n) \tag{2.7}$$

where N_x is the number of occurrences ξ_n less than x. Function $\theta(u)$:

$$\theta(u) = \begin{cases} 1, & u > 0 \\ 0, & u \leq 0 \end{cases}$$

The estimate (2.7) is unbiased:

$$M\{\hat{F}_N(x)\} = F(x)$$

and its variance is:

$$\sigma^2\left[\hat{F}(x)\right]=\frac{F(x)\left[1-F(x)\right]}{N}$$

A variety of other methods are used to estimate distribution laws that differ in complexity and accuracy: the method of histograms, the method based on expansion into series, and parametric and nonparametric methods [11, 12].

2.5 The Random Processes and RF Circuits

2.5.1 Linear Circuits

Definition. A *linear circuit is one for which output and input signals are linked by a linear differential equation*:

$$a_n(t)\frac{d^n}{dt^n}\cdot\eta(t)+...+a_1\frac{d}{dt}\eta(t)+a_0(t)\eta(t)$$
$$=b_m(t)\frac{d^m}{dt^m}\xi(t)+...+b_1(t)\frac{d}{dt}\xi(t)+b_0(t)\xi(t)$$

(2.8)

where $\eta(t)$ is the process (signal) at the circuit output, and $\xi(t)$ is the process (signal) at the circuit input.

To model how the linear circuit (system) transforms the random process $\xi(t)$ passing through it, one can use a description at the level of differential equations, impulse response or frequency response. In the general case, differential equations are preferable when both stationary and nonstationary modes of system operation are of interest while initial conditions are nonzero. When the initial conditions are zero ones, the impulse response may be a more convenient choice. And when only the stationary mode of system operation is considered, the frequency response is a reasonable choice.

When the impact of the random process on the linear circuit is simulated, typically there are two basic tasks of interest:

A. To determine mean, correlation function and spectral density of the output random signal $\eta(t)$ based on the corresponding parameters of the input random signal $\xi(t)$.

B. To determine the probability distribution function of the output signal $\eta(t)$ when the distribution function of the input signal $\xi(t)$ is known.

The second task is more generic and, if solved, covers task 1 also. But usually it is a far more complicated undertaking and no simple analytical models exist with the exception of the case when $\xi(t)$ is a stationary Gaussian process.

That is why the most common approach in linear circuit analysis is to define the correlation function of the output process based on the specified correlation function of the input process.

2.5.1.1 Method of Differential Equation

If $\xi(t)$ is the input random process with parameters:

$$m_\xi(t) = M\left[\xi(t)\right]$$

$$K_\xi(t_1,t_2) = M\left\{\left[\xi(t_1) - m_\xi(t_1)\right]\left[\xi(t_2) - m_\xi(t_2)\right]\right\},$$

and denoting $p = \dfrac{d}{dt}$,

$$A(p,t) = \sum_{i=0}^{n} a_i(t) \cdot p^i, \quad B(p,t) = \sum_{i=0}^{m} b_i(t) p^i$$

(2.8) becomes:

$$A(p,t) \cdot \eta(t) = B(p,t)\xi(t)$$

Then, formally:

$$\eta(t) = \frac{B(p,t)}{A(p,t)} \cdot \xi(t) = L(p,t) \cdot \xi(t)$$

where

$$L(p,t) = \frac{B(p,t)}{A(p,t)}$$

is called a linear operator of the system (circuit).

The output process parameters are modeled as:

$$m_\eta(t) = L(p,t) \cdot m_\xi(t)$$

$$K_\eta(t_1,t_2) = L(p,t_1)L(p,t_2)K_\xi(t_1 t_2)$$

2.5.1.2 The Method of Impulse Response

Definition. *Impulse response h(t) of the linear system (circuit) is the system output signal when input signal is a δ -function.*

Thus, the differential equation is:

$$a_n(t) \cdot \frac{d^n}{dt^n} h(t) + \ldots + a_1(t) \frac{d}{dt} h(t) + a_0(t) h(t)$$

$$= b_m(t) \frac{d^m}{dt^m} \delta(t) + \ldots + b_1 \frac{d}{dt} \delta(t) + b_0(t) \delta(t)$$

and $h(t)$ can be found when the equation is solved under zero initial conditions. In this case the output signal is:

$$\eta(t) = \int_0^t \xi(\tau) h(t-\tau) dt = \int_0^t h(x) \xi(t-x) dx$$

Models for the moments of the random process at the output of a linear system based on impulse response representation are given in Appendix 38. We should mention that the output signal $\eta(t)$ in the general case is a nonstationary process even if the input signal $\xi(t)$ is a stationary process. For passive linear circuits when the time interval considered is relatively large, the process $\eta(t)$ will approach a stationary one. In this case:

$$m_\eta = m_\xi \int_0^\infty h(x) dx$$

$$K_\eta(\tau) = \int_0^\infty h(x) dx \int_{-\infty}^{\tau+x} K_\xi(\tau) h(\tau + x - z) dz$$

The mutual correlation functions of the process $\eta(t)$ and a nonstationary $\zeta(t)$ or a stationary $\xi(t)$ processes correspondingly are:

$$K_{\zeta\eta}(t_1, t_2) = \int_0^{t_2} K_\zeta(t_1, \tau) h(t_2 - \tau) dt$$

$$K_{\xi\eta}(t_1, t_2) = \int_0^{t_2} K_\xi(t_1 - \tau) h(t_2 - \tau) dt$$

2.5.1.3 The Method of Frequency Response

Definition. *Complex frequency response of the system (circuit) is a function:*

$$H(t, j\omega) = |H(t, j\omega)| \cdot e^{j \arg [H(t, j\omega)]}$$

This representation is convenient when an estimate of the spectral density of the output process is required:

$$S_\eta(\omega, t) = S_\xi(\omega) \cdot |H(t, j\omega)|^2$$

If the input of the system is a harmonic oscillation $\chi(t, \omega) = e^{j\omega t}$, the differential equation is:

$$H_1(t, j\omega) = \frac{B(p, t)}{A(p, t)} \cdot e^{j\omega t}$$

The solution can be written as:

$$H_1(t, j\omega) = H(t, j\omega) \cdot e^{j\omega t}$$

For a system with constant parameters, frequency response does not depend on time: $H(t, j\omega) = H(j\omega)$. In this case frequency response $H(j\omega)$ and impulse response $h(t)$ are related by the Fourier transform pair:

$$h(t) = \frac{1}{2\pi} \int_{-\infty}^{\infty} H(j\omega) e^{j\omega t} \, d\omega$$

$$H(j\omega) = \int_{0}^{\infty} h(t) e^{-j\omega t} \, dt$$

The models of impulse and frequency responses for common linear circuits with constant parameters are given in Appendix 39. The correlation functions and spectral densities at the output of these circuits when the input signal is the stationary Gaussian white noise with zero mean are given in Appendix 40.

2.5.2 Nonlinear Circuits

The basic nonlinear transformation that is typically considered when nonlinear circuits are modeled is:

$$\eta(t) = \varphi[\xi(t)]$$

where $\varphi[\xi]$ is a nonlinear deterministic function.

In this case the output process in any arbitrary moment of time is completely defined by the input process at the same moment of time and function $\varphi[\xi]$. Such a transformation is called an inertialess or functional one. The circuit that performs the inertialess transformation is also called the inertialess circuit. Sometimes input signal $\xi(t)$ is additionally modified by linear circuits:

$$\zeta(t) = L_2 f[L_1 \xi(t)]; \quad \eta(t) = \varphi[\zeta(t)]$$

where L_1, L_2 are linear operators describing corresponding linear circuits.

2.5.2.1 General Approach

If n-dimensional probability density function $f_n(x_1,...x_n)$ of the random variables $\xi_1,...\xi_n$ is known and n-dimensional probability density function $f_n(y_1,...y_n)$ of the random variables $\eta_1,...,\eta_n$ has to be found:

$$y_1 = \varphi_1(x_1,...x_n)$$
$$. \quad . \quad . \quad . \quad .$$
$$y_n = \varphi_n(x_1,...,x_n)$$

and if the single-valued reciprocal function exists:

$$x_1 = q_1(y_1,...y_n)$$
$$. \quad . \quad . \quad .$$
$$x_1 = q_n(y_1,...y_n)$$

the probability density function of interest is:

$$f_n(y_1,...y_n) = f_n[q_1(y_1,...y_n)...q_n(y_1,...y_n)]\cdot|D_n|$$

$$D_n = \frac{\partial(x_1,...x_n)}{\partial(y_1,...y_n)}$$

If the reciprocal function is not single valued, the sum of all regions has to be taken.

2.5.2.2 Polynomial Transformation

If at some point C the signal can be expanded in terms of Taylor series:

$$\eta = f\,[\xi] = a_0 + a_1\,(\xi - C) + ... + a_n\,(\xi - C)^n$$

$$a_k = \frac{1}{k!}\frac{d^k\,f\,[\xi]}{d\xi^k}\bigg|_{\xi=C}$$

the moments can be found as:

$$m_\eta\,(t) = M\,\{\eta\,(t)\} = a_0 + a_1\,M\,\{\xi\,(t) - C\} + ... + M\,\{\xi\,(t) - C]^n\}$$

$$B_\eta\,(t_1,t_2) = a_0^2 + a_0\,a_1\,\big[m_\xi\,(t_1) + m_\xi\,(t_2) - 2C\big]$$

$$+ a_1^2\,\big[B_\xi\,(t_1,t_2) - C\,m_\xi\,(t_1) - C\,m_\xi\,(t_2) + C^2\big] + ...$$

$$+ a_n^2\,M\,\{[\xi\,(t_1) - C]^n\,[\xi\,(t_2) - C]^n\}$$

For the stationary Gaussian random process $\xi\,(t)$ with zero mean, variance $D_\xi = \sigma_\xi^2$ and normalized correlation function $R_\xi\,(\tau)$:

$$m_\gamma\,(t) = M\,\{\xi^\gamma\,(t)\} = \begin{cases} 1\cdot 3\cdot 5\cdot ... \cdot (\gamma - 1)\sigma_\xi^\gamma, & \gamma \text{ even} \\ 0, & \gamma \text{ odd} \end{cases}$$

$$m_{\mu\gamma}\,(\tau) = M\,\{\xi^\mu\,(t)\xi^\gamma\,(t+\tau)\} = \sigma_\xi^{\mu+\gamma} \times \sum_{k=o}^{\infty}\frac{1}{k!}N_{\mu k}\,N_{\gamma k}\,R_\xi^k\,(\tau)$$

$$N_{ik} = \int\limits_{-\infty}^{\infty}\xi^i\,\Phi^{(k+1)}\,(\xi)\,d\xi$$

$\Phi^{(n)}\,(z) = d\Phi\,(z)\,/\,dz^n$ is the nth-order derivative of the error function:

$$\Phi(z) = \frac{1}{\sqrt{2\pi}} \int_{-\infty}^{z} e^{-x^2/2} \, dx$$

2.5.2.3 The Piece-Discontinuous and Transcendental Transform

The nonlinear curve $\varphi[\xi]$ is often approximated by piece-discontinuous or transcendental functions when the impact of strong signals or interference on nonlinear circuit is considered. The moments of the output process can be modeled in two basic ways: by the direct method or by the method of characteristic functions.

The direct method employs nonlinear characteristics:

$$m_{\eta\gamma}(t) = M\left\{\eta^{\gamma}(t)\right\} = \int_{-\infty}^{\infty} \varphi^{\gamma}\left[\xi(t)\right] f_1(x;t) \, dx$$

$$m_{\eta 1,1}(t) = M\left\{\eta(t_1)\eta(t_2)\right\} = \int_{-\infty}^{\infty}\int_{-\infty}^{\infty} \varphi\left[\xi(t_1)\right]\varphi\left[\xi(t_2)\right] f_2(x_1,x_2;t_1,t_2) \, dx_1 \, dx_2$$

The method of characteristic functions employs a Laplace transform to model the nonlinear characteristic:

$$\eta = \varphi[\xi] = \frac{1}{2\pi} \int_{L} F(ju) e^{ju\xi} \, du$$

where L is an integration contour in the complex domain u. Integration contours and functions $F(ju)$

$$F(ju) = \int_{0}^{\infty} \varphi[\xi] e^{-ju\xi} \, d\xi$$

for some common nonlinear circuits are given in Appendix 41.

The function

$$m_{\eta_{1,1}}(t_1 \, t_2) = \frac{1}{4\pi^2} \int_{L}\int_{L} F(ju_1) F(ju_2) \theta_2(ju_1, ju_2;t_1,t_2) \, du_1 \, du_2$$

is the model for the case when θ_2 is a two-dimensional characteristic function of the input signal $\xi(t)$. When $\xi(t)$ is a stationary Gaussian process,

$\xi(t) \in N(0, \sigma_\xi^2)$, with correlation function $K_\xi(\tau) = \sigma_\xi^2 \cdot R_\xi(\tau)$ this model results in:

$$R_\eta(\tau) = \sigma_\xi^{2\gamma-2} \sum_{n=1}^{\infty} \left\{ \int_{-\infty}^{\infty} \varphi^{(\gamma)}[\xi] \Phi^{(n+1-\gamma)}\left(\frac{\xi}{\sigma_\xi}\right) d\xi \right\}^2 \frac{R_\varsigma^n(\tau)}{n!}$$

2.5.3 Random Narrowband Signal

In many radio engineering applications the signal can be considered to be narrowband. This means that the width of the frequency band Δf where the signal spectrum $S_\xi(f) \neq 0$ is small with respect to its central frequency f_0:

$$S_\xi(f) \neq 0, \quad f_0 - \frac{\Delta f}{2} \leq f \leq f_0 + \frac{\Delta f}{2}, \quad \Delta f \ll f_0$$

Often f_0 is chosen as:

$$f_0 = \frac{\omega_0}{2\pi} = \frac{\int_0^{\infty} f \, S_\xi(f) df}{\int_0^{\infty} S_\xi(f) df}$$

The correlation function of the narrowband stationary process:

$$K_\xi(\tau) = \sigma_\xi^2 \left[r_c(\tau) \cos \omega_0 \tau + r_s(\tau) \sin \omega_0 \tau \right] = \sigma_\xi^2 r(\tau) \cos \left[\omega_0 \tau + \gamma(\tau) \right]$$

where $r(t) = \sqrt{r_c^2(\tau) + r_s^2(\tau)}$;

$$\tan \gamma(t) = -\frac{r_s(\tau)}{r_c(\tau)} \; ;$$

σ_ξ^2 is a variance of $\xi(t)$.

If the spectral density $S_\xi(f)$ is a symmetrical function with respect to f_0, then:

$$R_\xi(\tau) = \sigma_\xi^2 \, r(\tau) \cos \omega_0 t$$

since $r_s(\tau) = 0, \quad \gamma(\tau) = 0$.

The sample of the narrowband signal looks like a modulated harmonic oscillation:

$$\xi(t) = A(t) \cos[\omega_0 t + \varphi(t)]$$

$$A(t) \geq 0, \quad -\pi \leq \varphi(t) \leq \pi$$

where amplitude $A(t)$ and phase $\varphi(t)$ variation in time is much slower compared to variation of the function $\cos \omega_0 \tau$. They are termed the *envelope* and *phase* of the narrowband random process. Another representation of the process $\xi(t)$:

$$\xi(t) = A_c(t) \cos \omega_0 t - A_s(t) \sin \omega_0 t$$

where $A_c(t) = A(t) \cos \varphi(t)$; $A_s(t) = A(t) \sin \varphi(t)$

Consequently,

$$A(t) = \sqrt{A_c^2(t) + A_s^2(t)} \qquad \varphi(t) = \arctan\left[\frac{A_s(t)}{A_c(t)}\right]$$

$$\omega(t) = \omega_0 + \frac{d\varphi(t)}{dt} = \omega_0 + \frac{\dot{A}_s(t) A_c(t) - \dot{A}_c(t) A_s(t)}{A_c^2(t) + A_s^2(t)}$$

$$\dot{A}(t) = dA(t)/dt$$

The common case in stochastic analysis of radio circuits is to consider the sum of the narrowband stationary Gaussian process with zero mean $\xi(t) = A(t) \cos[\omega_0 t + \varphi(t)]$ (random noise) and deterministic harmonic signal $s(t) = A_m \cos \omega_0 t$:

$$\eta(t) = s(t) + \xi(t) = V(t) \cos[\omega_0 t + \psi(t)]$$

$$V(t) \geq 0, -\pi \leq \psi(t) \leq \pi$$

and

$$V(t)\cos\psi(t) = A_m + A_c(t)$$

$$V(t)\sin\psi(t) = A_s(t)$$

Models of the basic statistical parameters of the resultant random process $\eta(t)$ when $\xi(t)$ has a correlation function $K_\xi(\tau) = \sigma_\xi^2 r(\tau)\cos\omega_0 t$ are given in Appendix 42.

Chapter 3

Methods of Modeling

3.1 Method of Carrier Frequency

This method is typically used to simulate RF and video-frequency radio circuits. The mathematical models can be devised both at the level of electrical diagrams and block diagrams. The signals and noise passing through the circuits are modeled by instantaneous voltages and currents (i.e., the wave structure is reproduced at the carrier frequency).

3.1.1 Narrowband Signal Modeling

Any deterministic narrowband signal can be modeled as:

$$u\left[t, \vec{\lambda}(t)\right] = A\left[t, \vec{\lambda}(t)\right] \cdot \sin\left\{\omega_0 t - \varphi\left[t, \vec{\lambda}(t)\right] - \varphi_0\right\}$$

where u is an instantaneous value of the signal voltage, A is an amplitude and φ is phase. Vector function $\vec{\lambda}(t)$ varies slowly in comparison with the carrier frequency ω_0. Thus, in terms of derivative of the function $\vec{\lambda}(t)$, it can be considered to be a constant parameter: $\vec{\lambda}(t) = \vec{\lambda}$, and the model becomes:

$$u = u\left(t, \vec{\lambda}\right) = A\left(t, \vec{\lambda}\right)\sin \Phi\left(t, \vec{\lambda}\right)$$

where

$$\Phi\left(t, \vec{\lambda}\right) = \omega_0 t - \varphi\left(t, \vec{\lambda}\right) - \varphi_0$$

is a complete phase.

The generic mathematical model describing a narrowband signal is:

$$\frac{d^2u}{dt^2} + a_1\left[t, \vec{\lambda}(t)\right]\frac{du}{dt} + a_0\left[t, \lambda(t)\right]u = 0$$

where

$$a_1 = -\left[\frac{\ddot{\Phi}(t, \vec{\lambda})}{\dot{\Phi}(t, \vec{\lambda})} + 2\frac{\dot{A}(t, \vec{\lambda})}{A(t, \vec{\lambda})}\right]$$

$$a_0 = \dot{\Phi}^2(t, \vec{\lambda}) + 2\frac{\dot{A}^2(t, \vec{\lambda})}{A^2(t, \vec{\lambda})} + \frac{\dot{A}(t, \vec{\lambda})\ddot{\Phi}(t, \vec{\lambda})}{A(t, \vec{\lambda})\dot{\Phi}(t, \vec{\lambda})} - \frac{\ddot{A}(t, \vec{\lambda})}{A(t, \vec{\lambda})}$$

The corresponding coefficients a_1, a_0 for different types of modulation are given in Appendix 43.

The solution of the differential equation for a narrowband signal with coefficients cited in Appendix 43 corresponds to the following models of the modulated signals:

A. Amplitude-modulated signal:

$$u\left[t, \vec{\lambda}(t)\right] = A_0\left[1 + m_{AM}\,\vec{\lambda}(t)\right]\sin\left(\omega_0 t - \varphi_0\right)$$

B. Frequency-modulated signal:

$$u\left[t, \vec{\lambda}(t)\right] = A_0\sin\left[\omega_0 t - m_{FM}\cdot\int_0^t \vec{\lambda}(t)dt - \varphi_0\right]$$

C. Phase-modulated signal:

$$u\left[t, \vec{\lambda}(t)\right] = A_0\sin\left[\omega_0 t - m_{PM}\cdot\vec{\lambda}(t) - \varphi_0\right]$$

Here m_{AM}, m_{FM}, m_{PM} are indices of amplitude, frequency and phase modulations that describe maximum deviation of the corresponding parameter from mean value.

3.1.2 Modeling of Common Devices and Circuits

The mathematical models of the basic devices used in radio engineering are given in Appendix 44. The linear radio-frequency (RF) and intermediate-frequency (IF) amplifiers controlled by an automatic gain control (AGC) circuit should be modeled as series connection of two elements: the first element is an element with transfer function $K_{RF}(u_{AGC})$ or $K_{IF}(u_{AGC})$ that is a nonlinear function of AGC voltage, and the second one is a stationary linear element with normalized impulse response $h_{RF}(t)$ or $h_{IF}(t)$. The mathematical modeling at this level is less accurate than modeling at the level of the circuit components and can be used in the early stages of system simulation when electrical diagrams are not yet available.

3.1.3 Simplified Algorithms

In general any component, circuit or device of a radio system can be described by the nonlinear differential equation:

$$\sum_{k=0}^{n} a_k(t) y^{(k)}(t) = \text{function} \left[y, y^{(1)}, \ldots y^{(n)}; \; x, x^{(1)}, \ldots x^{(m)} \right] \qquad (3.1)$$

where $x(t)$ and $y(t)$ are the input and output signals, respectively.

The first approach is applicable when time constant τ_0 of the circuit is much less than correlation interval τ_c of the input random signal $(\tau_0 \ll \tau_c)$. Then all derivatives in equation (3.1) are equal to zero, and the model takes the form:

$$a_0(t) \cdot y(t) = \text{function} \left[y(t), 0, \ldots 0 \right]$$

This equation has to be solved with respect to $y(t)$:

$$y(t) = F \left[x(t) \right]$$

to describe the output signal via specified input signal $x(t)$.

The second approach is applicable to the random signal $\xi(t) = x(t)$ and based on the approximation of the output signal as:

$$\eta(t) = m_\eta(t) + \eta^0(t)$$

i.e., by two components: mean $m_\eta(t)$ and fluctuation component $\eta^0(t)$. If nonlinear function $\eta = F(\xi)$ is determined, model (3.1) can be split into two differential equations: one equation is for $m_\eta(t)$ and the second one is for $\eta^0(t)$. If fluctuations $\eta^0(t)$ are small enough with respect to the mean $m_\eta(t)$, the second equation can be brought to a linear form.

3.2 Method of Complex Envelope

The essence of this method is to replace the RF element with a narrowband signal at its input by the model of an equivalent low-frequency element with the input signal represented by the complex envelope of the narrowband signal. The vector model of a signal, interference (other than noise) and noise can be represented as:

$$u_\Sigma(t) = \text{Re}\left\{\overline{A_s}(t)\cdot e^{j\omega_s t}\right\} + \text{Re}\left\{\overline{A_I}(t)\cdot e^{j\omega_I t}\right\} + \text{Re}\left\{\overline{A_n}(t)\cdot e^{j\omega_n t}\right\} = \text{Re}\left\{\overline{A(t)}\cdot e^{j\omega_0 t}\right\}$$

3.2.1 Modeling of the Stationary Linear Circuits

Any stationary linear RF element with resonance frequency ω_0 can be described by the impulse response $h(t)$, complex transfer function $H(j\omega)$ or linear differential equation with constant coefficients:

$$\sum_{i=0}^{n} a_i \cdot y^{(i)}(t) = \sum_{j=0}^{m} b_j \cdot x^{(j)}(t), \quad m < n \tag{3.2}$$

Based on any of these approaches the equivalent model can be devised by the method of complex envelope.

3.2.1.1 Time-Domain Method

If the stationary linear element with impulse response $h(t)$ is specified:

$$h(t) = \text{Re}\left\{\overline{H(t)}\cdot e^{j\omega_0 t}\right\} = \text{Re}\left\{H(t)\cdot e^{j[\omega_0 t - \varphi(t)]}\right\} = H(t)\cos[\omega_0 t - \varphi(t)],$$

where ω_0 is a resonant frequency for the specified element and $\overline{H(t)}$ is a complex envelope of impulse response:

$$\overline{H}(t) = H(t)\cdot e^{-j\varphi(t)} = h_c(t) - j h_s(t)$$

$$h_c(t) = \text{Re}\left\{\overline{H}(t)\right\} = H(t)\cos\varphi(t)$$

$$h_s(t) = \text{Im}\left\{\overline{H}(t)\right\} = H(t)\sin\varphi(t)$$

a convenient form to write down the models of the input and output signals is:

$$x(t) = \text{Re}\left\{A_x(t) \cdot e^{j[\omega_s t - \varphi_x(t)]}\right\}$$

$$y(t) = \text{Re}\left\{A_y(t) \cdot \exp j[\omega_s t - \varphi_y(t)]\right\}$$

where ω_s is a carrier frequency of the signal.

The convolution integral for complex envelope is:

$$\overline{A}_y(t) = \overline{A}_x(t) \otimes \overline{h}_{eq}(t) = \int_0^\infty \overline{A}_x(t-\tau)\overline{h}_{eq}(\tau)d\tau = \int_{-\infty}^t \overline{A}_x(\tau)\overline{h}_{eq}(t-\tau)d\tau$$

where $\overline{h}_{eq}(t) = h(t) \cdot e^{-j\omega_s t}$ is a complex impulse response of the model, i.e., equivalent low-frequency filter.

Typically, the following form is used:

$$\overline{A}_x = a_x - jb_x$$

$$h_{eq \cdot c} = \frac{1}{2}H(t) \cdot \cos[\Delta\omega_s t + \varphi(t)]$$

$$h_{eq \cdot s} = \frac{1}{2}H(t) \cdot \sin[\Delta\omega_s t + \varphi(t)]$$

where $\Delta\omega_s = \omega_s - \omega_0$.

The final model of the equivalent complex filter then takes the form:

$$\overline{A}_y(t) = a_y(t) - j \cdot b_y(t)$$

$$a_y(t) = \int_0^\infty a_x(t-\tau)h_{eq \cdot c}(\tau)d\tau - \int_0^\infty b_x(t-\tau)h_{eq \cdot c}(\tau)d\tau$$

$$b_y(t) - \int_0^\infty a_x(t-\tau)h_{eq \cdot s}(\tau)d\tau + \int_0^\infty b_x(t-\tau)h_{eq \cdot c}(\tau)d\tau$$

3.2.1.2 Frequency-Domain Method

The frequency-domain method is based on modeling of the transfer function $H(j\omega)$ rather than impulse response $h(t)$. For an equivalent low-frequency filter:

$$H_{eq}(j\Omega) = H[j(\omega_s + \Omega)] = H[j(\omega_0 + \Delta\omega_s + \Omega)]$$

This results in correspondence of the model $y(t) = H(j\omega) \cdot x(t)$ and the equivalent model $\overline{A_y}(t) = H_{eq}(j\Omega) \cdot \overline{A_x}(t)$. Analogously, the Laplace transform representations take the form:

$$y(t) = H(p) \cdot x(t)$$

$$\overline{A_y}(t) = H_{eq}(s) \cdot \overline{A_x}(t)$$

where $p = j\omega; \quad s = j\Omega$

$$H_{eq}(s) = H(p_0 + \Delta p_s + s)$$

$$p_0 = j\omega_0; \quad \Delta p_s = j\omega_s$$

Several different approaches are used to describe $H_{eq}(s)$ or $H_{eq}(j\Omega)$.

A. *The argument substitution.* If the transfer function of the simulated element is:

$$H(p) = \frac{B(p)}{A(p)} = \sum_{j=0}^{m} b_j \cdot p^j \bigg/ \sum_{i=0}^{n} a_i \cdot p^i$$

the equation can be written in terms of the following parameters:

$$p = s + p_s, \quad p_s = p_0 + \Delta p_s = j\omega_s = j(\omega_0 + \Delta\omega_s)$$

Then

$$H_{eq}(s) = \frac{B(s + p_s)}{A(s + p_s)} = \frac{B_{eq}(s)}{A_{eq}(s)}$$

or

$$H_{eq}(j\Omega) = \frac{\sum_{l=0}^{m} B_l \left[j(\omega_0 + \Delta\omega_s) \right](j\Omega)^l}{\sum_{k=0}^{n} A_k \left[j(\omega_0 + \Delta\omega_s) \right](j\Omega)^k} \tag{3.3}$$

The model $H_{eq}(j\Omega)$ is a fractional-rational function with complex coefficients.

B. *Simplification of the transfer function.* In this approach the factors in the transfer function of higher order are neglected (e.g., $1 + j\omega/\omega_0 + (j\omega/\omega_0)^2 \approx 1 + j\omega/\omega_0$).

The example of this approach for resonance amplifiers with a single parallel resonator and bandpass filter are given in Appendix 45.

C. *Method of shortened equation.* This method is applicable to linear circuits with the complex transfer function:

$$H(j\omega) = \delta \frac{P(j\omega,\delta)}{Q(j\omega,\delta)}$$

where δ is a small parameter and P,Q are the polynomials of any arbitrary type. The essence of the method is to expand the polynomials $P(j\omega,\delta), Q(j\omega,\delta)$ in terms of Taylor series around the point $j\omega_0$ and Maclaurin series around the point δ. After this expansion is done, the factors of the series higher than the second order are neglected.

D. *Method of shortened differential equations for complex envelopes.* The basic equation (3.2) can be rewritten in operator form:

$$\sum_{k=0}^{n} a_k (j\omega)^k \cdot y(t) = \sum_{l=0}^{m} b_l (j\omega)^l \cdot x(t), \quad m < n$$

The parameters $x(t)$ and $y(t)$ are modeled as:

$$x(t) = \mathrm{Re}\left\{ A_x(t) \cdot e^{j[\omega_s t - \varphi_x(t)]} \right\}$$

$$y(t) = \mathrm{Re}\left\{ A_y(t) \cdot e^{j[\omega_s t - \varphi_y(t)]} \right\} \tag{3.4}$$

and, using the Leibniz formula, this equation can be solved as:

$$\sum_{k=0}^{n} A_k \left[j(\omega_0 + \Delta\omega_s) \right] \cdot \overline{A}_y^{(k)}(t) = \sum_{l=0}^{m} B_l \left[j(\omega_0 + \Delta\omega_s) \right] \cdot \overline{A}_x^{(k)}(t) \qquad (3.5)$$

which corresponds exactly to the solution (3.3). The exact equation (3.5) can be shortened since slow-variation conditions for the complex envelope are met. This leads to:

$$\overline{A}^{(1)}(t) \approx \delta\omega_0 \overline{A}(t)$$

$$\overline{A}^{(2)}(t) \approx \delta^2 \omega_0^2 \overline{A}(t)$$

where δ is the first-order small value. The complex model can be represented by:

$$\overline{A}_x(t) = a_x - jb_x; \quad \overline{A}_y(t) = a_y - jb_y$$

When inserted into equation (3.5), such a representation gives two linear differential equations with real coefficients to find the solution $a_y(t), b_y(t)$ at the output of the complex filter.

3.2.2 Modeling of the Nonstationary Linear Circuits

Two methods are typically used to model nonstationary linear circuits: time-domain method and method of differential equations.

3.2.2.1 Time-Domain Method

In this case model (3.4) is inserted into the following equation:

$$\overline{A}_y(t) = \int_0^\infty \overline{A}_x(t-\tau)\overline{h}_{eq}(t,\tau)d\tau$$

with nonstationary complex impulse response

$$\overline{h}_{eq}(t,\tau) = h(t,\tau)e^{-j\omega_s\tau}$$

3.2.2.2 Method of Differential Equations

The equation for a nonstationary circuit is:

$$\sum_{k=0}^{n} A_k(t, j\omega_s)\overline{A}_y^{(k)}(t) = \sum_{l=0}^{m} B_l(t, j\omega_s)\overline{A}_x^{(l)}(t)$$

where A_k, B_l are complex coefficients:

$$A_k(t, j\omega_s) = \sum_{\mu=k}^{n} C_\mu^k (j\omega_s)^{\mu-k} a_\mu(t)$$

$$B_l(t, j\omega_c) = \sum_{\gamma=l}^{m} C_\gamma^l (j\omega_s)^{\gamma-l} b_\gamma(t)$$

3.2.3 Modeling of Inertialess Nonlinear Circuits

The inertialess nonlinear circuit is a circuit consisting of two stages connected in a series. The first stage is an inertialess circuit transforming the input signal according to some nonlinear law $y = G(x)$. The second stage is a formation circuit. It has a unity gain and passes beat frequencies centered around the kth harmonic $k\omega_s$ of the carrier frequency without any distortions (Figure 3.1).

3.2.3.1 Single-Channel Circuits

When $k = 0$ the inertialess nonlinear circuit is a detector, when $k = 1$ it is a nonlinear amplifier or bandpass limiter, and when $k = 2$ or higher it is a frequency multiplier. If the model of the input signal is:

$$x(t) = \mathrm{Re}\left\{\overline{A_x}(t) \cdot e^{j\omega_s t}\right\} = A_x(t) \cdot \cos\Phi_x(t)$$

then for $k \geq 1$ the output signal is:

$$y_k(t) = \mathrm{Re}\left\{\overline{A}_{y_k}(t) \cdot e^{jk\omega_s t}\right\} = A_{y_k}(t)\cos\Phi_y(t)$$

where

$$y(t) = G\{A_x(t)\cos\Phi_x(t)\} = G(A\cos\Phi)$$

$$\approx y_0(t) + \sum_{k=1}^{n} y_k(t) = L_0(A) + \sum_{k=1}^{n} L_k(A)\cos(k\Phi)$$

$$\Phi_x(t) = \omega_s t - \varphi_x(t)$$

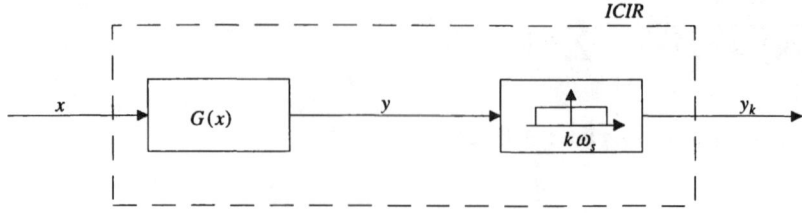

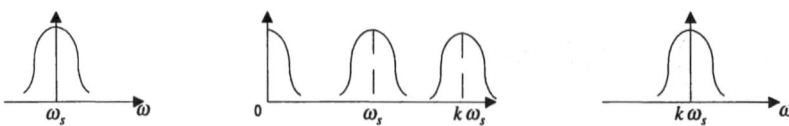

Figure 3.1 The inertialess nonlinear circuit model (from [13], p. 109)..

$$\Phi_{y_k}(t) = k\omega_s t - \varphi_{y_k}(t)$$

A. *Method of envelope and phase.* If the input of an inertialess nonlinear circuit is a narrowband signal, then the output signal can be expanded into Fourier series:

$$y(t) = G\{A_x(t)\cos\Phi_x(t)\} = G(A\cos\Phi)$$

$$\approx y_0(t) + \sum_{k=1}^{n} y_k(t) = L_0(A) + \sum_{k=1}^{n} L_k(A)\cos(k\Phi)$$

where

$$L_0(A) = \frac{1}{\pi} \int_0^\pi G(A\cos\Phi)d\Phi$$

$$L_k(A) = \frac{2}{\pi} \int_0^\pi G(A\cos\Phi)\cos k\Phi \, d\Phi$$

The model of the complex envelope transform is:

$$\overline{A}_{y_k}(t) = A_{y_k}(t) \cdot e^{-j\varphi_{y_k}(t)}$$

$$A_{y_k}(t) = L_k[A_x(t)], \quad k \geq 0$$

$$\varphi_{y_k}(t) = k\varphi_x(t), \quad k \geq 1$$

The coefficients $L_k(A)$ for some common nonlinear curves are given in Appendix 46.

B. *Method of contour integrals.* In this method nonlinear function $y = G(x)$ is represented as the Laplace transform of some image $g(p)$:

$$y = G(x) = \frac{1}{j2\pi} \int_{C-j\infty}^{C+j\infty} g(p)e^{px}\, dp, \quad x > 0, \quad p = \sigma + j\omega$$

If the input signal is $x = A\cos\Phi$

$$y(t) = \frac{1}{j2\pi} \int_{C-j\infty}^{C+j\infty} g(p)\exp[pA(t)\cos\Phi(t)]dp$$

It can be expanded in terms of the modified Bessel function series $I_k(z)$:

$$y(t) = \sum_{k=0}^{\infty} y_k(t) = \sum_{k=0}^{\infty} M_k[A(t)\cos k\Phi(t)]$$

where

$$M_k(A) = \frac{\varepsilon_k}{j2\pi} \int_{C-j\infty}^{C+j\infty} g(p)I_k(pA)dp$$

$$\varepsilon_k = \begin{cases} 1, & \text{if } k = 0 \\ 2 & \text{otherwise} \end{cases}$$

The model of the output signal takes the form:

$$\overline{A}_{y_k}(t) = A_{y_k}(t) \cdot e^{-j\varphi_{y_k}(t)}$$

$$A_{y_k}(t) = M_k[A_x(t)], \quad k \geq 0$$

$$\varphi_{y_k}(t) = k\varphi_x(t), \quad k \geq 1$$

If the curve $G(x)$ is a half-period nonlinear characteristic of γ th power:

$$y = G(x) = \begin{cases} k_0 x^{\gamma}, & x > 0 \\ 0, & x \leq 0 \end{cases}$$

then the image function is:

$$g(p) = \frac{k_0 \Gamma(\gamma + 1)}{p^{\gamma + 1}}$$

where $\Gamma(z)$ is a gamma function (see Appendix 1).

The coefficients are:

$$M_k(A) = k_0 \frac{\varepsilon_k \Gamma(\gamma + 1)}{2^{\gamma + 1} \Gamma\left(1 - \dfrac{k - \gamma}{2}\right) \Gamma\left(1 + \dfrac{k + \gamma}{2}\right)} A^{\gamma}$$

Thus, for detectors of the γ th power ($k = 0$):

$$M_k(A) = \frac{k_0 \Gamma(\gamma + 1) A^{\gamma}}{2^{\gamma + 1} \Gamma^2\left(1 + \dfrac{\gamma}{2}\right)}$$

3.2.3.2 Dual-Channel Circuits

Very common radio circuits are dual-channel circuits with a multiplier at the input (e.g., frequency converter, correlator). If both the input signal $x(t)$ and local oscillator signal $u_{LO}(t)$ are narrowband oscillations:

$$x(t) = \text{Re}\left\{ \overleftarrow{A}_x(t) \cdot e^{j\omega_s t} \right\}$$

$$u_{LO}(t) = \text{Re}\left\{ \overleftarrow{A}_{LO}(t) \cdot e^{j\omega_{LO} t} \right\}$$

and intermediate frequency $\omega_{IF} = \omega_s - \omega_{LO}$, then the model of the output signal $z(t)$ takes the form:

$$z(t) = \text{Re}\left\{ \overleftarrow{A}_z(t) \cdot e^{j\omega_{IF} t} \right\}$$

$$\overline{A}_z(t) = \overline{A}_x(t) \cdot \overline{K}_{IF}(t), \quad \overline{K}_{IF}(t) = \frac{1}{2} K_M \cdot \overline{A}_{LO}^*(t)$$

where K_M is a multiplier transfer coefficient.

For a correlator (phase discriminator) $\omega_{IF} \rightarrow 0$ that gives:

$$z(t) = \text{Re}\left\{\overline{A}_z(t)\right\}$$

$$\overline{A}_z(t) = \overline{A}_x(t) \cdot \overline{K}_{cor}(t) \quad \overline{K}_{cor}(t) = \frac{1}{2} K_M \cdot \overline{A}_{LO}(t) \cdot e^{j\omega_{IF}t}$$

3.2.4 Modeling of Inertial-Type Nonlinear Circuits

3.2.4.1 The Circuits That Can Be Reduced to the Inertialess

Two basic types of circuits can be reduced to inertialess ones:
1) The circuit CIR1 that is a series connection of any inertialess nonlinear circuits as described above and a linear circuit tuned to the kth frequency multiple $(k\omega_s)$ for additional filtering of the signal;
2) The circuit CIR2 that differs from CIR1 by including an additional resonant linear circuit at the input tuned to the frequency ω_s.

The mathematical description of CIR1 or CIR2 is done on a block-by-block basis by methods described in Section 3.2.3.

3.2.4.2 The Circuits That Cannot be Reduced to the Inertialess

The only way to devise the model for such a circuit is to compose a differential equation based on the circuit electrical diagram or block diagram. Then we attempt to shorten the equation and transfer it into complex envelope form. There are two basic approaches to get the shortened differential equations:

A. *Van der Pol method* (a method of slow-variation amplitudes). The essence of the method can be shown by taking a nonlinear circuit of the second order:

$$\ddot{y} + \omega_s^2 \cdot y = F(y, \dot{y}, x, \dot{x}) \tag{3.6}$$

The models of $x = x(t)$ and $y = y(t)$ are given by equation (3.4). The derivatives are:

$$\dot{x} = \mathrm{Re}\left\{\left(\overline{\dot{A}}_x + j\omega_s \overline{A}_x\right) \cdot e^{j\omega_s t}\right\}$$

$$\dot{y} = \mathrm{Re}\left\{\left(\overline{\dot{A}}_y + j\omega_s \overline{A}_y\right) \cdot e^{j\omega_s t}\right\}$$

$$\ddot{y} = \mathrm{Re}\left\{\left(\overline{\ddot{A}}_y + 2j\omega_s \overline{\dot{A}}_y - \omega_s^2 \overline{A}_y\right) \cdot e^{j\omega_s t}\right\}$$

By inserting the derivatives in equation (3.6) and multiplying both sides by $e^{-j\omega_s t}$, we can get:

$$\overline{\ddot{A}}_y + 2j\omega_s \overline{\dot{A}}_y = 2\exp(-j\omega_s t) \cdot F(y, \dot{y}, x, \dot{x}) + \left(\overline{\ddot{A}}_y + 2j\omega_s \overline{\dot{A}}_y\right)^* \cdot e^{-j2\omega_s t}$$

First, the high-frequency factors $e^{-j\omega_s t} = \cos\omega_s t - j\sin\omega_s t$ have to be eliminated. Both sides of the equation are averaged over the period $T = 2\pi/\omega_s$, and the complex envelopes $\overline{A}_x, \overline{A}_y$ and its derivatives $\overline{\dot{A}}_x, \overline{\dot{A}}_y, \overline{\ddot{A}}_y$ are considered to be constant over period T due to their slow variation in time. Thus, the equation becomes:

$$\overline{\ddot{A}}_y + 2j\omega_s \overline{\dot{A}}_y + \left(\omega_0^2 - \omega_s^2\right) \cdot \overline{A}_y = P\left(\overline{A}_y, \overline{\dot{A}}_y, \overline{A}_x, \overline{\dot{A}}_x\right) \qquad (3.7)$$

where

$$P\left(\overline{A}_y, \overline{\dot{A}}_y, \overline{A}_x, \overline{\dot{A}}_x\right) = \frac{1}{\pi} \int_0^{2\pi} F(y, \dot{y}, x, \dot{x}) \cdot e^{-j\Phi} d\Phi$$

After inserting $\overline{A}_x = a_x - jb_x$, $\overline{A}_y = a_y - jb_y$ into (3.7) the differential equations for $a_y(t), b_y(t)$ can be obtained.

The shortening of (3.7) can be done based on the conditions of slow variation:

$$\overline{A}^{(1)}(t) \approx \delta\omega_0 \overline{A}(t), \quad \overline{A}^{(2)}(t) \approx \delta^2 \omega_0^2 A(t), \ldots$$

For example, if we neglect the second derivative $\overline{\ddot{A}}_y$, the shortened differential equation of the first order has the form:

$$\overline{\ddot{A}}_y + a_0 \overline{A}_y = b_0 \cdot P\!\left(\overline{A}_y, \overline{\dot{A}}_y, \overline{A}_x, \overline{\dot{A}}_x \right)$$

Equation (3.6) can also be shortened in terms of amplitude A_y and phase φ_y if we use the expressions:

$$A\dot{A} = \dot{u}\!\left(\ddot{u} + \omega_0^2 u \right), \quad A = \left(u^2 + \omega_0^{-2} \cdot \dot{u}^2 \right)^{1/2}$$

$$\dot{\varphi} A^2 \omega_0 = u\!\left(\ddot{u} + \omega_0^2 u \right), \quad \varphi = \omega_0 t + \arctan(\dot{u}/\omega_0 u)$$

Thus:

$$\dot{A}_y = \dot{y}\!\left(\ddot{y} + \omega_s^2 y \right)\!/\omega_s^2 \cdot A_y = F(y, \dot{y}, x, \dot{x}) \cdot \dot{y}/\omega_s A_y$$

$$\dot{\varphi}_y' = y\!\left(\ddot{y} + \omega_s^2 y \right)\!/\omega_s \cdot A_y^2 = F(y, \dot{y}, x, \dot{x})/\omega_s \cdot A_y^2$$

where

$$y = A_y \cos\!\left(\omega_s t - \varphi_y \right)$$

$$\dot{y} = -\omega_s A_y \sin\!\left(\omega_s t - \varphi_y \right)$$

$$x = A_x \cos\!\left(\omega_s t - \varphi_x \right)$$

$$\dot{x} = -\omega_s A_x \sin\left(\omega_s t - \varphi_x \right)$$

Further shortening is done by averaging over $\Phi = \omega_s t$ on the right side of this model. This operation results in two nonlinear differential equations of the first order:

$$\dot{A}_y = F_1\!\left(A_y, A_x; \varphi_y, \varphi_x; \omega_s \right)$$

$$\dot{\varphi}_y = F_2\!\left(A_y, A_x; \varphi_y, \varphi_x; \omega_s \right)$$

B. *Method of linearization over fluctuations.* The essence of this method is as follows: let us consider the nonlinear circuit to be described by this equation:

$$\sum_{k=0}^{n} a_k y^{(k)}(t) = \sum_{l=0}^{m} b_l \frac{d^l}{dt^l} \{G[x(t) - Dy(t)]\}$$

or in operator representation:

$$A_y - BG(x - D_y) = 0$$

The input and output signals are modeled as oscillations with small amplitude and phase fluctuations:

$$x(t) = A[1 + \mu(t)]\cos[\omega_s t - \varphi(t)]$$

$$y(t) = y_k(t) = B_k[1 + \gamma_k(t)]\cos[k\omega_s t - \varphi_k(t)]$$

The assumption that fluctuations are small and come under the condition of slow variation in time means:

$$|\mu(t)| \ll 1, \quad \dot{\mu} \ll \omega_s \mu, \quad |\gamma_k(t)| \ll 1, \quad \dot{\gamma}_k \ll \omega_s \gamma_k, \quad \dot{\varphi} \ll \omega_s \varphi, \quad \dot{\varphi}_k \ll \omega_s \varphi_k$$

The next step is to substitute $x(t)$, $y(t)$ in the operator form of the equation, multiply both sides of the equation by $e^{jk\Phi} = e^{jk\omega_s t}$ and average over period $T = 2\pi/\omega_s$. This results in equation:

$$\frac{1}{2\pi}\int_0^{2\pi} \left[A\,\mathrm{Re}\{B_k(1+\gamma_k)e^{j(k\Phi-\varphi_k)}\} - BG\left[\mathrm{Re}\{A(1+\mu)e^{j(\Phi-\varphi)}\}\right.\right.$$

$$\left.\left. - D\,\mathrm{Re}\{B_k(1+\gamma_k)e^{j(k\Phi-\varphi_k)}\}\right]\right]e^{j\Phi}\,d\Phi = 0$$

or

$$\overline{F}(A, B_k, \delta^{(i)}) = \mathrm{Re}\{\overline{F}(A, B_k, \delta^{(i)})\} + j\,\mathrm{Im}\{\overline{F}(A, B_k, \delta^{(i)})\} = 0$$

where $\overline{F}$ is a complex function, and $\delta(i)$ are small values corresponding to small fluctuations $\mu, \gamma, \varphi, \varphi_k$. The latter equation may be converted to a linear one by expanding it into series over small fluctuations. To determine the linkage between small fluctuations the linear operator equation is used:

$$L_1\gamma_k + L_2\varphi_k = L_3\mu + L_4\varphi$$

The initial zero approximation for basic nonlinear equation is:

$$\overline{F}(A, B_k, 0) = 0$$

3.2.5 Modeling of Common Devices and Circuits

The mathematical models obtained by the method of complex envelope for the most common circuits are given in Appendix 47. Some comments apply:
1) Since the method of complex envelope retains no information about the carrier frequency, it is expedient to specify the carrier frequency (in parenthesis) related to the complex envelope in order to restore the instantaneous value of the signal;
2) For a multiplier complex envelopes at difference (d) and sum (s) frequencies can be described by the method of complex envelope only;
3) Two output effects are specified for a correlator (usually the factor at the sum frequency is neglected):

$$z_c(\tau, T) \approx z_{c_d}(\tau, T) = \frac{1}{2} \operatorname{Re}\left\{ \overline{A}_{c_d}(\tau, T) \cdot e^{\pm j\omega\tau} \right\}$$

$$\overline{A}_{c_d}(\tau, T) = \int_0^T \overline{A}_1(t)\overline{A}_2^*(t+\tau) e^{j\Delta\omega t} dt$$

$$\omega = \omega_2 \quad \Delta\omega = \omega_1 - \omega_2$$

4) To model the inertialess nonlinear circuits the method of contour integrals is typically used and function $y = G(x)$ is approximated by half-period nonlinearity of γ-th order. The special case of such a circuit is a logarithmic RF amplifier with logarithmic amplitude curve $A_y(t) = \log A_x(t)$ and nondistorted phase output $\varphi_y(t) = \varphi_x(t)$;
5) The models of ideal demodulators are dealing either with an amplitude of the input signal $A_x(t) = |\overline{A}_x(t)|$, or its phase $\varphi_x(t) = \arg \overline{A}_x(t)$, or the instantaneous frequency

$$\frac{d\varphi_x(t)}{dt} = \omega_0 \quad \omega(t) = \frac{d}{dt} \arg \overline{A}_x(t)$$

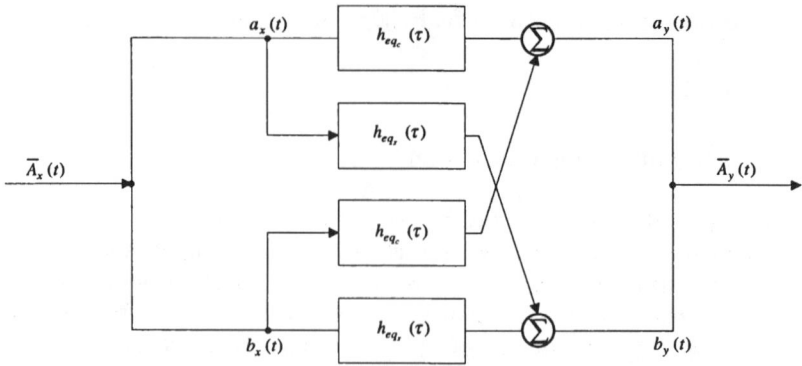

Figure 3.2 The complex filter model of the equivalent circuit (from [13], p. 103).

The general approach to model radio circuits by the method of complex envelope is to use the equivalent block diagram of the model represented as a complex filter (Figure 3.2). The components of the complex envelope are described as (see also Section 1.2):

$$ {a \atop b}(t) = {\text{Re} \atop \text{Im}}\{\overline{A}(t)\} = {\text{Re} \atop \text{Im}}\{A(t) \cdot e^{-j\varphi(t)}\} = A(t){\cos \atop \sin}\varphi(t) $$

When the complex filter model is used, the projections $a(t), b(t)$ should be used in the equations. The corresponding algorithms for demodulators are given in Table 3.1. If the circuit does not use information about amplitude (e.g., the circuit is a phase detector), the modeling algorithms can be obtained by taking into account the dependence:

$$ \cos\varphi = a/A = a/\sqrt{a^2 + b^2} \quad \sin\varphi = b/A = b/\sqrt{a^2 + b^2} $$

The spectral method is a good choice for modeling stationary linear circuits. In this case an RF circuit with the complex transfer function $H(jf)$ is transformed into an equivalent video-frequency circuit with the transfer function $H_{eq}(jF) = H[j(F + f_c)]$, where f_c is a carrier RF frequency. This approach is especially convenient for linear circuits with transfer functions approximated as:

$$ H(jf) = \frac{1}{1 + j\alpha}; \quad H(jf) = \frac{1 + \beta^2}{(1 + j\alpha)^2 + \beta^2} $$

where $\alpha = 2Q(f - f_0)/f_0$ is offset with respect to the resonance frequency f_0, and β is a coupling factor. Because

Table 3.1
Models of Demodulators (after [13], p.133)

Demodulator type	Model $\dfrac{z(t)}{K_{DM}}$
Amplitude (any)	$G\left[\sqrt{a^2(t)+b^2(t)}\right]$
Amplitude, linear	$\sqrt{a^2(t)+b^2(t)}$
Amplitude, square law	$a^2(t)+b^2(t)$
Phase	$arctg\left[b(t)/a(t)\right]$
Frequency	$\dfrac{a(t)\cdot \dot{b}(t)-b(t)\cdot \dot{a}(t)}{\sqrt{a^2(t)+b^2(t)}}$
Coherent	$a(t)\,\text{or}\,b(t)$
Amplitude-phase	$a_1(t)a_2(t)+b_1(t)b_2(t)$ *or* $b_1(t)a_2(t)-a_1(t)b_2(t)$

$$f = f_c + F = f_0 + \Delta f + F$$

the models of equivalent transfer functions are:

$$H_{eq}(jF) = \frac{1}{1+ j2Q(\Delta f + F)/f_0}$$

$$H_{eq}(jF) = \frac{1+\beta^2}{\left[1+ j2Q(\Delta f + F)/f_0\right]^2+\beta^2}$$

3.3 Method of Statistical Equivalents

The essence of the method is to substitute the real circuit (which has a random input signal) with the model that is equivalent to the original circuit in terms of statistical description only. The circuit equivalent ensures adequacy of the output signal only in terms of specified statistical parameters. Four basic approaches within this method are described below.

3.3.1 Method of Analytical Description

This method simulates output signal based on the analytical formula linking output $\eta(t)$ and input signals $\xi(t)$:

$$\eta(t) = F\{\xi(t)\}. \tag{3.8}$$

(See Chapter 2 about algorithms to transform random processes.)

The statistical characteristics of the output signal are always completely defined if model (3.8) and statistical characteristics of the process $\xi(t)$ are known.

3.3.2 Method of Statistical Linearization

The essence of the method is to model a low-frequency nonlinear circuit by substituting it with the linear statistical equivalent over mathematical expectation m_η and fluctuations $\eta^0 = \eta - m_\eta$. The model reproduces only mean and some simple moments (variance, correlation function) of the output fluctuations $\eta^0(t)$. For Gaussian random processes this is a very powerful method because its statistics are completely defined by the first two moments (mean and correlation function).

3.3.2.1 The Modeling of Inertialess Nonlinear Circuits

Let us consider the input and output signals of a low-frequency nonlinear circuit to be nonstationary random processes:

$$\xi(t) = x(t) = m_x(t) + x^0(t)$$

$$\eta(t) = y(t) = m_y(t) + y^0(t)$$

Typically, mean values $m_x(t)$, $m_y(t)$ are some deterministic signals, and $x^0(t)$, $y^0(t)$ are fluctuating interference (e.g., noise). The basic statistical parameters at the output are correlation function

$$K_y(t_1 t_2) = M\{y^0(t_1)y^0(t_2)\}$$

and variance

$$D_y(t) = \sigma_y^2(t) = K_y(t,t)$$

that can be found either analytically (see Chapter 2) or from experimental data.

The mathematical model of a nonlinear circuit is based on replacing it by the circuit that is linear (L) over fluctuations with output signal:

$$y_L(t) = m_{y_L}(t) + y_L^0(t)$$

and statistical characteristics:

$$K_{y_L}(t_1, t_2) = M\{y_L^0(t_1)y_L^0(t_2)\}$$

$$D_{y_L}(t) = K_{y_L}(t,t)$$

Various criteria can be used to check adequacy of the processes $y(t)$ and $y_L(t)$:

1) Mean values and variances are equal (CRITERION 1):

$$m_{y_L}(t) = m_y(t) \quad D_{y_L}(t) = D_y(t)$$

2) Mean values and correlation functions are equal (CRITERION 2):

$$m_{y_L}(t) = m_y(t) \quad K_{y_L}(t_1, t_2) = K_y(t,t)$$

3) Root-mean-square error is minimum (CRITERION 3):

$$\min \sigma^2 = \min M\{y(t) - y_L(t)\}^2\}$$

In practical applications the model of nonlinearity is typically defined as $y = G(x)$. In this case:

$$m_y(t) = M\{G(x)\} = \int_{-\infty}^{\infty} G(x) \cdot f_1(x,t) dx$$

$$D_y(t) = M\{G^2(x)\} - m_y^2(t) = \int_{-\infty}^{\infty} G^2(x) f_1(x,t) - m_y^2(t)$$

$$K_y(t_1, t_2) = M\{G(x_1)G(x_2)\} - m_y(t_1)m_y(t_2)$$

$$= \int_{-\infty}^{\infty} \int_{-\infty}^{\infty} G(x_1) \cdot G(x_2) f_2(x_1, x_2; t_1, t_2) dx_1 dx_2 - m_y(t_1)m_y(t_2)$$

A. *CRITERIA 1,3.* Two types of statistical equivalents are used:
1) Linear over fluctuations

$$m_{y_L} = G_0(m_x) \quad y_L^0(t) = C_1(t) \cdot x^0(t)$$

2) Linear over mathematical expectation and fluctuations

$$m_{y_L}(t) = C_0 \cdot m_x(t) \quad y_L^0(t) = C_1(t) \cdot x^0(t)$$

Parameters $C_0(t), C_1(t), G_0(m_x)$ are called the statistical linearization coefficients and are given in Appendix 48.

If the input signal is a Gaussian random process $x(t) \in N[m_x(t), \sigma_x^2(t)]$:

$$C_0(t) = C_0[m_x(t), \sigma_x(t), G]$$

$$C_1(t) = C_1[m_x(t), \sigma_x(t), G]$$

$$G_0(m_x) = G_0[m_x(t), \sigma_x(t), G]$$

B. *CRITERION 2.* The coefficients of statistical linearization $C_0(t), C_1(t)$ and impulse response are given in Appendix 48. The statistical parameters at the output of a nonlinear circuit are modeled as:

$$m_y(t) = C_0(t) = C_0(m_x, \sigma_x)$$

$$D_y(t) = K_y(t,t) = \sum_{n=1}^{\infty} \frac{C_n^2(t)}{\Gamma(n+1)}$$

$$D_{xy}(t) = -\sigma_x(t) \cdot C_1(t)$$

$$K_y(t_1,t_2) = \sum_{n=1}^{\infty} \frac{C_n(t_1)C_n(t_2)}{\Gamma(n+1)} \cdot R_x^n(t_1,t_2)$$

where $R_x(t_1,t_2)$ is the normalized correlation function of the input process;

$$C_n(t) = C_n(m_x,\sigma_x) = \int_{-\infty}^{\infty} G(\sigma_x z + m_x)\varphi^{(n)}(z)dz$$

$$\varphi^{(n)}(z) = \frac{d^n}{dz^n}\varphi(z) \quad \varphi(z) = \frac{1}{\sqrt{2\pi}}e^{-z^2/2}$$

If the input signal is a Gaussian process:

$$G_0(m_x,\sigma_x) = C_0(m_x,\sigma_x)$$

$$C_0^{(1)}(m_x,\sigma_x) = C_0^{(2)}(m_x,\sigma_x) = C_0(m_x,\sigma_x)/m_x$$

$$C_1^{(1)}(m_x,\sigma_x) = \pm\frac{1}{\sigma_x}\left\{\sum_{n=1}^{\infty}\frac{C_n^2(m_x,\sigma_x)}{\Gamma(n+1)}\right\}^{1/2}$$

$$C_1^{(2)}(m_x,\sigma_x) = \frac{1}{\sigma_x}\cdot C_1(m_x,\sigma_x)$$

3.3.2.2 The Modeling of Inertial-Type Nonlinear Circuits

The approaches differ for high-frequency and low-frequency circuits. We attempt to reduce high-frequency circuits to inertialess circuits and additional linear circuits. The former are modeled using approaches cited in Section 3.2.2.1, and the latter are modeled by the method of the carrier frequency (see Section 3.1). The following approach is common for low-frequency circuits:

1) All inertialess nonlinear circuits are substituted with its statistical equivalents using the method of statistical linearization, and two linear differential equations for m_y and y^0 are obtained;

2) Operator methods are used to solve these equations which brings differential equations to algebraic equations over m_y and σ_y;

3) Algebraic equations are solved by using special methods [14, 15] that make it possible to obtain expressions for parameters $m_y(t)$, $\sigma_y(t)$ at the circuit output.

3.3.2.3 The Harmonic Linearization

This method is used to devise statistical equivalents of nonlinear circuits when input and output signals are oscillatory processes. Let us consider an inertialess nonlinear circuit with wideband input and output signals:

$$x(t) = m_x(t) + x^0(t) = m_{x_{av}} + u_s(t) + x^0(t)$$

where $m_{x_{av}}$ is the constant component of a modulated signal averaged over time;

$$u_s(t) = A_s(t)\cos\Phi_s(t) = A_s(t)\cos[\omega_0 t - \varphi_c(t)] = m_x(t) - m_{x_{av}}$$

$x^0(t)$ is a stationary low-frequency noise with zero mean, variance $D_x = \sigma_x^2$ and correlation function $K_x(\tau) = D_x \cdot R_x(\tau)$.

The block diagram of the circuit and its statistical equivalent (model) are given in Figure 3.3.

The essence of the method is as follows: the block diagram in Figure 3.3(a) with the output signal

$$y(t) = m_y(t) + y^0(t)$$

has to be replaced with the block diagram in Figure 3.3(b), which is the linear equivalent with the output signal

$$y_L(t) = m_{y_L}(t) + y_L^0(t)$$

The signal $y(t)$ can be represented as:

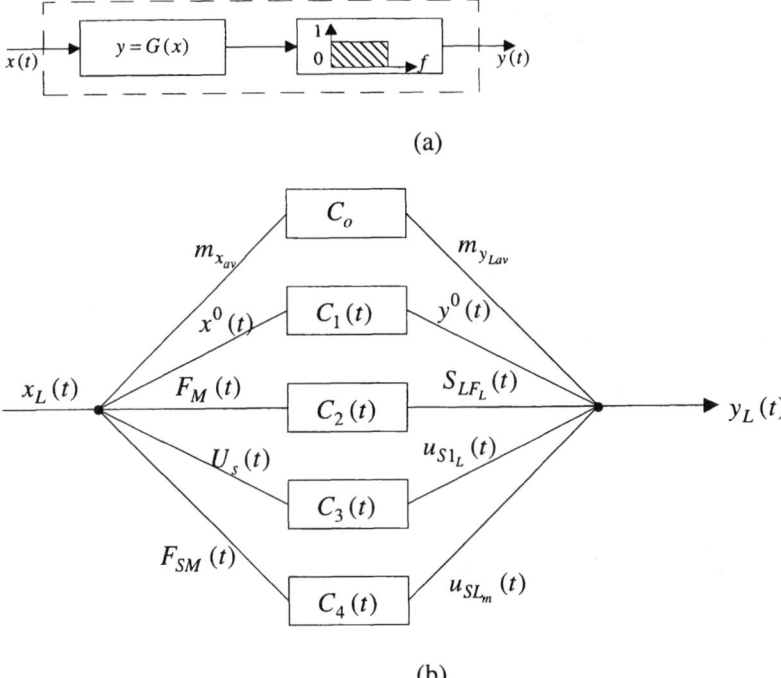

(a)

(b)

Figure 3.3 The block diagram of (a) the inertialess nonlinear ciruit and (b) its statistical eqivalent (after [13], p. 144).

$$y(t) = m_{y_{av}} + S_{LF}(t) + \sum_{m=1}^{\infty} u_{S_m}(t) + y^0(t)$$

where $m_{y_{av}}$ is an average (over time) constant component, $S_{LF}(t)$ is a low-frequency signal (demodulation effect), and

$$u_{S_m}(t) = A_{S_m}(t)\cos\{m[\omega_0 t - \varphi_s(t)]\}$$

is the mth harmonic of the output signal. Analogously:

$$y_L(t) = m_{y_{Lav}} + u_{S1_L}(t) + \sum_{m=2}^{\infty} u_{SL_m}(t) + S_{LF_l}(t) + y_L^0(t)$$

The third and fourth factors do not present at the input. In this case the following components have to be added to retain the linearity of the model at the input:

$$F_M(t) = A_S(t)$$

or depending on the modulation type:

$$F_{SM}(t) = A_S(t)\cos\{m[\omega_0 t - \varphi_s(t)]\}$$

Thus complete linear links in the equivalent model Figure 3.3(b) are:

$$m_{y_{Lav}} = C_0 \cdot m_{x_{av}}; \quad y_L^0(t) = C_1(t) \cdot x^0(t)$$

$$S_{LF_L}(t) = C_2(t) \cdot F_M(t)$$

$$u_{S1_L}(t) = C_3(t) \cdot u_s(t) = C_3(t) A_s(t)\cos\Phi_s(t)$$

$$u_{sm_L}(t) = C_4(t) \cdot F_{sm}(t)$$

The expressions to calculate coefficients $C_i(t)$, $i = \overline{0,4}$ are given in Appendix 48. There:

$$h_{mn}(t) = \frac{j^{m+n}}{2\pi} \int_C z^n g(jz) J_m[zA_s(t)] \times e^{jm_{xav}z} \times e^{-\frac{1}{2}\sigma_x^2 z^2} dz$$

$J_m(u)$ is a Bessel function, and g is the Laplace image of transform $y = G(x)$:

$$y = G(x) = \frac{1}{j2\pi} \int_{C-j\infty}^{C+j\infty} g(p)\exp(px)dp, \quad x > 0, \quad p = \sigma + j\omega$$

3.3.3 Method of Generation

Typically, this method uses the Monte-Carlo approach to determine statistical characteristics of the signal and noise in an equivalent model. If the input signal is the result of a combination of the signal and noise:

$$u_\Sigma = u_s + u_n$$

the output voltage $\eta(t)$ can be modeled as a combination of the signal $C(t)$ and noise $\xi(t)$ components:

$$\eta(t) = C(t) + \xi(t)$$

To create a model by the method of generation, the following algorithm has to be devised:

$$C(t) = F_1\left\{t, \vec{P}_s(t), \vec{P}_n(t), \vec{P}_\Sigma(t)\right\}$$

$$\vec{P}_\xi(t) = F_2\left\{t, \vec{P}_s(t), \vec{P}_n(t), \vec{P}_\Sigma(t)\right\}$$

where $\vec{P}$ is the set of statistical parameters of the signal (s), noise (n) and its sum (Σ) determined by stochastic simulation (Monte-Carlo method) (see Chapter 2). The block-diagram of the equivalent model is given in Figure 3.4.

3.3.4 Method of Filtration

This method is often used to model the circuits containing a discriminator as a basic element. The discriminator is a circuit in which the output signal depends upon how the input signal differs from a reference signal or from any other signal.

The generic block diagram of a discriminator (D) is given in Figure 3.5.

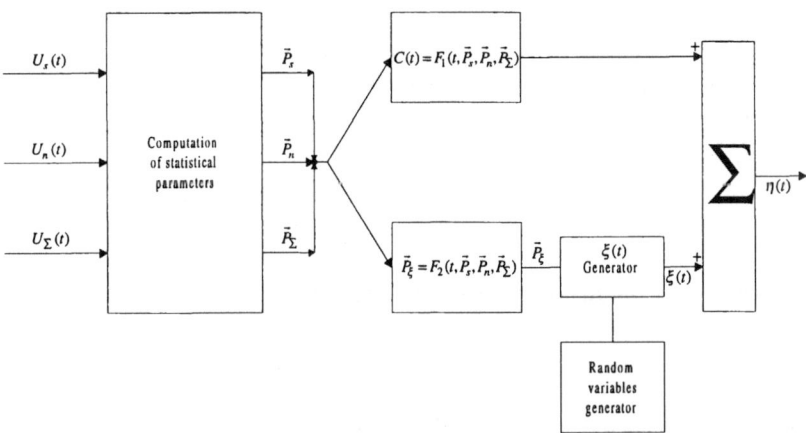

Figure 3.4 The block-diagram of a statistical equivalent for the method of generation (after [13], p. 149).

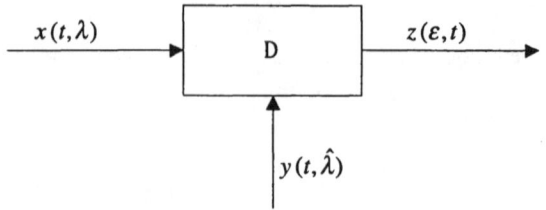

Figure 3.5 The block diagram of a discriminator (after [13], p. 150).

When voltage $x(t,\lambda)$ of the input signal and voltage $y(t,\hat{\lambda})$ of the reference signal are processed, it results in the low-frequency voltage

$$z(\varepsilon,t)=L_{LF}(\varepsilon)=F\{x(t,\lambda),y(t,\hat{\lambda})\}$$

This voltage is an even function of the offset signal $L_{LF}(\varepsilon)$, which is a difference between the parameter of interest $\lambda(t)$ and its estimate $\hat{\lambda}(t)$ generated by the voltage-controlled oscillator (VCO):

$$\varepsilon(t)=\lambda(t)-\hat{\lambda}(t)$$

There are two basic modes of operation of tracking circuits with a discriminator: when the interference does not contain information and when it does. In the former case

$$x(t,\lambda)=u_s(t,\lambda)+u_I(t)$$

and the single parameter $\hat{\lambda}\to\lambda$ is tracked. In the latter case the interference contains some information λ_I (common case for imitation jamming) and the model of the input signal is:

$$x(t,\lambda_s,\lambda_I)=u_s(t,\lambda_s)+u_I(t,\lambda_I)$$

Thus the discriminator tracks two parameters: λ_s,λ_I.

3.3.4.1 The Statistical Equivalent of a Discriminator

The statistical equivalent of a discriminator is based on the following assumptions:

A. The output voltage is divided into mean and fluctuation components:

$$m_z(\varepsilon,t) = M\{z(\varepsilon,t)\} \quad \xi(t) = z^0(\varepsilon,t)$$

$$z(\varepsilon,t) = m_z(\varepsilon,t) + \xi(t)$$

B. The slow-fluctuating parameter $\xi(t)$ is statistically averaged over noise fluctuations, considered to be constant and approximated as white noise with correlation function $K_z(\tau,\varepsilon,t) = N_0(\varepsilon,t) \cdot \delta(\tau)$, where

$$N_0(\varepsilon,t) = N_z(0,\varepsilon,t) = \int_{-\infty}^{\infty} K_z(\tau,\varepsilon,t) d\tau$$

In general,

$$N_z(f,\varepsilon,t) = \int_{-\infty}^{\infty} K_z(\tau,\varepsilon,t) \cdot e^{-j2\pi ft} d\tau$$

is the nonstationary spectral density of white noise.
   With these assumptions:

$$z_{eq}(\varepsilon,t) = m_z(\varepsilon,t) + \sqrt{N_0(\varepsilon,t)} \cdot \xi_N(t) \tag{3.9}$$

where $\xi_N(t)$ is white noise with unity spectral density.
   The model (3.9) is represented by a block diagram of an equivalent discriminator (Figure 3.6).

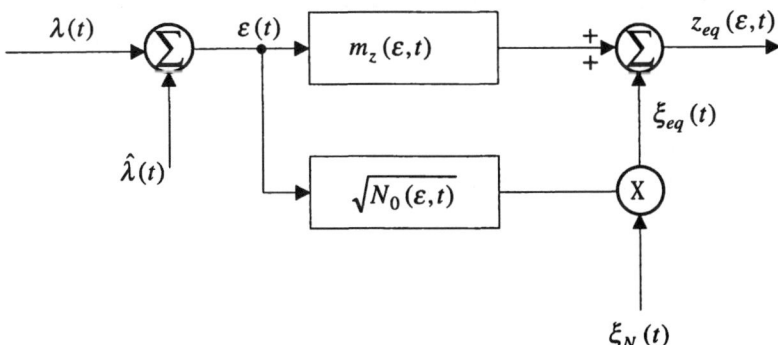

Figure 3.6 The block diagram of an equivalent discriminator (after [13], p. 153).

C. If $z(\varepsilon,t)$ is a stationary random process, then:

$$z_{eq}(\varepsilon,t) = m_z(\varepsilon) + \sqrt{N_0(\varepsilon)} \cdot \xi_N(t) = a(\varepsilon) + b(\varepsilon) \cdot \xi_N(t)$$

where $a(\varepsilon) = m(\varepsilon)$ is a discriminator curve (odd function of parameter ε) , and $b(\varepsilon) = \sqrt{N_0(\varepsilon)}$ is a fluctuation curve (even function of parameter ε).

D. If only the linear portion of the discrimination curve is considered $(|\varepsilon| \ll 1)$, the model becomes:

$$z_{eq}(\varepsilon,t) = K_D \cdot \varepsilon + \sqrt{N_{00} + N_{02} \cdot \varepsilon^2} \cdot \xi_N(t)$$

where

$$K_D = \left. \frac{da(\varepsilon)}{d\varepsilon} \right|_{\varepsilon=0}$$

is a discrimination response slope.

3.3.5 Method of Information Parameter

This method is widely used to simulate tracking RF systems involved in transformation of any information parameter: amplitude, frequency, phase, delay time, and so on. A variety of such systems exist: tracking, track-while-scan and guidance radars, phase-locked loops in communication receivers, optimum and quasioptimum demodulators, and others.

The single-channel variant of a tracking RF system can be represented by the block diagram of Figure 3.7. Here all notations are the same as those used in Section 3.3.4. The essence of the method is to substitute the real system in Figure 3.7 by the automatic control loop in Figure 3.8 with the input signal equal to the information parameter $\lambda(t)$. Typically, the following stages are involved:

A. Input parameter $x(t, \lambda)$ is substituted with parameter $\lambda(t)$, or if there are two parameters of interest, with $\lambda_s(t), \lambda_I(t)$.

B. The discriminator is substituted with its statistical equivalent.

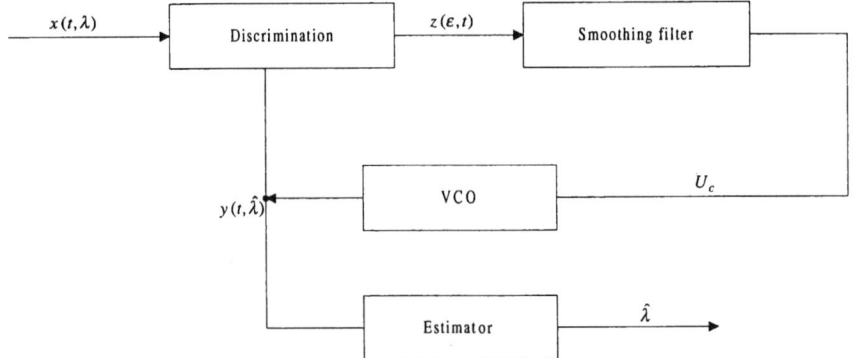

Figure 3.7 The block diagram of a single-channel tracking system (after [13], p. 162).

C. The voltage-controlled oscillator (VCO) is substituted with the low-frequency model.

D. The mathematical model to simulate the loop is devised (Figure 3.8). Typically, one of the approaches that is well devised in the theory of automatic control systems is applied: loop simulation based on integral equations, differential equations or some simplified approaches [14, 15].

3.4 Mathematical Description of Circuit Components

The current flow through an electrical circuit is described by the system of equations based on Ohm's law and Kirchhoff's laws:

$$i = (u + e)/R \qquad \text{Ohm's law}$$

$$\sum_{k=1}^{n} i_k = 0 \qquad \text{the first Kirchhoff's law}$$

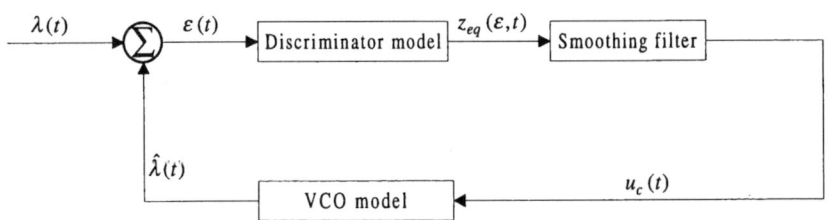

Figure 3.8 The block diagram of the automatic control loop equivalent to the single-channel tracking system (after [13], p. 167).

$$\sum_{k=1}^{n} u_k = \sum_{k=1}^{n} e_k \qquad \text{the second Kirchhoff law}$$

Here i is current flowing through the section of the circuit, u is voltage between the beginning and the end of this section, e is the resultant electromotance, R is a complete active resistance of the section, i_k are currents coming $(+i_k)$ or leaving $(-i_k)$ some point of the circuit. The second Kirchhoff law is valid for any closed-loop circuits, u_k and e_k are voltages and electromotances at the specified section of the circuit. The model for power is:

$$P = i \cdot u = \frac{u^2}{R} = i \cdot R$$

The basic circuit components: resistors (R), capacitors (C) and inductances (L) are often termed RCL components. All components of electrical circuits have some resistance. If, while current flows through the component, the energy is irrevocably dissipated, it is resistive impedance; otherwise it is reactance. Theoretically, a resistor has a resistive impedance, and capacitor and inductance are reactances. In practice, there are no pure reactances since real capacitors and inductances always have some loss resistances R_c or R_L.

All real circuit components are nonlinear (its impedance depends on value of passing current or voltage). In the majority of cases this variation of impedance can be considered relatively small. That is why in terms of mathematical models RCL components are typically considered to be linear ones (the impedance is independent from value of current or voltage). In this case the mathematical description of circuit components is simplified considerably and can be represented by equations based on Kirchhoff's laws. The models of the basic RCL components based on assumptions of impedance linearity are given in Appendix 49.

When current is modeled as a harmonic oscillation:

$$i = I_m \cdot \sin \omega t$$

the following equations for resistor u_R, inductance u_L and capacitor u_C voltages apply:

$$u_R = R \cdot I_m \sin \omega t = U_{m_R} \cdot \sin \omega t$$

$$u_L = L \cdot \frac{di}{dt} = \omega L \cdot I_m \cos \omega t = U_{m_L} \cdot \cos \omega t$$

$$u_C = \frac{1}{C} \int i \, dt = -\frac{I_m}{\omega C} \cdot \cos \omega t = -U_{m_C} \cdot \cos \omega t$$

where

$$U_{m_R} = R \cdot I_m$$

$$U_{m_L} = X_L \cdot I_m, \quad X_L = \omega L \text{ is an inductance}$$

$$U_{C_L} = X_C \cdot I_m, \quad X_C = 1/(\omega C) \text{ is a capacitance}$$

These models can be also represented in the form:

$$U_R = U_{m_R} \cdot \sin \omega t$$

$$U_L = U_{m_L} \cdot \sin (\omega t + \pi/2)$$

$$U_C = U_{m_C} \cdot \sin (\omega t - \pi/2)$$

which shows how voltage phase changes when current flows through reactances. The model for instantaneous power is:

$$p(t) = iu = i_m U_m \sin \omega t \cdot \sin (\omega t + \varphi)$$

where phase φ depends on impedance of the component. The average (over period T) power is:

$$P_{av} = \frac{1}{T} \int_0^T p(t) \, dt = \frac{I_m U_m}{2} \cos \varphi = I \cdot U \cdot \cos \varphi$$

where $I = I_m / \sqrt{2}$, $U = U_m / \sqrt{2}$ are effective current and voltage correspondingly. Obviously, $P_{av_R} = I \cdot U$, $P_{av_L} = P_{av_C} = 0$.

References

[1] Goldenberg, L. M., B. D. Matyushkin, and M. I. Polyak, *Digital Signal Processing (A Handbook)*, (in Russian), Moscow: Radio I Svyaz, 1985.

[2] Gonorovskiy, I. S., *Radio Circuits and Signals*, (in Russian), Moscow: Sovetskoe Radio, 1977.

[3] Kontorovich, M. I., *The Processes in Electrical Circuits*, (in Russian), Moscow: Radio I Svyaz, 1975.

[4] Goryainov, V. T., A. G. Juravlev, and V. I. Tikhonov, *Stochastic Radio Engineering*, (in Russian), Moscow: Sovetskoe Radio, 1980.

[5] Abegauz, G. G., et al., *Calculation of Probabilistic Data (A Handbook)*, (in Russian), Moscow: Voenizdat, 1970.

[6] Bratley, P., B. L. Fox, and E. L. Scharge, *A Guide to Simulation*, New York: Springer-Verlag, 1983.

[7] Knuth, D. E., Seminumerical Algorithms, in *The Art of Computer Programming*, Vol. 2, 2nd ed., Reading, MA: Addison-Wesley, 1981.

[8] Devroye, L., *Non-Uniform Random Variate Generation*, New York: Springer-Verlag, 1986.

[9] Press, W. H., et al., "Numerical Recipes in C," in *The Art of Scientific Computing*, Cambridge, England: Cambridge University Press, 1992.

[10] Bikov, V. V., *Digital Simulation in Stochastic Radio Engineering*, (in Russian), Moscow: Sovetskoe Radio, 1971.

[11] Pollyak, Yu. G., *Stochastic Computer Simulation*, (in Russian), Moscow: Sovetskoe Radio, 1971.

[12] Shaligin, A. S., and Yu. I. Palagin, *The Applied Methods of Stochastic Simulation*, (in Russian), Moscow: Mashinostroenie, 1986.

[13] Borisov, Yu. P., and V. V. Tsvetnov, *The Mathematical Modeling of Radio Systems and Devices*, (in Russian), Moscow: Radio I Svyaz, 1985.

[14] Pugachev, V. S., *The Theory of Random Functions and Its Application to Automatic Control Systems*, (in Russian), Moscow: Fizmatgiz, 1960.

[15] Kazakov, I. B., and B. G. Dostupov, *Statisitical Dynamics of Nonlinear Automatic Systems*, (in Russian), Moscow: Fizmatgiz, 1962.

Selected Bibliography

Brigham, E. O., *The Fast Fourier Transform*, Englewood Cliffs, NJ: Prentice Hall, 1974.

Burdic, W. S., *Radar Signal Analysis*, Englewood Cliffs, NJ: Prentice Hall, 1968.

Cook, C. E., and M. Bernfeld, *Radar Signals*, Norwood, MA: Artech House, 1993.

Dahlquist, G., and A. Bjorck, *Numerical Methods*, Englewood Cliffs, NJ: Prentice Hall, 1974.

Davenport, W. B., and W. L. Root, *An Introduction to the Theory of Random Signals and Noise*, New York: McGraw-Hill, 1980.

Feller, W., *An Introduction to Probability Theory and Its Application*, New York: John Wiley and Sons, 1968.

Forsythe, G. E., M. A. Malcom, and C. B. Moler, *Computer Methods for Mathematical Computation*, Englewood Cliffs, NJ: Prentice Hall, 1977.

Gray, L. M., and L. D. Davisson, *Random Process: A Mathematical Approach for Engineers*, Englewood Cliffs, NJ: Prentice Hall, 1986.

Haykin, S., *Communication Systems*, New York: John Wiley and Sons, 1988.

Helstrom, C. W., *Probability and Stochastic Processes for Engineers*, London: Macmillan, 1984.

Korn, G. A., and T. M. Korn, *Mathematical Handbook*, New York: McGraw-Hill, 1968.

Kunt, M., *Digital Signal Processing*, Norwood, MA: Artech House, 1986.

Leonov, A. I., (ed.), *Modeling in Radar*, (in Russian), Moscow: Sovetskoe Radio, 1979.

Levin, B. R., *Theoretical Fundamentals of Stochastic Radio Engineering,* (in Russian), Moscow: Sovetskoe Radio, 1974.

Lewis, B. L., F. F. Kretschmer, Jr., and W. W. Shelton, *Aspects of Radar Signal Processing*, Norwood, MA: Artech House, 1986.

Malvar, H. S., *Signal Processing with Lapped Transforms*, Norwood, MA: Artech House, 1992.

Nathanson, F. E., *Radar Design Principles: Signal Processing and the Environment*, New York: McGraw-Hill, 1990.

Nitzberg, R., *Adaptive Signal Processing for Radar*, Norwood, MA: Artech House, 1992.

Nussbaumer, G., *The Fast Fourier Transform and Convolution Evaluation Algorithms*, (in Russian), Moscow: Radio I Svyaz, 1985.

Oppenheim, A. V., and A. S. Willsky, *Signals and System*, Englewood Cliffs, NJ: Prentice Hall, 1983.

Oppenheim, A. V., and R. W. Schaefer, *Digital Signal Processing*, Englewood Cliffs, NJ: Prentice Hall, 1975.

Peebles, P. Z., Probability, *Random Variables and Random Signals Principles*, New York: McGraw-Hill, 1980.

Pettai, R., *Noise in Receiving Systems*, New York: John Wiley and Sons, 1984.

Picinbonbo, B., *Principles of Signals and Systems: Deterministic Signals*, Norwood, MA: Artech House, 1988.

Ppoulis, A., *Probability, Random Variables and Random Signal Principles*, New York: McGraw-Hill, 1980.

Schwartz, M., *Signal Processing*, New York: McGraw-Hill, 1975.

Stark, H., and J. W. Woods, *Probability, Random Processes and Estimation Theory for Engineers*, Englewood Cliffs, NJ: Prentice Hall, 1986.

Tikhonov, V. I., *Stochastic Radio Engineering*, (in Russian), Moscow: Radio i Svaz, 1982.

Urkowitz, H., *Signal Theory and Random Processes*, Norwood, MA: Artech House, 1983.

Van Trees, H. L., *Detection, Estimation and Modulation Theory*, (3 vols.), New York: McGraw-Hill, 1968–1971.

Vizmuller, P., *RF Design Guide: Systems, Circuits, and Equations*, Norwood, MA: Artech House, 1995.

Wong, E., *Introduction to Random Processes*, New York: Springer-Verlag, 1983.

Wozencraft, J. M., and I. M. Jacobs, *Principles of Communication Engineering*, New York: John Wiley and Sons, 1965.

PART 2

RADAR SYSTEM SIMULATION

The software package with the programs cited in Part 2 is available as an optional supplement to the Handbook.

Chapter 4

Radar Cross Section

4.1 Effective RCS

4.1.1 DESCRIPTION

Models the effective RCS of a point target modified by the pattern-propagation factor which makes a correction for the effect of interference lobes in the case of multipath propagation. *Assumption*: a target is located within the reflection lobe region over a smooth, flat earth whose reflection coefficient is ≈ 1.

4.1.2 INPUT DATA

===

Parameter	Dimension and Description

===

$\sigma_0 := 1$	m^2, free-space target RCS
$\lambda := 0.2$	m, wavelength
$R := 1$	nm, target range
$h_a := 50$	ft, antenna height
$h_t := 10$	ft, target height

===

4.1.3 MODEL

$R_m := R \cdot 1852$	$R_m = 1.852 \times 10^3$	m, target range
$h_{a\_m} := h_a \cdot 0.3048$	$h_{a\_m} = 15.24$	m, antenna height
$h_{t\_m} := h_t \cdot 0.3048$	$h_{t\_m} = 3.048$	m, target height

Note: over a reflecting surface, interference nulls occur for targets at elevation angle increments $\theta_n = \lambda/2h_a$. Because the target elevation angle (for $R \gg h_t \gg h_a$) is $\theta_t \approx h_t/R$, the propagation factor F can be expressed as below.

153

$$F := 2 \cdot \sin\left(\frac{2 \cdot \pi \cdot h_{a\_m} \cdot h_{t\_m}}{R_m \cdot \lambda} \right)$$ propagation factor

$$\sigma_{eff} := \sigma_0 \cdot F^4$$ m², effective RCS

$$\sigma_{eff\_dB} := 10 \cdot \log\left(\sigma_{eff} \right)$$ dBsm, effective RCS with respect to 1 m²

4.1.4 OUTPUT DATA

RCS is $\sigma_{eff} = 4.0$ m² or $\sigma_{eff\_dB} = 6.1$ dBsm for input data specified

4.1.5 REFERENCES
[1] Barton, D. K., and S. A. Leonov, (eds.), *Radar Technology Encyclopedia*, Norwood, MA: Artech House, 1997, p. 365.
[2] Long, M. W., (ed.), *Airborne Early Warning System Concepts*, Norwood, MA: Artech House, 1992, p. 120.

4.1.6 EXAMPLES

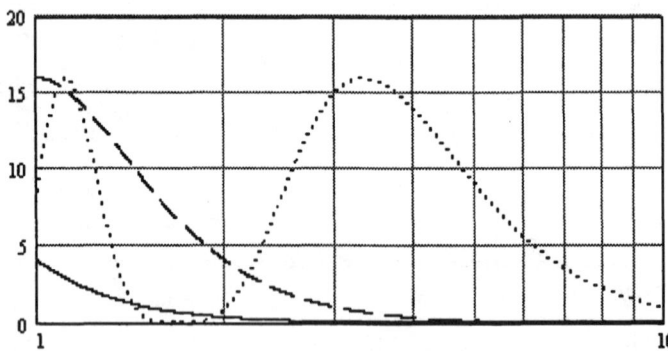

Figure 4.1 Effective RCS (m²) vs. target range (nm) for a low-flying target (height = 10 ft, antenna height =50 ft, free-space RCS = 1 m²) for various frequency bands: L band (0.23 m) - solid line, S band (0.1 m) - dashed line, X band (0.03m) - dotted line.

4.2 RCS Fluctuations

4.2.1 DESCRIPTION
Models RCS of a fluctuating target as a function of observation time.

4.2.2 INPUT DATA

==

Parameter	Dimension and Description

==

$\sigma_0 := 1$	m^2, average RCS
case $:= 1$	Swerling cases (1 to 4)
$T := 0.25$	s, RCS observation time
PRI $:= 0.01$	s, pulse repetition interval
$t_c := 3 \cdot PRI \qquad t_c = 0.03$	s, RCS fluctuation correlation time

Note: for Swerling cases 1 and 3 $t_c >$ PRI; for Swerling case 2 and 4 $t_c <$ PRI.

==

4.2.3 MODEL

$\Delta t := t_c$ s, sampling interval

$$N := \mathrm{ceil}\left(\frac{T}{\Delta t}\right) \qquad N = 9$$ number of points in RCS sample

$n := 0..N$ cycle

$$\sigma := \left| \begin{array}{l} \text{for } n \in 0..N \\ \quad \left| \begin{array}{l} \text{for } k \in 0..1 \\ \quad z_{u_k} \leftarrow \mathrm{rnd}(1) \\ z_n \leftarrow \left| \begin{array}{ll} -\ln\left(z_{u_0}\right) & \text{if} \quad \left| \begin{array}{l} \text{case} = 1 \\ \text{case} = 2 \end{array} \right. \\ -\ln\left(z_{u_0} \cdot z_{u_1}\right) & \text{if} \quad \left| \begin{array}{l} \text{case} = 3 \\ \text{case} = 4 \end{array} \right. \end{array} \right. \end{array} \right. \\ \sigma \leftarrow \sigma_0 \cdot z \end{array} \right.$$

m^2, RCS sample

sample of uniform distribution between 0 and 1

the exponential distribution

the chi-square distribution with 4 degrees of freedom

$\sigma_{av_n} := \sigma_0$ m^2, average RCS

4.2.4 OUTPUT DATA

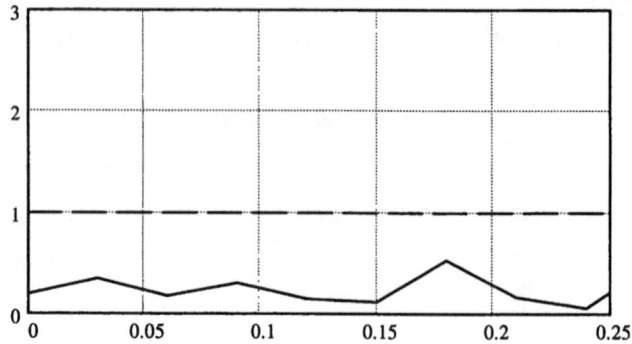

Figure 4.2 The RCS of fluctuating target (m²) vs. RCS observation time (s) for

average RCS (m²): $\sigma_0 = 1$ Swerling model: case = 1   (a low

value of the slowly fluctuating RCS sample persists throughout
the time of this plot).

4.2.5 REFERENCES

[1] Barton, D. K., and S. A. Leonov, (eds.), *Radar Technology Encyclopedia*, Norwood,
MA: Artech House, 1997, p. 366.
[2] Skolnik, M., *Introduction to Radar Systems*, New York: McGraw-Hill, 1980, p. 46.

4.2.6 EXAMPLES

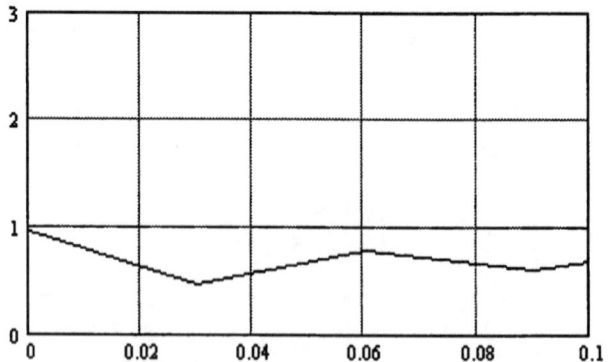

Figure 4.3 The RCS of fluctuating target (m²) vs. RCS observation time (s) for

average RCS (m²): $\sigma_0 = 1$ Swerling model: case = 1 (another run

from the random number generator shows a greater value of RCS
in this plot than in Figure 4.2).

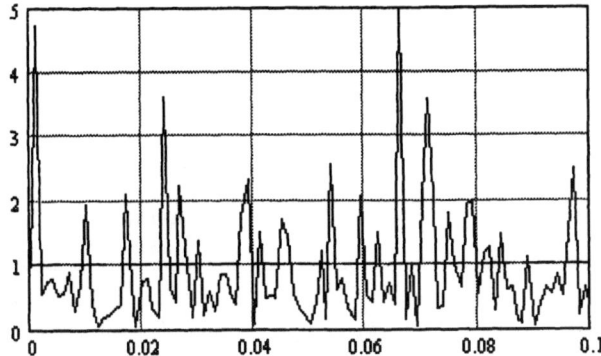

Figure 4.4 The RCS of fluctuating target (m²) vs. RCS observation time (s) for average RCS (m²): $\sigma_0 = 1$ Swerling model: case = 2   (the more rapid fluctuations are apparent for Swerling case 2).

4.3 RCS of a Target for a Bistatic Radar

4.3.1 DESCRIPTION

Calculates the RCS of the target for a bistatic radar as a function of the monostatic RCS and bistatic angle

4.3.2 INPUT DATA

==

Parameter	Dimension and Description

==

$\sigma_0 := 5$ m^2, RCS of monostatic target

$K := 3.5$ empirical coefficient defined by target configuration and complexity

Note: when unknown the coefficient K can be derived based on Babinet's principle for the forward-scatterer path ($\beta = 180$) as

$$K := \frac{\ln\left(4 \cdot \pi \cdot \dfrac{A^2}{\lambda^2 \sigma_0}\right)}{\pi - 2.4}$$

where A is the area of the target projected normal to the radar beam, λ is the wavelength.

input_type := 1 $1 \rightarrow$ input bistatic angle β
$2 \rightarrow$ input radar - target geometry

if input_type = 1

$\beta := 180$ deg, bistatic angle

Note: bistatic angle is the angle between directions from the target to the transmitting and receiving positions of the multistatic radar.

if input_type = 2

$R_1 := 28000$ m, the transmitter - target range

$R_2 := 12000$ m, the receiver - target range

$B := 20000$ m, baseline

==

4.3.3 MODEL

4.3.3.1 Constants

$z := \dfrac{\pi}{180}$ coefficient to transform degrees to radians

4.3.3.2 Bistatic Radar Cross Section

$$\gamma := \mathrm{acos}\left[\frac{1}{2 \cdot R_1 \cdot R_2} \cdot \left(R_1{}^2 + R_2{}^2 - B^2\right)\right]$$ rad, bistatic angle for input_type = 2

$$\alpha := \begin{vmatrix} \beta \cdot z & \text{if input\_type} = 1 \\ \gamma & \text{if input\_type} = 2 \end{vmatrix} \qquad \alpha = 3.142$$ rad, bistatic angle

$$\frac{\alpha}{z} = 180$$ deg, bistatic angle

$$\sigma := \sigma_0 \cdot \left[\left[1 + e^{(K \cdot |\alpha| - 2.4 \cdot K - 1)}\right]\right] \qquad \sigma = 29.656$$ m², radar cross section

$$\sigma_{dB} := 10 \cdot \log\left(\sigma + 10^{-10}\right)$$ dBsm, radar cross section with respect to 1 m²

4.3.4 OUTPUT DATA

RCS is: $\sigma = 29.7$ m² or $\sigma_{dB} = 14.7$ dBsm for input data specified

4.3.5 REFERENCES

[1] Barton, D. K., and S. A. Leonov, (eds.), *Radar Technology Encyclopedia*, Norwood, MA: Artech House, 1997, p. 368.
[2] Aver'yanov, V. Ya., *Radar Stations and Systems with Space Diversity,* (in Russian), Minsk: Nauka i Tekhnika, 1978, p. 21.

4.3.6 EXAMPLES

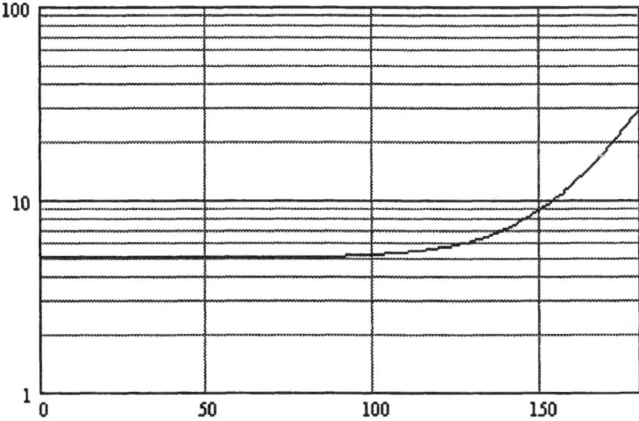

Figure 4.5 The RCS of a target (m²) vs. bistatic angle (deg) for monostatic RCS (m²):
$\sigma_0 = 5$ and $K = 3.5$

4.4 RCS of Aircraft and Ships

4.4.1 DESCRIPTION
Calculates the RCS of an aircraft or a maritime vessel as a function of its length and radar frequency.

4.4.2 INPUT DATA

===

Parameter	*Dimension and Description*

===

$type := 2$

$1 \rightarrow$ aircraft
$2 \rightarrow$ ship

if type = 1

$L_a := 20$

m, aircraft length

if type = 2

$angle := 1$

$1 \rightarrow$ low elevation angles   (close to grazing angle)
$2 \rightarrow$ high elevation angles (other than grazing ones)

$f := 2.8$

GHz, radar frequency

$L_s := 200$

m, ship length

===

4.4.3 MODEL

$\sigma_a := 0.01 \cdot L_a^2 \qquad \sigma_a = 4$

m2, radar cross section of an aircraft

$D := 2.5 \cdot 10^{-6} \cdot L_s^3 \qquad D = 20$

kilotons, displacement of a ship

$\sigma_s := \begin{vmatrix} 1644 \cdot f^{\frac{1}{2}} \cdot D^{\frac{3}{2}} & \text{if } angle = 1 \\ D \cdot 10^3 & \text{if } angle = 2 \end{vmatrix}$

m2, radar cross section of a ship

$\sigma := \begin{vmatrix} \sigma_a & \text{if } type = 1 \\ \sigma_s & \text{if } type = 2 \end{vmatrix}$

m2, radar cross section of a target specified

$\sigma_{dB} := 10 \cdot \log(\sigma)$

dBsm, radar cross section with respect to 1 m2

4.4.4 OUTPUT DATA

RCS is: $\sigma = 2.461 \times 10^5$ m2 or $\sigma_{dB} = 53.91$ dBsm for input data specified

4.4.5 REFERENCES

[1] Barton, D. K., and S. A. Leonov, (eds.), *Radar Technology Encyclopedia*, Norwood, MA: Artech House, 1997, pp. 362, 367.

[2] Morchin, W., *Radar Engineers Sourcebook*, Norwood, MA: Artech House, 1993, pp. 110, 119.

[3] Skolnik, M., *Introduction to Radar Systems*, New York: McGraw-Hill, 1980, p. 42.

4.4.6 EXAMPLES

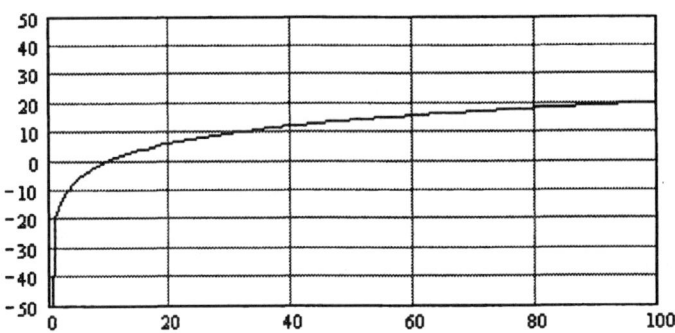

Figure 4.6 The RCS of an aircraft (dB) vs. its length (meters).

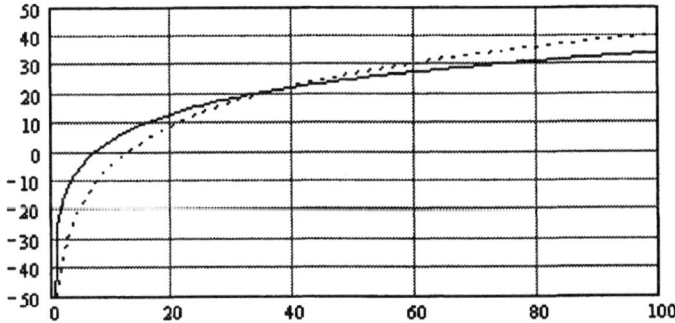

Figure 4.7 The RCS of a ship (dB) vs. its length (meters) (solid line is RCS for high elevation angles, dotted line is RCS for low elevation angles).

4.5 RCS of Antennas

4.5.1 DESCRIPTION

Calculates the RCS of an antenna modeled as a flat plate with the area of the antenna. *Assumption*: only "antenna mode" is modeled. The "structure" component is typically obtained based on empirical or experimentally measured data. The RCS of an antenna is then composed of two portions: the "antenna mode" and the "structure".

4.5.2 INPUT DATA

===

| *Parameter* | *Dimension and Description* |

===

$A_e := 40$ m², antenna effective aperture

$\lambda := 0.3$ m, wavelength

$\Gamma := 1$ reflection coefficient

===

4.5.3 MODEL

$$\sigma := 4 \cdot \pi \cdot \frac{A_e^2}{\lambda^2} \cdot \Gamma$$
     m², radar cross section

$$\sigma_{dB} := 10 \cdot \log(\sigma)$$
     dBsm, radar cross section with respect to 1 m²

4.5.4 OUTPUT DATA

RCS is: $\sigma = 2.234 \times 10^5$ n² or $\sigma_{dB} = 53.5$ dBsm for input data specified

4.5.5 REFERENCES

[1] Barton, D. K., and S. A. Leonov, (eds.), *Radar Technology Encyclopedia*, Norwood, MA: Artech House, 1997, p. 364.
[2] Morchin, W., *Radar Engineers Sourcebook*, Norwood, MA: Artech House, 1993.

4.5.6 EXAMPLES

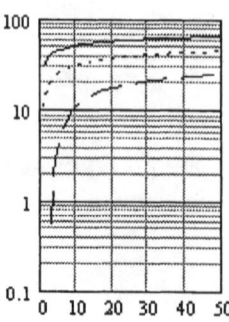

Figure 4.8 The RCS of antenna (dB) vs. the area of its aperture (m²) for $\lambda = 0.1$ m (solid line), $\lambda = 1$ m (dotted line) and $\lambda = 10$ m (dashed line).

4.6 RCS of Birds and Insects

4.6.1 DESCRIPTION
Calculates the RCS of birds and insects as a function of their weight.

4.6.2 INPUT DATA

==

Parameter   *Dimension and Description*

==

W := 20   grams, the weight

4.6.3 MODEL

$\sigma_{dB} := -46 + 5.8 \cdot \log(W)$ dBsm, radar cross section with respect to 1 m²

Note: typically there is 5.5 dB standard deviation about the values predicted
with this model.

4.6.4 OUTPUT DATA
RCS is: $\sigma_{dB} = -38.5$ dBsm for input data specified

4.6.5 REFERENCES

[1] Barton, D. K., and S. A. Leonov, (eds.), *Radar Technology Encyclopedia*, Norwood,
   MA: Artech House, 1997, p. 364.
[2] Morchin, W., *Radar Engineers Sourcebook*, Norwood, MA: Artech House, 1993,
   p. 121.

4.6.6 EXAMPLES

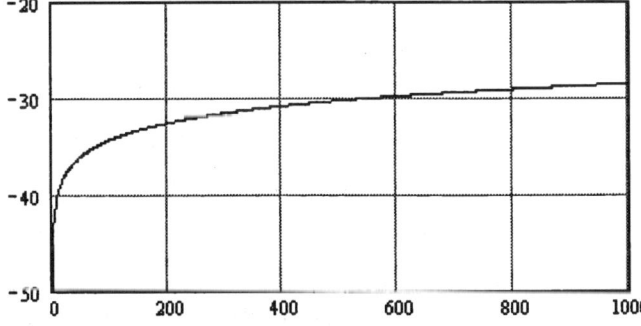

Figure 4.9 The RCS of a bird or an insect (dBsm) vs. the weight (grams).

4.7 RCS of Simple Shapes

4.7.1 DESCRIPTION

Calculates the RCS of simple shapes (those that permit the use of the analytic (exact) methods of RCS prediction). The following shapes are covered: a cone, a cylinder, a disk, an edge, an ellipsoid of revolution, an ogive, a paraboloid, a plate, a reflector (corner), a sphere, and a wire.

4.7.2 INPUT DATA

===

Parameter	Dimension and Description

===

shape := 12	$1 \rightarrow$ a cone or an ogive
	$2 \rightarrow$ a cylinder (circular)
	$3 \rightarrow$ a disk (circular)
	$4 \rightarrow$ an edge (curved)
	$5 \rightarrow$ an edge (straight)
	$6 \rightarrow$ an ellipsoid of revolution
	$7 \rightarrow$ a paraboloid (circular)
	$8 \rightarrow$ a plate (flat)
	$9 \rightarrow$ a reflector (corner, dihedral)
	$10 \rightarrow$ a reflector (corner, trihedral)
	$11 \rightarrow$ a sphere
	$12 \rightarrow$ a wire
$\lambda := 1$	m, wavelength

if shape = 1

$\alpha := 20$	deg, half-angle of a cone or half-angle formed by the ogive axis and the tangent line at the ogive apex

if shape = 2

$R_{cyl} := 1$	m, radius of a cylinder
$L_{cyl} := 3$	m, length of a cylinder
$\theta_2 := 90$	deg, angle between the cylinder axis and radar line-of-sight

if shape = 3

disk_type := 2	$1 \rightarrow$ electrically large disk
	$2 \rightarrow$ electrically small circular disk at normal incidence
$R_{disk} := 0.2$	m², radius of a disk

if disk_type = 1

$\theta_3 := 10$	deg, angle relative to the normal to the disk

if shape = 4

$R_{edg} := 0.2$ m, radius of a curved edge contour

if shape = 5

$L_{edg} := 1$ m, length of a straight edge

if shape = 6

$a_{ell} := 0.5$ m, semimajor axis

$b_{ell} := 0.1$ m, semiminor axis

if shape = 7

$a_{par} := 1.5$ m, radius of the apex curvature

if shape = 8

$x := 2$ m, plate length

$y := 2$ m, plate width

$\theta_8 := 20$ deg, angle relative to the normal to the plate

$\phi_8 := 45$ deg, angle between the plane that contains the normal and radar line-of-sight and the x dimension

if shape = 9

$w_{di} := 0.5$ m, height and width of each plate

$\theta_9 := 45$ deg, the angle relative to the bisector angle of the dihedral

if shape = 10

$type := 1$ 1 → nonintersecting sides are square
2 → intersecting planes are circular
3 → nonintersecting sides form triangles rather than squares

$x_{tri} := 0.5$ m, length of each intersecting side

if shape = 11

$R_{sph} := 0.2$ m, radius of a sphere

if shape = 12

$R_w := 0.01$ m, radius of a wire

$L_w := 5$ m, length of a wire

$polarization := 2$ 1 -> parallel polarization
2 -> perpendicular polarization

==

4.7.3 MODEL

4.7.3.1 Constants

$$z := \frac{\pi}{180}$$ coefficient to transform degrees to radians

$$k := \frac{2 \cdot \pi}{\lambda} \qquad k = 6.283$$ wave number

4.7.3.2 RCS of a Cone or an Ogive (incidence: nose-on)

$$\sigma_1 := \frac{\lambda^2}{16 \cdot \pi} \cdot \tan(z \cdot \alpha)^4 \qquad \sigma_1 = 3.491 \times 10^{-4}$$ m², RCS

Note: the model applies to an infinite cone or electrically large circular ogive (dimensions are large relative to a wavelength).

4.7.3.3 RCS of a Cylinder

$$\sigma_2 := \frac{R_{cyl} \cdot \lambda \cdot \sin(z \cdot \theta_2)}{2 \cdot \pi} \cdot \left[\frac{\sin\left[\left(2 \cdot \pi \cdot \frac{L_{cyl}}{\lambda} \right) \cdot \cos(z \cdot \theta_2) \right]}{\cos(z \cdot \theta_2)} \right]^2$$

$$\sigma_2 = 56.549$$ m², RCS

Note: the model applies to a right circular electrically large cylinder. The expression is valid when $\dfrac{4 \cdot \pi \cdot R_{cyl} \cdot \sin(z \cdot \theta_2)}{\lambda} \gg 1.$

4.7.3.4 RCS of a Disk

$$J_1(u) := \frac{1}{\pi} \cdot \int_0^\pi \cos(u \cdot \sin(t) - t) \, dt$$ Bessel function of the first kind and order 1

$$\sigma 1_3 := \frac{4 \cdot \pi \cdot \left(\pi \cdot R_{disk}^2 \right)^2}{\lambda^2} \cdot \left(2 \cdot \frac{J_1\left(2 \cdot k \cdot R_{disk} \cdot \sin(z \cdot \theta_3) \right)}{2 \cdot k \cdot R_{disk} \cdot \sin(z \cdot \theta_3)} \right)^2 \cdot \cos(z \cdot \theta_3)^2$$

$$\sigma 1_3 = 0.183$$ m², RCS of an electrically large disk

$$\sigma 2_3 := 1125\pi \cdot R_{disk}^2 \cdot \left(\frac{R_{disk}}{\lambda} \right)^4 \qquad \sigma 2_3 = 0.226$$ m², RCS of a small disk

$$\sigma_3 := \begin{vmatrix} \sigma 1_3 & \text{if } \text{disk\_type} = 1 \\ \sigma 2_3 & \text{if } \text{disk\_type} = 2 \end{vmatrix} \qquad \sigma_3 = 0.226 \qquad\qquad \text{m}^2, \text{ RCS of a disk}$$

4.7.3.5 RCS of a Curved Edge (incidence: perpendicular to the edge)

$$\sigma_4 := \frac{R_{edg} \cdot \lambda}{2} \qquad\qquad \sigma_4 = 0.1 \qquad\qquad\qquad \text{m}^2, \text{ RCS}$$

4.7.3.6 RCS of a Straight Edge (incidence: perpendicular to the edge)

$$\sigma_5 := \frac{L_{edg}^2}{\pi} \qquad\qquad \sigma_5 = 0.318 \qquad\qquad\qquad \text{m}^2, \text{ RCS}$$

4.7.3.7 RCS of an Ellipsoid of Revolution (incidence: along the major axis)

$$\sigma_6 := \frac{\pi \cdot b_{ell}^4}{a_{ell}^2} \qquad\qquad \sigma_6 = 1.257 \times 10^{-3} \qquad\qquad \text{m}^2, \text{ RCS}$$

Note: the model is valid for an electrically large ellipsoid (a_{ell}, $b_{ell} \gg \lambda$).

4.7.3.8 RCS of a Paraboloid (nose-on incidence)

$$\sigma_7 := \pi \cdot a_{par}^2 \qquad\qquad \sigma_7 = 7.069 \qquad\qquad\qquad \text{m}^2, \text{ RCS}$$

4.7.3.9 RCS of a Plate

$$u := k \cdot x \cdot \sin(z \cdot \theta_8) \cdot \cos(z \cdot \phi_8)$$

$$v := k \cdot y \cdot \sin(z \cdot \theta_8) \cdot \cos(z \cdot \phi_8)$$

$$\sigma_8 := \frac{4 \cdot \pi \cdot (x \cdot y)^2}{\lambda^2} \cdot \left(\frac{\sin(u)}{u} \cdot \frac{\sin(v)}{v} \right)^2 \cdot \cos(z \cdot \theta_8)^2 \qquad \sigma_8 = 2.28 \times 10^{-4} \quad \text{m}^2, \text{ RCS}$$

Note: the model is valid for an electrically large flat plate.

4.7.3.10 RCS of a Corner Reflector, Dihedral

$$\sigma_9 := \begin{vmatrix} \dfrac{16 \cdot \pi \cdot w_{di}^4}{\lambda^2} \cdot \sin(z \cdot \theta_9)^4 & \text{if } 0 < \theta_9 \leq 45 \\[20pt] \dfrac{16 \cdot \pi \cdot w_{di}^4}{\lambda^2} \cdot \sin\left[z \cdot (90 - \theta_9) \right]^4 & \text{if } 45 < \theta_9 < 90 \end{vmatrix} \qquad \sigma_9 = 0.785 \quad \text{m}^2, \text{ RCS}$$

Note: the model applies to a corner reflector, consisting of two square planes, viewed in the plane normal to both planes.

4.7.3.11 RCS of a Corner Reflector, Trihedral

$$\sigma_{10} := \begin{vmatrix} 3 \cdot \dfrac{4 \cdot \pi \cdot x_{tri}^{4}}{\lambda^{2}} & \text{if type = 1} \\[2em] 1.202 \cdot \dfrac{4 \cdot \pi \cdot x_{tri}^{4}}{\lambda^{2}} & \text{if type = 2} \\[2em] \dfrac{1}{3} \cdot \dfrac{4 \cdot \pi \cdot x_{tri}^{4}}{\lambda^{2}} & \text{if type = 3} \end{vmatrix} \qquad \sigma_{10} = 2.356 \qquad \text{m}^2\text{, RCS}$$

Note: the model applies to a corner reflector, consisting of three mutually perpendicular, centered and equal planes, viewed from the direction of maximum triple-bounce area.

4.7.3.12 RCS of a Sphere

$$\rho := k \cdot R_{sph} \qquad \rho = 1.257 \qquad \text{the ratio of sphere circumference to wavelength}$$

Coefficients in series for RCS calculation, where js and ys are spherical Bessel functions of the first and second kind of order n.

$$a(n,\rho) := \frac{js(n,\rho)}{js(n,\rho) - j \cdot ys(n,\rho)}$$

$$b(n,\rho) := \frac{-\left[\dfrac{d}{d\rho}(\rho \cdot js(n,\rho)) \right]}{\dfrac{d}{d\rho}(\rho \cdot js(n,\rho) - j \cdot \rho \cdot ys(n,\rho))}$$

RCS of a sphere, m^2

$$\sigma_{11} := \pi \cdot R_{sph}^{2} \cdot \left[\frac{1}{\rho^{2}} \cdot \left[\left| \sum_{n=1}^{2 \cdot \text{ceil}(\rho)} (-1)^{n} \cdot (2 \cdot n + 1) \cdot (a(n,\rho) + b(n,\rho)) \right| \right]^{2} \right]$$

Note: the summation should actually extend to $n = \infty$, but it is sufficient to set the upper limit to $n > \rho$ with some margin.

$\sigma_{11} = 0.346$ m², RCS

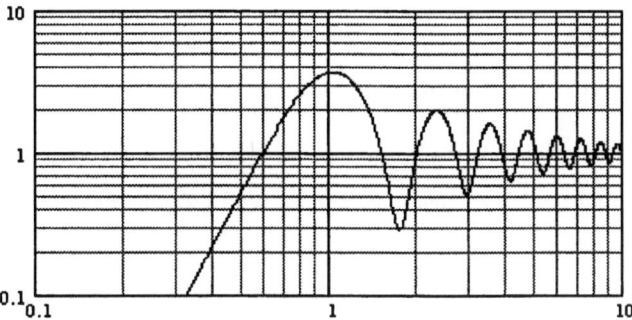

Figure 4.10 The RCS of a conducting sphere normalized to its projected area πR_{sph}^{2} vs. the ratio of sphere circumference to wavelength ρ.

4.7.3.13 RCS of a Wire (normal incidence)

polarization = 2 polarization selected

$$\sigma_{12} := \begin{vmatrix} \dfrac{\pi \cdot L_w^{\ 2}}{\left(\dfrac{\pi}{2}\right)^2 + \ln\left(\dfrac{\lambda}{1.78 \cdot \pi \cdot R_w}\right)^2} & \text{if polarization} = 1 \\[2em] \dfrac{9}{4} \cdot \pi \cdot L_w^{\ 2} \cdot \left(k \cdot R_w\right)^4 & \text{if polarization} = 2 \end{vmatrix}$$

$\sigma_{12} = 2.754 \times 10^{-3}$ m², RCS

Note: the model is valid when $kR_w \ll 1$. For parallel polarization it also excludes wire lengths near odd multiples of $\lambda/2$ where resonance occurs.

4.7.3.14 RCS of a Specified Simple Shape

k := shape   k = 12 specified shape

$$\sigma := \begin{vmatrix} \sigma_k & \text{if shape} = k \\ 0 & \text{otherwise} \end{vmatrix} \qquad \sigma = 2.754 \times 10^{-3}$$ m², RCS

$\sigma_{dB} := 10 \cdot \log\left(\sigma + 10^{-100}\right)$ $\sigma_{dB} = -25.6$ dBsm, RCS

4.7.4 OUTPUT DATA

RCS is: $\sigma = 2.8 \times 10^{-3}$ m^2 or $\sigma_{dB} = -25.6$ dBsm for input data specified.

4.7.5 REFERENCES

[1] Barton, D. K., and S. A. Leonov, (eds.), *Radar Technology Encyclopedia*, Norwood, MA: Artech House, 1997, p. 369.

[2] Morchin, W., *Radar Engineers Sourcebook*, Norwood, MA: Artech House, 1993, pp. 93–110.

[3] Kerr, D. E., *Propagation of Short Radio Waves*, New York: McGraw-Hill, 1951, pp. 451, 452.

Chapter 5

Antennas and Propagation

5.1 Antenna Patterns and Gain

5.1.1 DESCRIPTION

Simulates the theoretical normalized voltage antenna patterns: 1) omnidirectional, 2) sin(x)/x type, 3) Gaussian, and 4) cosecant-squared. Calculates antenna gain as a function of its area and wavelength. Antenna pattern simulation and antenna gain calculation portions of the program are indpendent segments and input data entered are not related to each other.

5.1.2 INPUT DATA

==

Parameter	Dimension and Description

==

Antenna Pattern Simulation

antenna_type := 3

$1 \rightarrow$ omnidirectional pattern
$2 \rightarrow$ sin(x)/x type pattern
$3 \rightarrow$ Gaussian pattern
$4 \rightarrow$ cosecant-squared pattern

BW := 5   deg, half-power (3 dB) beamwidth

θ_{max} := 10   deg, beam axis angle

θ := 8   deg, current angle

Antenna Gain Calculation

λ := 0.1   m, wavelength

A := 20   m², antenna physical aperture

η_{ap} := 1   aperture efficiency

Note: aperture efficiency is defined as in [1, p. 246].

==

171

5.1.3 MODEL

5.1.3.1 Constants

$$z := \frac{\pi}{180}$$
coefficient to transform degrees to radians

$$c1 := \frac{1.39157}{\sin\left(\dfrac{z \cdot BW}{2}\right)}$$ $c1 = 31.903$ normalization coefficient for sin(x)/x pattern

$$c2 := \frac{\sqrt{2 \cdot \ln(2)}}{\sin\left(\dfrac{z \cdot BW}{2}\right)}$$ $c2 = 26.993$ normalization coefficient for Gaussian pattern

5.1.3.2 Normalized Voltage Patterns

$$f := \begin{vmatrix} 1 \quad \text{if antenna\_type} = 1 \\[4pt] \text{if antenna\_type} = 2 \\[2pt] \quad \begin{vmatrix} f \leftarrow 1 \quad \text{if } \theta - \theta_{max} = 0 \\[6pt] \dfrac{\sin\left[c1 \cdot z \cdot (\theta - \theta_{max})\right]}{c1 \cdot z \cdot (\theta - \theta_{max})} \quad \text{otherwise} \end{vmatrix} \\[10pt] f \leftarrow \exp\left[\left[-c2^2 \cdot \dfrac{\left[z \cdot (\theta - \theta_{max})\right]^2}{4}\right]\right] \quad \text{if antenna\_type} = 3 \\[10pt] \text{if antenna\_type} = 4 \\[2pt] \quad \begin{vmatrix} \left[\exp\left[\left[-c2^2 \cdot \dfrac{\left[z \cdot (\theta - \theta_{max})\right]^2}{4}\right]\right]\right] \quad \text{if } \theta \le \theta_{max} + \dfrac{BW}{2} \\[10pt] \dfrac{1}{\sqrt{2}} \cdot \dfrac{\sin\left[z \cdot \left(\theta_{max} + \dfrac{BW}{2}\right)\right]}{\sin(z \cdot \theta)} \quad \text{if } \theta > \theta_{max} + \dfrac{BW}{2} \end{vmatrix} \end{vmatrix}$$

5.1.3.3 Antenna Gain

$$G := \frac{4 \cdot \pi \cdot A \cdot \eta_{ap}}{\lambda}$$ $G = 2.513 \times 10^3$ antenna power gain

$$G\_db := 10 \cdot \log(G)$$ $G\_db = 34.002$ dB, antenna power gain

5.1.4 OUTPUT DATA

The normalized voltage pattern f = 0.801 and gain   G_db = 34.002 dB
for input data specified.

5.1.5 REFERENCE

[1] Barton, D. K., and S. A. Leonov, (eds.), *Radar Technology Encyclopedia*, Norwood,
MA: Artech House, 1997, pp. 30, 40, 245, 297.

5.1.6 EXAMPLES

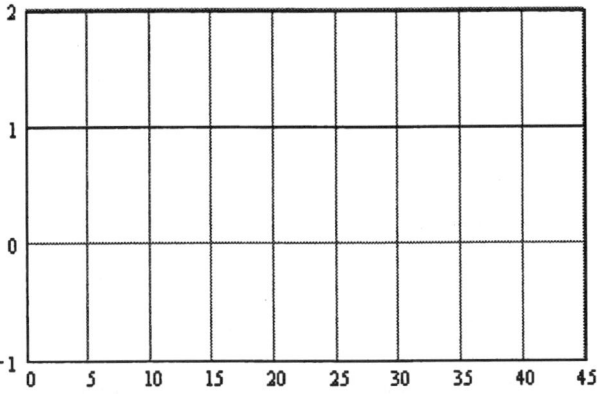

Figure 5.1 Normalized voltage gain vs. angle (deg) for omnidirectional pattern.

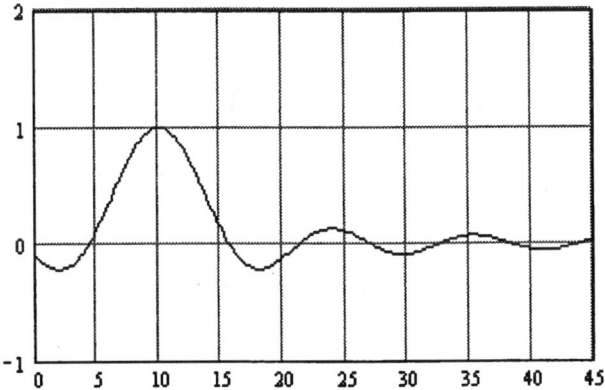

Figure 5.2 Normalized voltage gain vs. angle (deg) for sin(x)/x pattern.

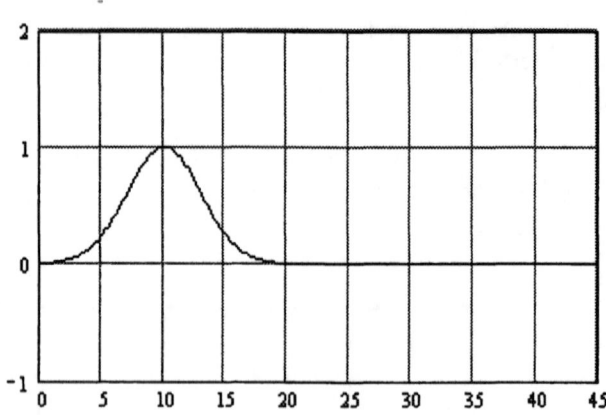

Figure 5.3 Normalized voltage gain vs. angle (deg) for a Gaussian pattern.

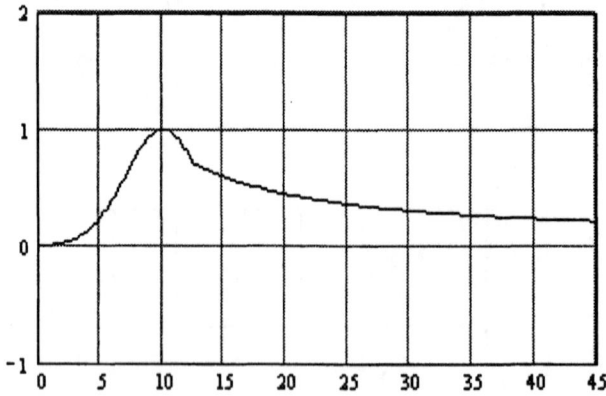

Figure 5.4 Normalized voltage gain vs. angle (deg) for csc-squared pattern.

5.2 Phased-Array Patterns and Gain

5.2.1 DESCRIPTION
Simulates the typical theoretical patterns for one-dimensional (1D) and two-dimensional (2D) phased arrays. Calculates array gain as a function of the number of array elements and incidence angle.

5.2.2 INPUT DATA

==

Parameter	*Dimension and Description*

==

Array Pattern Simulation

array_type := 1

$1 \rightarrow$ linear (1D) array
$2 \rightarrow$ planar (2D) array

distribution := 1

$1 \rightarrow$ uniform
$2 \rightarrow$ cosine-squared on a pedestal

$a_0 := 0.2$ pedestal amplitude

$b_0 := 1 - a_0$

parameters of cosine-squared distribution if distribution = 2

$\lambda := 0.1$

m, wavelength

Linear Array

$P := 100$

number of elements

$d_1 := 0.01$

m, distance between array elements

$\Delta\psi := 0.36$

deg, phase shift between adjacent elements

$\theta_1 := 0$

deg, angle from normal to array

Planar Array

$N := 100$

number of elements in X plane

$M := 100$

number of elements in Y plane

$d_2 := 0.01$

m, distance between array elements

$\Delta_c := \dfrac{\pi}{8}$ $\Delta_{ps} := \dfrac{\pi}{4}$

rad, increments of phase control algorithm

$\upsilon_2 := 0$

deg, angle from normal to array

$\theta_2 := 0$

deg, angle in the plane of the array between X axis and projection of an incident wave

Gain Calculation

$\eta_{ar} := 1$ array radiation efficiency

$c := \dfrac{3}{2}$ correction coefficient to raise cosine function
(depends on matching elements to space: for
many practical arrays k = 3/2)

==

5.2.3 MODEL

5.2.3.1 Constants

$z := \dfrac{\pi}{180}$ coefficient to transform degrees to radians

$k := \dfrac{2 \cdot \pi}{\lambda}$ $k = 62.832$ wave number

5.2.3.2 Normalized Linear Array Pattern

$p := 0 .. P - 1$ cycle

$x_p := (p + 0.5) \cdot d_1$ m, x coordinate of the nth element for
uniformly spaced array

$L := (P - 1) \cdot d_1$ $L = 0.99$ m, length of the array

Amplitude distribution of the illuminating field of the array nth element

$$A_1(p) := \begin{vmatrix} 1 & \text{if distribution} = 1 \\[2mm] a_0 + b_0 \cdot \cos\left[\pi \cdot \left(\dfrac{x_p}{L} + 0.5\right)\right]^2 & \text{if distribution} = 2 \end{vmatrix}$$

$\phi(p) := -p \cdot z \cdot \Delta\psi$ rad, phase distribution of the illuminating field
for the array nth element

$I_1(p) := A_1(p) \cdot e^{j \cdot \phi(p)}$ amplitude-phase distribution of the illuminating
field for the array nth element

$$f_{a1}(\theta_1) := \dfrac{\left| \sum_p I_1(p) \cdot \exp\left(j \cdot k \cdot x_p \cdot \sin(z \cdot \theta_1)\right) \right|}{\sum_p A_1(p)}$$ linear array factor

$$f_{a1}(\theta_1) = 0.984 \qquad \text{normalized array factor for} \quad \theta_1 = 0 \quad \text{deg}$$

$$f_e(\theta_1) := \left(\cos(z \cdot \theta_1)\right)^{\frac{c}{2}} \qquad\qquad \text{normalized element pattern}$$

$$f_e(\theta_1) = 1 \qquad\qquad \text{normalized element pattern for} \quad \theta_1 = 0 \quad \text{deg}$$

$$f_1(\theta_1) := f_{a1}(\theta_1) \cdot f_e(\theta_1) \qquad\qquad \text{normalized array pattern}$$

$$f_1(\theta_1) = 0.984 \qquad \text{normalized array pattern for} \qquad \theta_1 = 0 \quad \text{deg}$$

$$G_1 := \pi \cdot \eta_{ar} \cdot P \cdot \cos(z \cdot \theta_1)^c \qquad G_1 = 314.159 \qquad \text{array power gain}$$

$$G\_db_1 := 10 \cdot \log(G_1) \qquad G\_db_1 = 24.971 \qquad \text{dB, array power gain}$$

5.2.3.3 Normalized Planar Array Pattern

$$n := 0 .. N - 1 \qquad\qquad\qquad\qquad\qquad \text{cycle}$$

$$m := 0 .. M - 1 \qquad\qquad\qquad\qquad\qquad \text{cycle}$$

$$u_0 := \cos(z \cdot \upsilon_2) \cdot \sin(z \cdot \theta_2)$$
$$v_0 := \sin(z \cdot \upsilon_2) \cdot \sin(z \cdot \theta_2) \qquad \text{sine-space coordinates for a planar array}$$

$$x_n := n \cdot d_2 \qquad\qquad\qquad \text{m, } x \text{ coordinate of the } nm\text{th element}$$

$$y_m := m \cdot d_2 \qquad\qquad\qquad \text{m, } y \text{ coordinate of the } nm\text{th element}$$

Amplitude distribution of the illuminating field of the array nth element

$$A_2(n,m) := \begin{vmatrix} 1 & \text{if distribution} = 1 \\[2em] a_0 + b_0 \cdot \cos\left[\pi \cdot \dfrac{\sqrt{(x_n)^2 + (y_m)^2}}{d_2 \cdot \sqrt{N^2 + M^2}}\right]^2 & \text{if distribution} = 2 \end{vmatrix}$$

$$\phi(n) := \Delta_c \cdot \text{floor}(k \cdot u_0 \cdot x_n + 0.5) \qquad \text{phase distribution for } x \text{ coordinate}$$

$$\phi(m) := \Delta_c \cdot \text{floor}(k \cdot v_0 \cdot y_m + 0.5) \qquad \text{phase distribution for } y \text{ coordinate}$$

Phase distribution of the illuminating field of the array nth element

$$\phi 0(n,m) := \Delta_{ps} \cdot \text{floor}\left(\frac{\phi(n) + \phi(m)}{\Delta_{ps}} + 0.5\right)$$

Amplitude-phase distribution of the illuminating field of the array nth element

$$I_2(n,m) := A_2(n,m) \cdot e^{j \cdot \phi 0(n,m)}$$

Normalized planar array pattern

$$f_2(\theta_2, \upsilon_2) := \frac{\left| \sum\limits_n \sum\limits_m I_2(n,m) \cdot \exp\left[j \cdot k \cdot (x_n \cdot u_0 + y_m \cdot v_0)\right] \right|}{\sum\limits_n \sum\limits_m A_2(n,m)}$$

$f_2(\theta_2, \upsilon_2) = 1$　　　　normalized array pattern for　$\theta_2 = 0$　　$\upsilon_2 = 0$　　deg

$G_2 := \pi \cdot \eta_{ar} \cdot N \cdot M \cdot \cos(z \cdot \theta_2)^c$　　$G_2 = 3.142 \times 10^4$　　　array power gain

$G\_db_2 := 10 \cdot \log(G_2)$　　　$G\_db_2 = 44.971$　　dB, array power gain

5.2.3.4 Specified Array Data

array_type = 1

　　　　　　　　　　　　　　　　　　array type specified

$$f := \begin{vmatrix} f_1(\theta_1) & \text{if array\_type} = 1 \\ f_2(\theta_2, \upsilon_2) & \text{if array\_type} = 2 \end{vmatrix}$$

normalized array voltage pattern

$$G\_db := \begin{vmatrix} G\_db_1 & \text{if array\_type} = 1 \\ G\_db_2 & \text{if array\_type} = 2 \end{vmatrix}$$

dB, array power gain

5.2.4 OUTPUT DATA

The normalized voltage pattern　$f = 0.984$　and gain　$G\_db = 24.971$　dB for input data specified.

5.2.5 REFERENCE

[1] Barton, D. K., and S. A. Leonov, (eds.), *Radar Technology Encyclopedia*, Norwood, MA: Artech House, 1997, pp. 30, 96, 297.

5.2.6 EXAMPLES

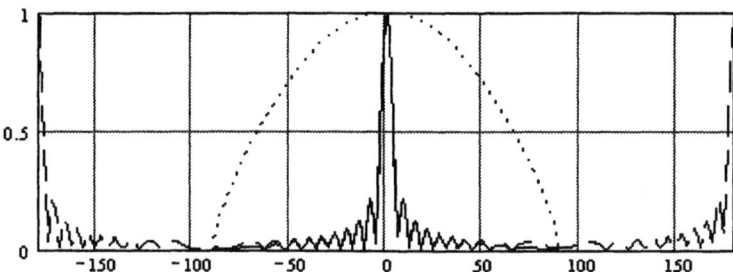

Figure 5.5 Normalized voltage gain vs. angle (deg) for the linear array of   N = 100
elements with 1-cm spacing between the elements and uniform amplitude
distribution (dashed line is array factor, dotted line is an element pattern,
solid line is array pattern).

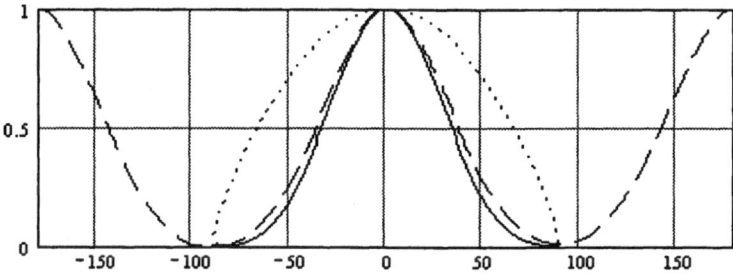

Figure 5.6 Normalized voltage gain vs. angle (deg) for the linear array of   N = 10
elements with 1-cm spacing between the elements and uniform amplitude
distribution (dashed line is array factor, dotted line is an element pattern,
solid line is array pattern).

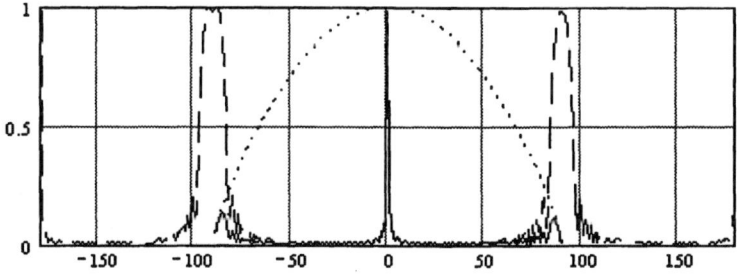

Figure 5.7 Normalized voltage gain vs. angle (deg) for the linear array of N = 100
elements with 10-cm spacing between the elements and uniform amplitude
distribution (dashed line is array factor, dotted line is an element pattern,
solid line is array pattern).

5.3 Pattern-Propagation Factor

5.3.1 DESCRIPTION

Calculates the pattern-propagation factor for the following three regions: 1) interference (optical) region (reflection PPF) where lobing structure of electromagnetic field exists; 2) transition (intermediate) region (combined PPF) where lobing field structure is corrected by diffraction PPF; 3) diffraction region (diffraction PPF) where no lobing structure is present (attenuation only) for smooth-sphere or knife-edge diffraction cases. *Assumptions:* for optical region: a) absence of absorption losses and abnormal refractive effects; b) path difference between direct and reflected wave is small compared to target distance; c) target distance is much greater than the distance from antenna to reflection point, so the rays are parallel (grazing angle is equal to elevation pointing angle); for diffraction region: a target range is longer than the radar horizon and the path is blocked.

5.3.2 INPUT DATA

===

Parameter	Dimension and Description

===

Target Parameters

$R_t := 25000$ m, target range

$h_t := 10$ m, target height above ground level

Radar Parameters

$f := 1500$ MHz, radar frequency

polarization := 1 1 -> horizontal
2 -> vertical

$h_a := 25$ m, antenna focal point height above surface level

Antenna Pattern

antenna_type := 3 1 → omnidirectional
2 → sin(x)/x type
3 → Gaussian

$BW := 2$ deg, half-power (3-dB) beamwidth

$\theta_{max} := 5.0$ deg, beam axis elevation angle

$\theta := 0.5$ deg, elevation pointing angle

Underlying Surface

surface_type := 3

$0 \rightarrow$ absolutely refractive surface ($\rho_0 = 1$)

$1 \rightarrow$ good soil (wet)

$2 \rightarrow$ average soil

$3 \rightarrow$ bad soil (dry)

$4 \rightarrow$ salt water

$5 \rightarrow$ fresh water

$6 \rightarrow$ snow or ice

$\sigma_h := 1$ m, rms roughness of the surface

$0 \rightarrow$ no vegetation

vegetation := 0 $1 \rightarrow$ thin grass

$2 \rightarrow$ dense weeds or brush

$3 \rightarrow$ dense trees

===

5.3.3 MODEL

5.3.3.1 Constants

$z := \dfrac{\pi}{180}$ coefficient to transform degrees to radians

$a_e := 8493333$ m, effective earth radius (4/3 approximation)

$c := 2.997925 \cdot 10^8$ m/s, velocity of light

$c1 := \dfrac{1.39157}{\sin\left(\dfrac{z \cdot BW}{2}\right)}$ $c1 = 79.735$ normalization coefficient for sin(x)/x pattern

$c2 := \dfrac{\sqrt{2 \cdot \ln(2)}}{\sin\left(\dfrac{z \cdot BW}{2}\right)}$ $c2 = 67.464$ normalization coefficient for Gaussian pattern

5.3.3.2 Path Difference and Propagation Region

$\lambda := \dfrac{c}{f \cdot 10^6}$ $\lambda = 0.2$ m, wavelength

Parameters to calculate the path difference

$p := \sqrt{\dfrac{4 \cdot a_e \cdot (h_a + h_t) + R_t^2}{3}}$ $p = 2.459 \times 10^4$

$$\Phi := \text{acos}\left[\frac{2 \cdot a_e \cdot (h_t - h_a) \cdot R_t}{p^3}\right] \qquad \Phi = 2.014$$

Distance from radar point of specular reflection

$$d_1 := \frac{R_t}{2} - p \cdot \cos\left(\frac{\Phi + \pi}{3}\right) \qquad d_1 = 1.612 \times 10^4$$

Height of radar antenna above plane tangent at point of specular reflection

$$H_1 := h_a - \frac{d_1^2}{2 \cdot a_e} \qquad H_1 = 9.711$$

Height of target above plane tangent at point of specular reflection

$$H_2 := \frac{(R_t - d_1) \cdot H_1}{d_1} \qquad H_2 = 5.353$$

Path difference: the extra one-way distance traveled by the specular reflected ray as compared to the direct path to the target (meters)

$$\delta\!0 := \frac{2 \cdot H_1 \cdot H_2}{R_t} \qquad \delta\!0 = 4.159 \times 10^{-3}$$

$$Z := \frac{\lambda}{8} \qquad Z = 0.025 \qquad \text{m, threshold to compare against the path difference}$$

$$\text{region} := \left| \begin{array}{l} \text{if } H_1 > 0 \\ \quad \left| \begin{array}{l} 1 \quad \text{if } \delta\!0 > Z \\ 2 \quad \text{if } \delta\!0 \le Z \\ \end{array} \right. \\ 3 \quad \text{if } \delta\!0 \le Z \text{ if } H_1 \le 0 \end{array} \right. \qquad \text{region} = 2 \qquad \text{PPF region}$$

5.3.3.3 Interference Region

A. Pattern Factors

$\psi := \theta$ $\qquad$ $\psi = 0.5$ $\qquad\qquad\qquad\qquad\qquad$ deg, grazing angle

$\theta_{p\_d} := \theta - \theta_{max}$ $\qquad$ $\theta_{p\_d} = -4.5$ $\qquad$ deg, angle of direct ray (target) relative to beam axis

$\theta_{p\_r} := -\theta - \theta_{max}$ $\qquad$ $\theta_{p\_r} = -5.5$ $\qquad$ deg, angle of reflected ray relative to beam axis

$\qquad\qquad$ Normalized voltage pattern factor for direct beam

$$f\_d := \begin{cases} 1 & \text{if } \text{antenna\_type} = 1 \\[2mm] \text{if } \text{antenna\_type} = 2 \\ \qquad \begin{cases} f \leftarrow 1 & \text{if } \theta_{p\_d} = 0 \\[2mm] \dfrac{\sin\left[c1\cdot z\cdot\left(\theta_{p\_d}\right)\right]}{c1\cdot z\cdot\left(\theta_{p\_d}\right)} & \text{otherwise} \end{cases} \\[6mm] f \leftarrow \exp\left[-c2^2\cdot\dfrac{\left[z\cdot\left(\theta_{p\_d}\right)\right]^2}{4}\right] & \text{if } \text{antenna\_type} = 3 \end{cases}$$

$f\_d = 8.949 \times 10^{-4}$

$\qquad\qquad$ Normalized voltage pattern factor for reflected beam

$$f\_r := \begin{cases} 1 & \text{if } \text{antenna\_type} = 1 \\[2mm] \text{if } \text{antenna\_type} = 2 \\ \qquad \begin{cases} f \leftarrow 1 & \text{if } \theta_{p\_r} = 0 \\[2mm] \dfrac{\sin\left[c1\cdot z\cdot\left(\theta_{p\_r}\right)\right]}{c1\cdot z\cdot\left(\theta_{p\_r}\right)} & \text{otherwise} \end{cases} \\[6mm] f \leftarrow \exp\left[\left[-c2^2\cdot\dfrac{\left[z\left(\theta_{p\_r}\right)\right]^2}{4}\right]\right] & \text{if } \text{antenna\_type} = 3 \end{cases}$$

$f\_r = 2.795 \times 10^{-5}$

B. Fresnel Reflection Coefficient

$\varepsilon_r :=$ | 0 if surface_type $= 0$
| 25 if surface_type $= 1$
| 15 if surface_type $= 2$
| 3 if surface_type $= 3$
| if surface_type $= 4$
| | 80 if $f \leq 1500$
| | $80 - 0.00733 \cdot (f - 1500)$ if $1500 < f \leq 3000$
| | $69 - 0.00243 \cdot (f - 3000)$ if $3000 < f \leq 10000$
| if surface_type $= 5$
| | 80 if $f \leq 1500$
| | $80 - 0.00733 \cdot (f - 1500)$ if $1500 < f \leq 3000$
| | $69 - 0.00243 \cdot (f - 3000)$ if $3000 < f \leq 10000$
| if surface_type $= 6$
| | 3.2 if $f \leq 2000$
| | 3.2 if $2000 < f \leq 10000$

relative dielectric constant of the surface material

$\sigma_e :=$ | 0 if surface_type $= 0$
| 0.02 if surface_type $= 1$
| 0.005 if surface_type $= 2$
| 0.001 if surface_type $= 3$
| if surface_type $= 4$
| | 4.3 if $f \leq 1500$
| | $4.3 + 0.00148 \cdot (f - 1500)$ if $1500 < f \leq 3000$
| | $6.52 + 0.001314 \cdot (f - 3000)$ if $3000 < f \leq 10000$
| if surface_type $= 5$
| | 1 if $f \leq 1500$
| | $1 + 0.00106 \cdot (f - 1500)$ if $1500 < f \leq 3000$
| | $1 + 0.00106 \cdot (f - 3000)$ if $3000 < f \leq 10000$
| if surface_type $= 6$
| | 0.000057 if $f \leq 2000$
| | $0.000057 + 6.79 \cdot 10^{-8} \cdot (f - 2000)$ if $2000 < f \leq 10000$

mho/m, conductivity of the surface material

$$\varepsilon_c := \varepsilon_r - j \cdot 60 \cdot \frac{c}{f \cdot 10^6} \cdot \sigma_e \qquad \text{complex dielectric constant}$$

Complex Fresnel reflection coefficient for horizontal polarization

$$\Gamma_H := \begin{vmatrix} 1 & \text{if } surface\_type = 0 \\ \dfrac{\sin(z \cdot \psi) - \sqrt{\varepsilon_c - \cos(z \cdot \psi)^2}}{\sin(z \cdot \psi) + \sqrt{\varepsilon_c - \cos(z \cdot \psi)^2}} & \text{otherwise} \end{vmatrix}$$

Complex Fresnel reflection coefficient for vertical polarization

$$\Gamma_V := \begin{vmatrix} 1 & \text{if } surface\_type = 0 \\ \dfrac{\varepsilon_c \cdot \sin(z \cdot \psi) - \sqrt{\varepsilon_c - \cos(z \cdot \psi)^2}}{\varepsilon_c \cdot \sin(z \cdot \psi) + \sqrt{\varepsilon_c - \cos(z \cdot \psi)^2}} & \text{otherwise} \end{vmatrix}$$

$$\rho_0 := \begin{vmatrix} |\Gamma_H| & \text{if } polarization = 1 \\ |\Gamma_V| & \text{if } polarization = 2 \end{vmatrix} \qquad \text{magnitude of Fresnel reflection coefficient}$$

$$\rho_0 = 0.988$$

$$\phi := \begin{vmatrix} 180 & \text{if } surface\_type = 0 \\ \text{otherwise} \\ \quad \begin{vmatrix} \dfrac{\arg(\Gamma_H)}{z} & \text{if } polarization = 1 \\ \dfrac{\arg(\Gamma_V)}{z} & \text{if } polarization = 2 \end{vmatrix} \end{vmatrix} \qquad \text{phase of Fresnel reflection coefficient, deg}$$

$$\phi = 179.998$$

C. Specular Scattering Coefficient

$$\rho_s := \exp\left[-2 \cdot \left(2 \cdot \pi \cdot \frac{\sigma_h}{\lambda} \cdot \sin(z \cdot \psi) \right)^2 \right] \qquad \rho_s = 0.86$$

D. Vegetation Coefficient

$$K := \begin{vmatrix} 1 & \text{if vegetation} = 0 \\ 1 & \text{if vegetation} = 1 \\ 3 & \text{if vegetation} = 2 \\ 10 & \text{if vegetation} = 3 \end{vmatrix}$$

multiplication factor depending
on vegetation type

$$\rho_v := \begin{vmatrix} 1 & \text{if vegetation} = 0 \\ \exp\left(-\frac{K}{\lambda} \cdot \sin(z \cdot \psi)\right) & \text{otherwise} \end{vmatrix}$$

$\rho_v = 1$ vegetation coefficient

E. Divergence Factor

$$u := \sqrt{\frac{a_e}{2 \cdot h_a}} \cdot \tan(z \cdot \theta)$$

parameter

$$D := \sqrt{\frac{1}{3} \cdot \left(1 + 2 \cdot \frac{u}{\sqrt{u^2 + 3}}\right)}$$

$D = 0.966$ divergence factor accounting for
sphericity of the earth

F. Reflection PPF

$$\delta := 2 \cdot h_a \cdot \sin(z \cdot \theta) \qquad \delta = 0.436$$

m, path difference between direct
and reflected beams

$$\alpha := \frac{2 \cdot \pi}{\lambda} \cdot \delta + z \cdot \phi \qquad \alpha = 16.859$$

rad, PPF phase angle

PPF in interference region

$$F_r := f\_d \cdot \left| 1 + \frac{f\_r}{f\_d} \cdot D \cdot \rho_0 \cdot \rho_s \cdot \rho_v \cdot e^{-j \cdot \alpha} \right|$$

$F_r = 8.858 \times 10^{-4}$

or equivalent expression

$$F_r := f\_d \cdot \sqrt{1 + \left(\frac{f\_r}{f\_d} \cdot D \cdot \rho_0 \cdot \rho_s \cdot \rho_v\right)^2 + 2 \cdot \left(\frac{f\_r}{f\_d} \cdot D \cdot \rho_0 \cdot \rho_s \cdot \rho_v\right) \cdot \cos(\alpha)}$$

$F_r = 8.858 \times 10^{-4}$ reflection pattern-propagation factor

$F_{r\_db} := 20 \cdot \log(F_r)$ dB, reflection pattern-propagation factor

$F_{r\_db} = -61.054$

5.3.3.4 Diffraction Region

A. Pattern Factors

$R_h := \sqrt{2 \cdot a_e} \cdot \left(\sqrt{h_a} + \sqrt{h_t} \right)$ $R_h = 3.364 \times 10^4$ m, horizon range

$\theta_0 := -\theta_{max}$ $\theta_0 = -5$ deg, angle of antenna pattern with respect to maximum radiation angle at the horizon

Normalized voltage pattern factor

$$f\_0 := \begin{vmatrix} 1 & \text{if antenna\_type} = 1 \\ & \text{if antenna\_type} = 2 \\ & \quad \begin{vmatrix} f \leftarrow 1 & \text{if } \theta_0 = 0 \\ \dfrac{\sin\left[c1 \cdot z \cdot (\theta_0) \right]}{c1 \cdot z \cdot (\theta_0)} & \text{otherwise} \end{vmatrix} \\ f \leftarrow \exp\left[\left[-c2^2 \cdot \dfrac{\left[z \cdot (\theta_0) \right]^2}{4} \right] \right] & \text{if antenna\_type} = 3 \end{vmatrix}$$

$f\_0 = 1.725 \times 10^{-4}$

B. Smooth-Sphere Diffraction

Height-gain factors

$$b := \begin{vmatrix} 0 & \text{if polarization} = 1 \\ 1 & \text{if polarization} = 2 \end{vmatrix} \qquad b = 0$$

$$h_{min} := \frac{\lambda}{2 \cdot \pi} \cdot \frac{\left[\varepsilon_r^2 + \left(60 \cdot \sigma_e \cdot \lambda \right)^2 \right]^{\frac{b}{2}}}{\left[\left(\varepsilon_r^2 - 1 \right)^2 + \left(60 \cdot \sigma_e \cdot \lambda \right)^2 \right]^{\frac{1}{4}}} \qquad h_{min} = 0.011$$

$$h1 := \begin{vmatrix} h_{min} & \text{if } h_a < h_{min} \\ h_a & \text{otherwise} \end{vmatrix}$$

$h1 = 25$

$$h2 := \begin{vmatrix} h_{min} & \text{if } h_t < h_{min} \\ h_t & \text{otherwise} \end{vmatrix}$$

$h2 = 10$

$$h_c := 30 \cdot \lambda^{\frac{2}{3}}$$

$h_c = 10.255$

$$g1 := \begin{vmatrix} 1 & \text{if } h_a \le h_c \\ 0.1356 \cdot \left(\dfrac{h_a}{h_c}\right)^{-0.904} \cdot 10^{0.948 \cdot \sqrt{\frac{h_a}{2 \cdot h_c}}} & \text{otherwise} \end{vmatrix}$$

$g1 = 0.675$

$$g2 := \begin{vmatrix} 1 & \text{if } h_t \le h_c \\ 0.1356 \cdot \left(\dfrac{h_t}{h_c}\right)^{-0.904} \cdot 10^{0.948 \cdot \sqrt{\frac{h_t}{2 \cdot h_c}}} & \text{otherwise} \end{vmatrix}$$

$g2 = 1$

Smooth-sphere diffraction propagation factor

$$F_{sm} := 9.29 \cdot 10^{-6} \cdot \lambda^{-\frac{3}{2}} \cdot R_t^{\frac{1}{2}} \cdot \exp\left(-7.12 \cdot 10^{-5} \cdot R_t \cdot \lambda^{-\frac{1}{3}}\right) \cdot g1 \cdot h1 \cdot g2 \cdot h2$$

$F_{sm} = 0.132$

C. Knife-Edge Diffraction

$$a := \frac{2}{\lambda} \cdot \left(\frac{1}{R_h} + \frac{1}{R_t - R_h}\right)$$

$a = -8.606 \times 10^{-4}$ parameter

$$par := \frac{R_h}{R_t} \cdot \left[\frac{(R_t - R_h)^2}{2 \cdot a_e} + \sigma_h\right] \cdot \sqrt{a}$$

$par = 0.213i$ parameter

$$p := \begin{vmatrix} par & \text{if } a > 0 \\ 1 & \text{otherwise} \end{vmatrix}$$

$p = 1$

$$F_{ke\_db} := \begin{vmatrix} -(6 + 8 \cdot p) & \text{if } p \le 1 \\ -\left(6.4 + 20 \cdot \log\left(\sqrt{p^2 + 1} + p\right)\right) & \text{otherwise} \end{vmatrix}$$

dB, knife-edge diffraction propagation factor

$$F_{ke\_db} = -14$$

$$F_{ke} := 10^{\frac{F_{ke\_db}}{10}}$$

knife-edge diffraction propagation factor

$$F_{ke} = 0.04$$

D. Diffraction PPF

$$\sigma_h = 1$$

m, rms roughness of the surface specified

$$\sigma_{h\_cr} := \frac{\sqrt{\lambda \cdot R_h}}{2} \qquad \sigma_{h\_cr} = 40.998$$

m, the required rms roughness to support knife-edge diffraction

$$F := \begin{vmatrix} F_{sm} & \text{if } \sigma_h \le \sigma_{h\_cr} \\ F_{ke} & \text{otherwise} \end{vmatrix}$$

diffraction propagation factor

$$F_d := f\_0 \cdot F$$

diffraction pattern-propagation factor

$$F_d = 2.277 \times 10^{-5}$$

$$F_{d\_db} := 20 \cdot \log(F_d)$$

dB, diffraction pattern-propagation factor

$$F_{d\_db} = -92.852$$

5.3.3.5 Transition Region

$$F_{t\_db} := F_{r\_db} \cdot \frac{80}{Z} + F_{d\_db} \cdot \left(1 - \frac{80}{Z}\right)$$

dB, transition PPF

$$F_{t\_db} = -87.558$$

5.3.3.6 Pattern-Propagation Factor for Specified Conditions

$$region = 2$$

region specified

$$F\_db := \begin{vmatrix} F_{r\_db} & \text{if } region = 1 \\ F_{t\_db} & \text{if } region = 2 \\ F_{d\_db} & \text{if } region = 3 \end{vmatrix} \qquad F\_db = -87.558$$

dB, PPF

5.3.4 OUTPUT DATA

The pattern-propagation factor   F_db = −87.558   dB for input data specified.

5.3.5 REFERENCES

[1] Barton, D. K., and S. A. Leonov, (eds.), *Radar Technology Encyclopedia*, Norwood, MA: Artech House, 1997, p. 312.
[2] Barton, D. K., *Modern Radar System Analysis*, Norwood, MA: Artech House, 1989, p. 291.
[3] Barton, D. K., and W. F. Barton, *Modern Radar System Analysis: Software and Users Manuals*, Norwood, MA: Artech House, pp. 150, 165.
[4] Blake, L. V., *Machine Plotting of Radio/Radar Vertical Plane Coverage Diagrams*, NRL Report 7098, June 25, 1970.

Chapter 6

Waveforms and Signal Processing

6.1 Pulsed Waveform Generation

6.1.1 DESCRIPTION

Generates the following common pulsed waveforms: unmodulated pulse, frequency-modulated pulse, and phase-coded pulse.

6.1.2 INPUT DATA

===

Parameter	*Dimension and Description*

===

$f\_s := 200$	MHz, digital sampling rate
$\tau := 3$	μs, pulsewidth
$f_0 := 5$	MHz, carrier frequency
waveform := 1	1 → unmodulated waveform 2 → frequency-modulated waveform 3 → phase-modulated waveform

Unmodulated Waveform

$\psi_c := 0$	rad, initial phase

Frequency-Modulated Waveform

modulation := 1	1 → linear frequency modulation 2 → nonlinear frequency modulation
LINEARITY := 0.7 index := 3	parameters to change frequency linearity law

Note: to obtain the waveform with desired bandwidth, enter repeated values under parameters to change frequency linearity law and run the program until parameter dSP in the final section reaches the correct value.

191

Phase-Modulated Waveform

LENGTH := 13   phase code length

Note: Barker code with LENGTH = 13 is used as an example. For more information about common codes, and limitations applied to the length and type of codes, see [2, pp. 29–117].

===

6.1.3 MODEL

6.1.3.1 Constants

y_th := −3   dB, level to calculate spectrum width

$I := 2^{12}$ $I = 4.096 \times 10^3$ number of samples to calculate spectrum

6.1.3.2 Time-Domain Simulation

$\Delta t := \dfrac{1}{f\_s}$ $\Delta t = 5 \times 10^{-3}$ μs, sampling interval

$N := \text{floor}\left(\dfrac{\tau}{\Delta t}\right)$ $N = 600$ number of samples in the waveform

$n := 0 .. N - 1$ cycle

A. Unmodulated Pulse

$\theta\_UM_n := \psi_c$ unmodulated waveform phase

B. Frequency-Modulated Pulse

$T_n := \dfrac{n}{f\_s} - \dfrac{1}{2 \cdot f\_s}$ μs, time samples

$t_n := \dfrac{T_n - \dfrac{\tau}{2}}{\left(\dfrac{\tau}{2}\right)}$ normalized and centered time samples

Phase-modulation law for a linear frequency modulation

$\theta 1_n := (-1)^{\text{index}} \cdot \text{LINEARITY} \cdot \tau \cdot (t_n)^2$

Phase-modulation law for a nonlinear frequency modulation

$\theta 2_n := \text{LINEARITY} \cdot \tau \cdot \left[(-1)^{\text{index}} \cdot (t_n)^2 + \left[\text{acos}\left[(t_n)^2\right]^{\frac{1}{\text{index}+1}} \right] \right]$

$$\theta\_FM_n := \begin{vmatrix} \theta 1_n & \text{if} & \text{modulation} = 1 \\ \theta 2_n & \text{if} & \text{modulation} = 2 \end{vmatrix}$$ frequency-modulated waveform phase

C. Phase-Modulated Pulse

$$\Delta\tau := \frac{\tau}{\text{LENGTH}}$$ $\Delta\tau = 0.231$ μs, phase modulation increment

$$K := \text{floor}\left(\frac{\Delta\tau}{\Delta t}\right)$$ $K = 46$ number of samples within modulation increment

Phase-modulated waveform phase (Barker code LENGTH = 13 case)

$$\theta\_PM_n := \begin{vmatrix} 0 & \text{if} & n \leq K \\ 0 & \text{if} & K < n \leq 2 \cdot K \\ 0 & \text{if} & 2 \cdot K < n \leq 3 \cdot K \\ 0 & \text{if} & 3 \cdot K < n \leq 4 \cdot K \\ 0 & \text{if} & 4 \cdot K < n \leq 5 \cdot K \\ \pi & \text{if} & 5 \cdot K < n \leq 6 \cdot K \\ \pi & \text{if} & 6 \cdot K < n \leq 7 \cdot K \\ 0 & \text{if} & 7 \cdot K < n \leq 8 \cdot K \\ 0 & \text{if} & 8 \cdot K < n \leq 9 \cdot K \\ \pi & \text{if} & 9 \cdot K < n \leq 10 \cdot K \\ 0 & \text{if} & 10 \cdot K < n \leq 11 \cdot K \\ \pi & \text{if} & 11 \cdot K < n \leq 12 \cdot K \\ 0 & \text{if} & n > 12 \cdot K \end{vmatrix}$$

D. Pulse with the Specified Modulation Law

waveform = 1   waveform type specified

$$\theta_n := \begin{vmatrix} \theta\_UM_n & \text{if} & \text{waveform} = 1 \\ \theta\_FM_n & \text{if} & \text{waveform} = 2 \\ \theta\_PM_n & \text{if} & \text{waveform} = 3 \end{vmatrix}$$ phase of the specified waveform

$$WF\_comp_n := \exp\left(-j \cdot \theta_n\right)$$ complex baseband waveform

$$a_n := \text{Re}\left(WF\_comp_n\right)$$
$$b_n := \text{Im}\left(WF\_comp_n\right)$$ complex baseband waveform coefficients

$a_0 := 0$ $b_0 := 0$ set first coefficients to zero

Complex Waveform Coefficients

Real   *Imaginary*

$a_0 = 0$ $b_0 = 0$

$a_1 = 1$ $b_1 = 0$

....

$a_{N-2} = 1$ $b_{N-2} = 0$

$a_{N-1} = 1$ $b_{N-1} = 0$

$WF\_bsb := a + j \cdot b$ representation of the complex baseband waveform via its coefficients

$V1 := Re(WF\_bsb)$ baseband waveform voltage

$WF\_f0_n := WF\_bsb_n \cdot exp\left(-j \cdot 2 \cdot \pi \cdot f_0 \cdot \Delta t \cdot n\right)$ complex waveform at the frequency f0

$V2 := Re(WF\_f0)$ waveform voltage

Writing complex waveforms to files

$WRITEPRN("wfmbsb.prn") := WF\_bsb$

$WRITEPRN("wfmf0.prn") := WF\_f0$

6.1.3.3 Frequency-Domain Simulation

A. Waveform Spectrum Computation

$I = 4.096 \times 10^3$ number of samples in extended waveform

$i\_s := ceil\left(\dfrac{I - N}{2}\right)$ $i\_s = 1.748 \times 10^3$ first nonzero sample

$i\_f := i\_s + N - 1$ $i\_f = 2.347 \times 10^3$ last nonzero sample

$i := 0 .. I - 1$ cycle

$EWF1_i := 0$

$EWF1_{i\_s+n} := WF\_bsb_n$ extended baseband waveform

$EV1_i := Re(EWF1_i)$ extended waveform voltage

$EWF2_i := 0$

$EWF2_{i\_s+n} := WF\_f0_n$ extended waveform at the frequency f0

$EV2_i := Re(EWF2_i)$ extended waveform voltage

$SP\_EWF := cfft(EWF2)$ extended waveform spectrum

$I := last(SP\_EWF)$ $I = 4.095 \times 10^3$ last sample in the waveform spectrum

$\Delta f := \dfrac{f\_s}{I}$ $\Delta f = 0.049$ MHz, frequency sampling interval

$i\_0 := \dfrac{I + 1}{2}$ $i\_0 = 2.048 \times 10^3$ central sample

$f_i := (i - i\_0) \cdot \Delta f$ MHz, current frequency

$f_0 = 5$ MHz, carrier frequency

$f_0 = -100.024$

$f_I = 99.976$ MHz, frequency range

Note: the carrier frequency f0 specified in the input data menu must be within f_0 - f_M range. To increase this range, increase sampling rate f_s.

B. The Procedure to Center Spectrum Around Zero Frequency

$i1\_s := 0$ $i1\_s = 0$

$i1\_f := i\_0 - 1$ $i1\_f = 2.047 \times 10^3$

$i2\_s := 0$ $i2\_s = 0$

$i2\_f := i\_0 - 1$ $i2\_f = 2.047 \times 10^3$

$i1 := i1\_s .. i1\_f$ $i2 := i2\_s .. i2\_f$ cycles i1 and i2

$SP0\_EWF_{i1} := SP\_EWF_{i\_0+i1}$

$SP0\_EWF_{i2+i\_0} := SP\_EWF_{i2}$ spectrum centered around zero frequency

$$SP_i := 20 \cdot \log\left(\left|SP0\_EWF_i + 10^{-10}\right|\right) \qquad \text{dB, waveform spectrum}$$

$$SP\_MAX := \max(SP) \qquad \text{spectrum maximum}$$

$$SPN := SP - SP\_MAX \qquad \text{dB, normalized waveform spectrum}$$

C. The Spectrum Width Calculation

MHz, frequency where the spectrum curve crosses the specified level upward

$$f\_b(SPC) := \begin{vmatrix} \text{for } i \in 1..I-2 \\ \qquad \text{break if } SPC_{i-1} \le y\_th \text{ if } SPC_i > y\_th \\ \qquad df\_b \leftarrow \dfrac{SPC_i - y\_th}{SPC_i - SPC_{i-1}} \cdot \Delta f \\ \qquad f\_b \leftarrow f_i - df\_b \\ \qquad f\_b \end{vmatrix}$$

MHz, frequency where the spectrum curve crosses the specified level downward

$$f\_e(SPC) := \begin{vmatrix} \text{for } i \in 1..I-1 \\ \qquad \text{break if } SPC_{i-1} > y\_th \text{ if } SPC_i \le y\_th \\ \qquad df\_e \leftarrow \dfrac{SPC_{i-1} - y\_th}{SPC_{i-1} - SPC_i} \cdot \Delta f \\ \qquad f\_e \leftarrow f_i + df\_e \\ \qquad f\_e \end{vmatrix}$$

$$dSP := f\_e(SPN) - f\_b(SPN) \qquad \text{MHz, waveform spectrum width}$$
$$\text{at} \quad y\_th = -3 \quad \text{dB level}$$
$$dSP = 0.341$$

Note: the specrum width calculation procedure works only for monotonous spectrum functions at levels higher than y_th dB. Be cautious when used for oscillating spectrum shapes with peak-to-valley spans more than y_th dB, such as those for linear frequency-modulated waveforms.

6.1.4 REFERENCES

[1] Barton, D. K., and S. A. Leonov, (eds.), *Radar Technology Encyclopedia*, Norwood, MA: Artech House, 1997, pp. 89, 472, 475.
[2] Lewis, B. L., F. F. Kretschmer, and W. W. Shelton, *Aspects of Radar Signal Processing*, Norwood, MA: Artech House, 1986, pp. 29–117.

6.1.5 OUTPUT DATA

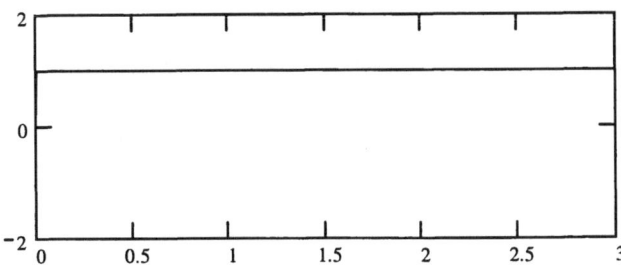

Figure 6.1 Baseband   waveform = 1   voltage vs. time (μs), pulsewidth (μs): $\tau = 3$

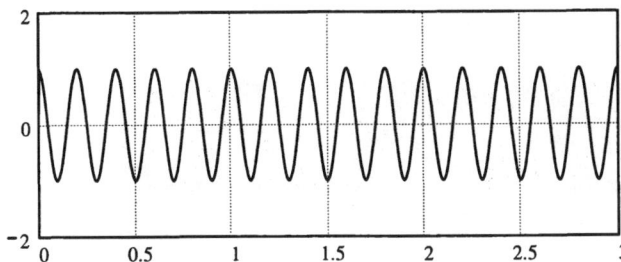

Figure 6.2 Voltage of the waveform = 1   vs. time (μs) at the specified frequency
$f_0 = 5$ MHz.

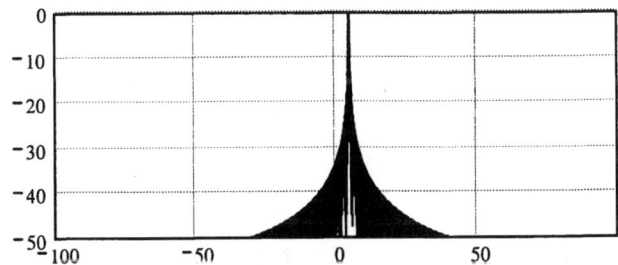

Figure 6.3 Spectrum of the   waveform = 1   (dB) with the carrier frequency

$f_0 = 5$ MHz vs. frequency (MHz),   spectrum width (MHz):

dSP = 0.341.

6.1.6 EXAMPLES

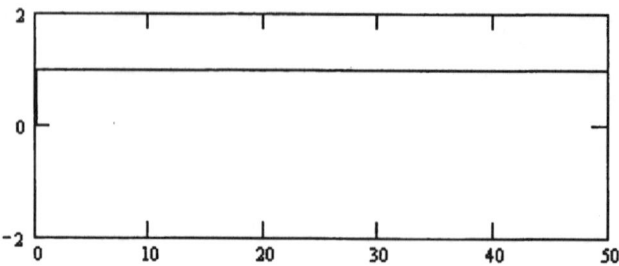

Figure 6.4 Baseband waveform voltage vs. time (μs) for 50-μs unmodulated waveform.

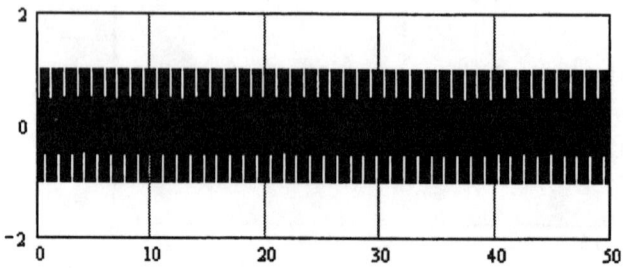

Figure 6.5 Voltage of the waveform at the specified frequency 5 MHz vs. time (μs) for 50-μs unmodulated waveform.

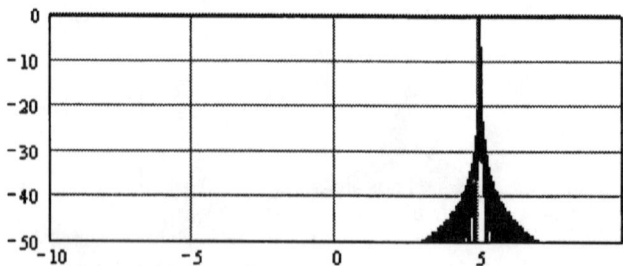

Figure 6.6 Spectrum (dB) of the waveform at the specified frequency 5 MHz vs. frequency (MHz) for 50-μs unmodulated waveform.

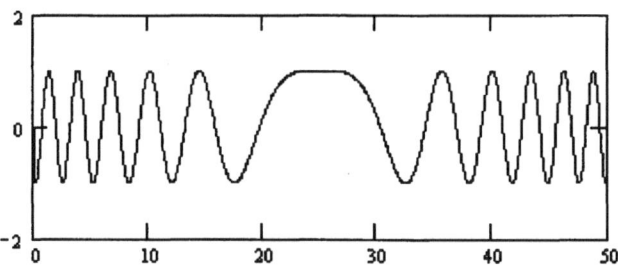

Figure 6.7 Baseband waveform voltage vs. time (µs) for 50-µs linear frequency-modulated waveform.

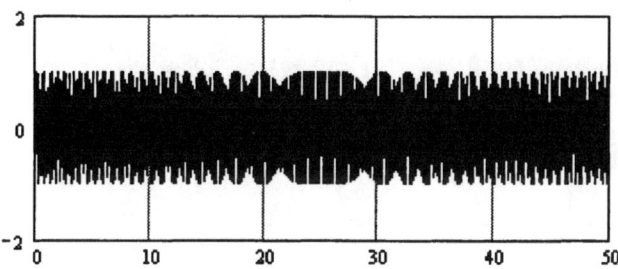

Figure 6.8 Voltage of the waveform at the specified frequency 5 MHz vs. time (µs) for 50-µs linear frequency-modulated waveform.

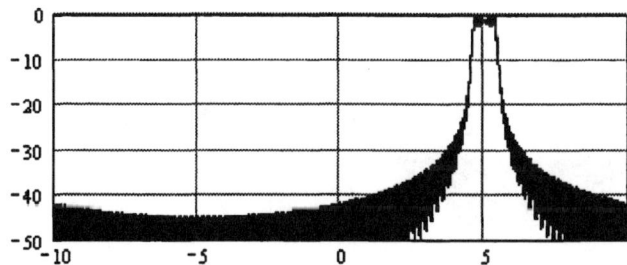

Figure 6.9 Spectrum (dB) of the waveform at the specified frequency 5 MHz vs. frequency (MHz) for 50-µs linear frequency-modulated waveform.

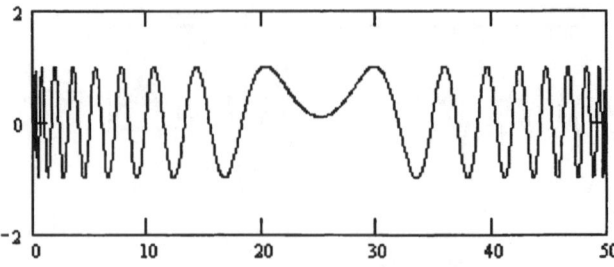

Figure 6.10 Baseband waveform voltage vs. time (μs) for 50-μs nonlinear frequency-modulated waveform.

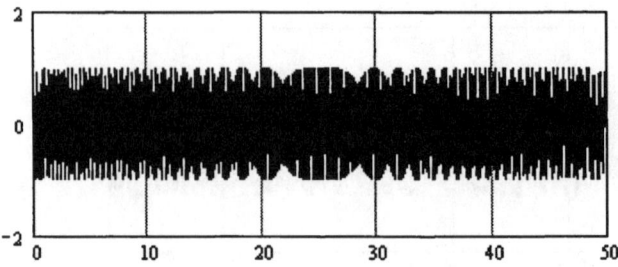

Figure 6.11 Voltage of the waveform at the specified frequency 5 MHz vs. time (μs) for 50-μs nonlinear frequency-modulated waveform.

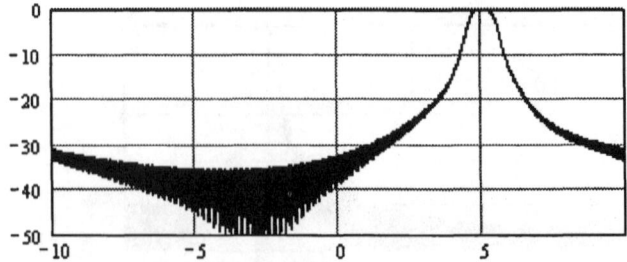

Figure 6.12 Spectrum (dB) of the waveform at the specified frequency 5 MHz vs. frequency (MHz) for 50-μs nonlinear frequency-modulated waveform.

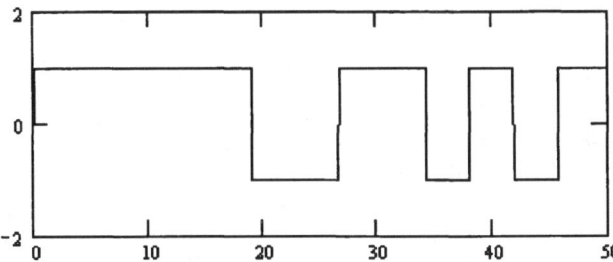

Figure 6.13 Baseband waveform voltage vs. time (μs) for 50-μs phase-coded waveform.

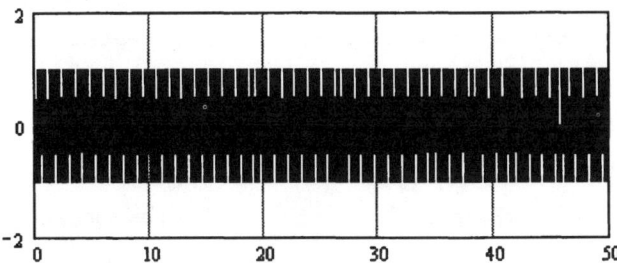

Figure 6.14 Voltage of the waveform at the specified frequency 5 MHz vs. time (μs) for 50-μs phase-coded waveform.

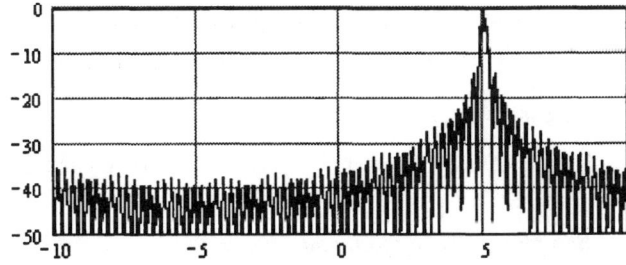

Figure 6.15 Spectrum (dB) of the waveform at the specified frequency 5 MHz vs. frequency (MHz) for 50-μs phase-coded waveform.

6.2 Digital Pulse Compression

6.2.1 DESCRIPTION

Simulates digital pulse compression of the modulated uncompressed waveform. Pulse compression is simulated in the frequency domain, i.e., multiplication of the spectra or the uncompressed waveform and pulse compression coefficients is employed to obtain the compressed waveform. The impact of radar circuits on the uncompressed waveform is simulated by filtering with the receiver filter that has the narrowest bandwidth, decimation (since the digital sampling rate of the analog-to-digital converter is typically smaller than the sampling rate of waveform generation) and doppler frequency shift produced by a moving target. Weighting in the frequency domain may be applied to control the range sidelobes of the compressed waveform.

6.2.2 INPUT DATA

===

Parameter   ***Dimension and Description***

===

Input Signal

Note: generate the uncompressed waveform by a routine cited in Section 6.1 (Pulsed Waveform Generation) and enter the file here.

S := READPRN("wfmbsb0.prn")   input baseband waveform

$N := last(S) + 1$ $N = 1 \times 10^3$ number of samples in the input file

Pulse Compression Coefficients

$K := 251$ number of complex pulse compression coefficients

Note: the number of complex pulse compression coefficients is typically defined by the number of taps of a radar pulse compressor.

compression := 1 $1 \rightarrow$ generate PC coefficients by this program

                        $2 \rightarrow$ input PC coefficients from previously generated file

if compression = 2

PC_COEF_FILE_K := READPRN("pccoef.prn")   PC coefficients file

Sampling, Decimation, and Filtering

f_s := 20   MHz, waveform generation sampling rate

Note: the sampling rate entered here must correspond to the sampling rate used to generate input uncompressed waveform in Section 6.1.

decimation := 10   decimation ratio

Note: the decimation ratio is the ratio of the waveform generation digital sampling rate to the analog-to-digital converter sampling rate.

H0_FIL := READPRN("filter5.prn")   frequency response used to simulate
                                            the receiver filter with the narrowest
                                                              bandwidth

Note: generate the filter frequency response by a routine cited in Section 6.3 (Frequency-Selective Filtering) and enter the file here.

Doppler Shift

doppler := 0 $0 \rightarrow$ no doppler shift
                                    $1 \rightarrow$ doppler shift corresponding to f_d

$f_d := 1000$ Hz, doppler frequency if doppler = 1

Weighting

weight := 1 $0 \rightarrow$ no weighting
                                             $1 \rightarrow$ cosine weighting
                                             $2 \rightarrow$ Kaiser weighting
                if weight = 1

$\Delta f\_w\_c := 3$ MHz, 3-dB bandwidth of weighting function

n_w := 1   order of power in cos^n_c function

Note: more weighting is applied if parameter Δf_w_c is decreased and n_w is increased, resulting in lower level of range sidelobes but wider main lobe and higher pulse compression loss.
                if weight = 2

$\alpha := 1.5$ Kaiser weighting parameter

Note: more weighting is applied if parameter α is increased, resulting in a lower level of range sidelobes but wider main lobe and higher pulse compression loss.
==
6.2.3 MODEL

6.2.3.1 Constants
y_th := −3   dB, level to calculate spectrum width

$M := 2^{12}$ $M = 4.096 \times 10^3$ number of waveform samples for spectrum
                                                              calculations
6.2.3.2 Sampling Rates and Intervals
$\Delta t := \dfrac{1}{f\_s}$ $\Delta t = 0.05$ μs, sampling interval before decimation

$$f\_d := \frac{f\_s}{\text{decimation}} \quad f\_d = 2 \qquad \text{MHz, sampling rate after decimation}$$

$$\Delta t\_d := \frac{1}{f\_d} \qquad \Delta t\_d = 0.5 \qquad \text{μs, sampling interval after decimation}$$

$$M1 := \text{floor}\left(\frac{M}{\text{decimation}} - 1\right) \qquad \text{number of samples after decimation}$$

$$M1 = 408$$

$n := 0..N-1$ \hspace{5cm} cycle n

$m := 0..M-1$ \hspace{5cm} cycle m

$m1 := 0..M1-1$ \hspace{4.8cm} cycle m1

$\dim(a,b) := b - a + 1$ \qquad number of sample calculations: a = first sample; b = last sample

6.2.3.3 Uncompressed Waveform Before Filtering and Decimation (N Samples)

$WF1\_UNC := S$ \hspace{3cm} complex baseband uncompressed waveform

$V1 := \text{Re}(WF1\_UNC)$ \hspace{4cm} waveform voltage

$\tau_{unc} := N \cdot \Delta t \qquad \tau_{unc} = 50$ \hspace{2cm} μs, uncompressed waveform duration

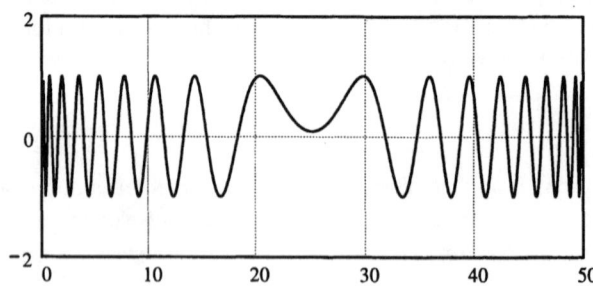

Figure 6.16 Baseband waveform before filtering and decimation.

6.2.3.4 Uncompressed Waveform Before Filtering and Decimation Expanded for Better Transformation into Frequency Domain (M Samples)

$$n\_s := \text{ceil}\left(\frac{M-N}{2}\right) \qquad n\_s = 1.548 \times 10^3 \qquad \text{first nonzero sample (start)}$$

$n\_f := n\_s + N - 1$ $n\_f = 2.547 \times 10^3$ last nonzero sample (finish)

$n\_0 := \dfrac{M}{2}$ $n\_0 = 2.048 \times 10^3$ central sample

$dim(n\_s, n\_f) = 1 \times 10^3$ number of samples

$EWF1\_UNC_m := 0$ sets all values to zero

$EWF1\_UNC_{n\_s+n} := WF1\_UNC_n$ expanded complex baseband waveform

$EV1 := Re(EWF1\_UNC)$ waveform voltage

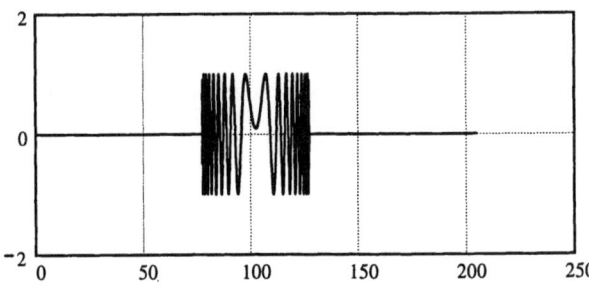

Figure 6.17 Expanded baseband waveform before filtering and decimation.

6.2.3.5 Spectrum of the Uncompressed Waveform Before Filtering

$SP\_F\_in := cfft(EWF1\_UNC)$ Complex spectrum of the waveform at the filter input

$\Delta f := \dfrac{f\_s}{M}$ $\Delta f = 4.883 \times 10^{-3}$ MHz, frequency sampling interval

$f_m := (m - n\_0) \cdot \Delta f$ MHz, frequency counts

$f_{n\_0} = 0$ MHz, central frequency

$f_0 = -10$ $f_{M-1} = 9.995$ MHz, frequency limits

$p1 := 0 .. n\_0 - 1$ $p2 := 0 .. n\_0 - 1$ cycles p1 and p2

$$SP0\_F\_in_{p1} := SP\_F\_in_{n\_0+p1}$$

Complex spectrum of EWF1_UNC: centered around zero frequency

$$SP0\_F\_in_{p2+n\_0} := SP\_F\_in_{p2}$$

$$SP0_m := 20 \cdot \log\left(\left|SP0\_F\_in_m + 10^{-10}\right|\right)$$ dB, waveform spectrum at the filter input

$$SP0\_MAX := \max(SP0)$$ spectrum maximum

$$SP0N1 := SP0 - SP0\_MAX$$ dB, normalized waveform spectrum at the filter input

6.2.3.6 Filter Frequency Response

$$H_m := 20 \cdot \log\left(\left|H0\_FIL_m + 10^{-10}\right|\right)$$ dB, filter frequency response

$$HN := H - \max(H)$$ dB, normalized filter frequency response

6.2.3.7 Spectrum of the Uncompressed Waveform After Filtering

$$filtering(SPC, H) := \begin{vmatrix} spec \leftarrow \overrightarrow{SPC \cdot H} \\ spec \end{vmatrix}$$ filtering algorithm

$$SP0\_F\_out := filtering(SP0\_F\_in, H0\_FIL)$$ spectrum of the filtered waveform

$$SP0_m := 20 \cdot \log\left(\left|SP0\_F\_out_m + 10^{-10}\right|\right)$$ dB, spectrum of the filtered waveform

$$SP0N2 := SP0 - \max(SP0)$$ dB, normalized spectrum of the filtered waveform

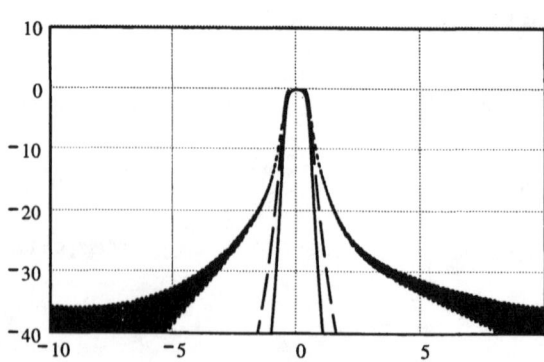

Figure 6.18 Spectra of the filter input waveform (dotted line), filter frequency response (dashed line), and filter output waveform (solid line).

6.2.3.8 Expanded Uncompressed Waveform at the Filter Output (M Samples)

$SP\_F\_out_{p1} := SP0\_F\_out_{n\_0+p1}$ transformation of the spectrum centered around zero frequency

$SP\_F\_out_{n\_0+p2} := SP0\_F\_out_{p2}$

Note: Mathcad functions *icfft* and *cfft* work with the spectra denoted SP but not with spectra denoted SP0 that are centered around zero frequency.

$EWF2\_UNC := icfft(SP\_F\_out)$ expanded waveform at the filter output

$EV2 := Re(EWF2\_UNC)$ expanded waveform voltage

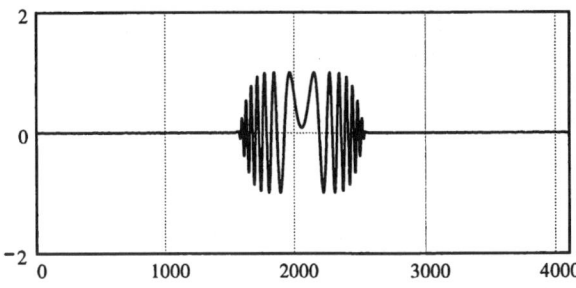

Figure 6.19 Expanded uncompressed waveform at the filter output.

6.2.3.9 Filtered and Decimated Waveform (K Samples)

$K\_half := floor\left(\dfrac{K}{2}\right)$ $K\_half = 125$ the middle sample

$nT\_s := n\_0 - K\_half \cdot decimation$ $nT\_s = 798$ the first nonzero count

$nT\_f := n\_0 + K\_half \cdot decimation$ $nT\_f = 3.298 \times 10^3$ the last nonzero count

$k := 0 .. K - 1$ cycle

$EWFK\_UNC_k := EWF2\_UNC_{nT\_s+k \cdot decimation}$ filtered and decimated waveform before doppler shift

$DOPPLER := \begin{vmatrix} \text{for } k \in 0 .. K - 1 \\ \quad \Theta_k \leftarrow exp\left(-j \cdot 2 \cdot \pi \cdot f_d \cdot \Delta t\_d \cdot 10^{-6} \cdot k\right) \\ \Theta \end{vmatrix}$ additional doppler shift phase doppler = 0   doppler shift mode specified

Fitered and decimated waveform with doppler shift superimposed

$$\text{EWF3\_UNC\_K} := \begin{vmatrix} \text{EWFK\_UNC} & \text{if } \text{doppler} = 0 \\[1em] \overrightarrow{\text{EWFK\_UNC} \cdot \text{DOPPLER}} & \text{if } \text{doppler} = 1 \end{vmatrix}$$

EV3 := Re(EWF3_UNC_K)   waveform voltage

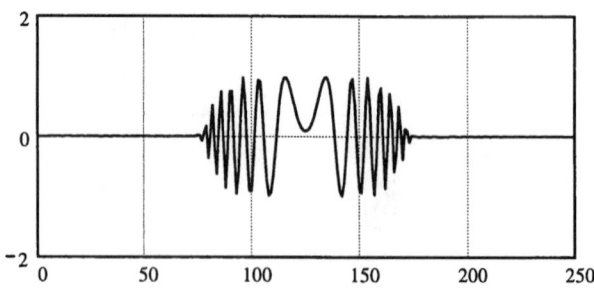

Figure 6.20 Filtered and decimated uncompressed waveform.

6.2.3.10 Expanded Filtered and Decimated Waveform (M1 Samples)

$$m1\_0 := \text{floor}\left(\frac{M1}{2}\right)$$ $m1\_0 = 204$ the middle sample

$mT\_s := m1\_0 - K\_half$ $mT\_s = 79$ the first nonzero count

$mT\_f := m1\_0 + K\_half$ $mT\_f = 329$ the last nonzero count

$\text{EWF3\_UNC\_M1}_{m1} := 0$ sets all values to zero

$\text{EWF3\_UNC\_M1}_{mT\_s+k} := \text{EWF3\_UNC\_K}_k$ expanded waveform

6.2.3.11 Spectrum of the Filtered and Decimated Waveform (M1 Samples)

SP_EWF3_UNC_M1 := cfft(EWF3_UNC_M1)   waveform spectrum

$$\Delta f1\_d := \frac{f\_d}{M1}$$ $\Delta f1\_d = 4.902 \times 10^{-3}$ MHz, frequency sampling interval

$$f1_{m1} := (m1 - m1\_0) \cdot \Delta f1\_d$$

                                                   MHz, frequency counts

$$f1_{m1\_0} = 0$$ MHz, central frequency

$$f1_0 = -1 \qquad f1_{M1-1} = 0.995$$ MHz, frequency limits

$$r1 := 0 .. \, m1\_0 - 1$$ cycle r1

$$SP0\_EWF3\_UNC\_M1_{r1} := SP\_EWF3\_UNC\_M1_{m1\_0+r1}$$ spectrum centered around zero frequency

$$SP0\_EWF3\_UNC\_M1_{m1\_0+r1} := SP\_EWF3\_UNC\_M1_{r1}$$

$$SP0_{m1} := 20 \cdot \log\left(\left| SP0\_EWF3\_UNC\_M1_{m1} + 10^{-10} \right| \right)$$ dB, spectrum

$$SP0N3 := SP0 - \max(SP0)$$ dB, normalized spectrum

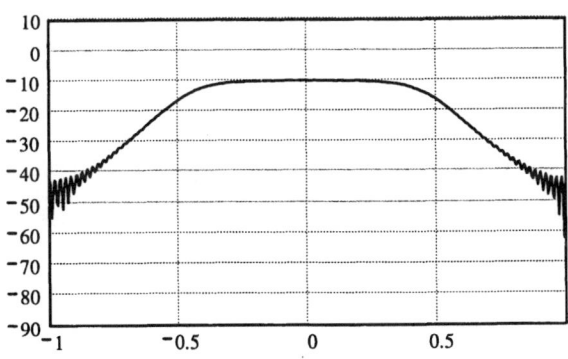

Figure 6.21 Spectrum of the filtered and decimated waveform (dB) vs. frequency (MHz).

6.2.3.12 Uncompressed Waveform at Pulse Compressor Input (M1 samples)

$$mT\_s = 79$$ the first nonzero sample

$$mT\_f = 329$$ the last nonzero sample

$$t1 := 0 .. \, mT\_s - 1 \qquad t2 := mT\_f + 1 .. \, M1 - 1$$ cycles t1 and t2

$$EWF3\_UNC\_M1_{t1} := 0$$ setting waveform to zero beyond mTs and mT_f

$$EWF3\_UNC\_M1_{t2} := 0$$

$$WF\_UNC := EWF3\_UNC\_M1$$ uncompressed waveform

$$EV4 := Re(WF\_UNC)$$ waveform voltage

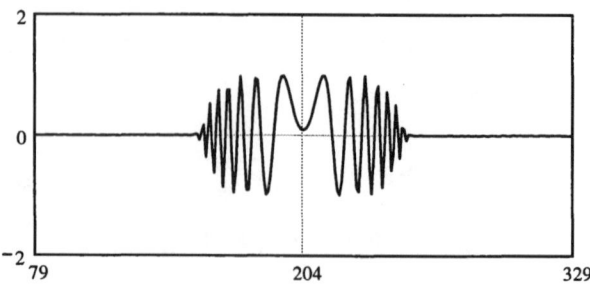

Figure 6.22 The uncompressed waveform at pulse compressor input.

6.2.3.13 Unweighted Pulse Compression Coefficients (M1 Samples)

Zero-centered spectrum of the pulse compression coefficients

SP0_PC_COEF_UNW_M1 := SP0_EWF3_UNC_M1

Transformation of the spectrum for the *icfft* function

$$SP\_PC\_COEF\_UNW\_M1_{rl} := SP0\_PC\_COEF\_UNW\_M1_{ml\_0+rl}$$

$$SP\_PC\_COEF\_UNW\_M1_{ml\_0+rl} := SP0\_PC\_COEF\_UNW\_M1_{rl}$$

Pulse compression coefficients before weighting

PC_COEF_UNW_M1 := icfft(SP_PC_COEF_UNW_M1)

PC_r := Re(PC_COEF_UNW_M1)   real coefficients

PC_i := Im(PC_COEF_UNW_M1)   imaginary coefficients

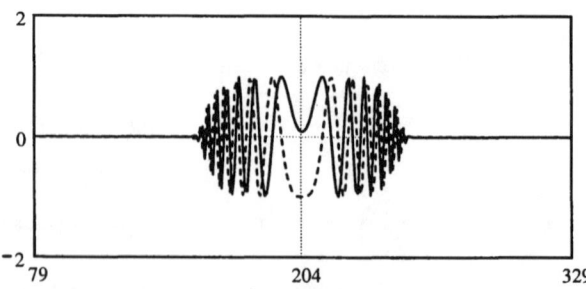

Figure 6.23 The unweighted pulse compression coefficients (solid line is a real part, dotted line is an imaginary part).

6.2.3.14 Frequency Weighting Function (M1 Samples)

$$f\_w_{m1} := \frac{m1 - m1\_0}{M1} \cdot f\_d \qquad \text{MHz, frequency counts}$$

$$I_0(z) := \frac{1}{2 \cdot \pi} \cdot \int_0^{2 \cdot \pi} \exp(z \cdot \cos(u))\, du \qquad \text{Bessel function}$$

$$WF_{m1} := \begin{cases} 1 & \text{if weight} = 0 \\ \cos\left(\pi \cdot \dfrac{f\_w_{m1}}{\Delta f\_w\_c}\right)^{n\_w} & \text{if weight} = 1 \\ \dfrac{I_0\left[\pi \cdot \alpha \cdot \sqrt{1 - \left(\dfrac{f\_w_{m1}}{f\_w_0}\right)^2}\right]}{I_0(\pi \cdot \alpha)} & \text{if weight} = 2 \end{cases} \qquad \text{weighting function}$$

$$\text{weight} = 1 \qquad \text{weighting function specified}$$

$$WC_{m1} := 20 \cdot \log\left[WF_{m1} \cdot \left(WF_{m1\_0}\right)^{-1}\right] \qquad \text{dB, weighting coefficients}$$

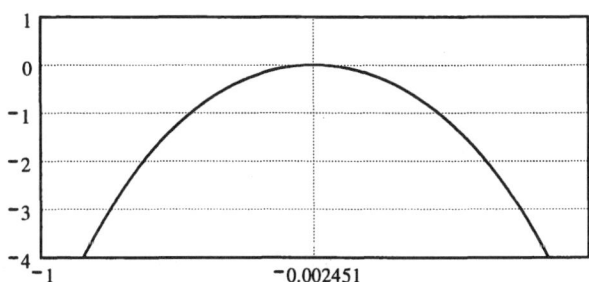

Figure 6.24 The weighting coefficients (dB) vs. frequency (MHz).

6.2.3.15 Spectrum of the Weighted Pulse Compression Coefficients (M1 Samples)

Zero-centered spectrum of the weighted pulse compression coefficients

$$SP0\_PC\_COEF\_W\_M1_{m1} := SP0\_PC\_COEF\_UNW\_M1_{m1} \cdot WF_{m1}$$

$$SP0_{m1} := 20 \cdot \log\left(\left|SP0\_PC\_COEF\_W\_M1_{m1} + 10^{-10}\right|\right) \qquad \text{dB, spectrum}$$

$$SP0N4 := SP0 - \max(SP0) \qquad \text{dB, normalized spectrum}$$

Spectrum of the weighted pulse compression coefficients

$$SP\_PC\_COEF\_W\_M1_{m1} := 0$$

$$SP\_PC\_COEF\_W\_M1_{r1} := SP0\_PC\_COEF\_W\_M1_{m1\_0+r1}$$

$$SP\_PC\_COEF\_W\_M1_{m1\_0+r1} := SP0\_PC\_COEF\_W\_M1_{r1}$$

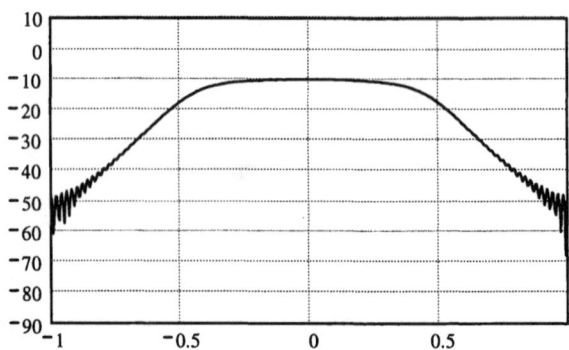

Figure 6.25 Spectrum of the weighted pulse compression coefficients (dB) vs. frequency (MHz).

6.2.3.16 Weighted Pulse Compression Coefficients (K Samples)

$$PC\_COEF\_W\_M1 := \text{icfft}(SP\_PC\_COEF\_W\_M1) \qquad \text{coefficients, M1 samples}$$

$$PC\_COEF\_W\_M1_{t1} := 0$$

$$PC\_COEF\_W\_M1_{t2} := 0 \qquad \text{setting coefficients to zero beyond mTs and mT\_f}$$

$$PC\_COEF\_W\_K_k := PC\_COEF\_W\_M1_{mT\_s+k} \qquad \text{coefficients, K samples}$$

Writing pulse compression coefficients to a file

$$\text{WRITEPRN}("pccoef.prn") := PC\_COEF\_W\_K$$

6.2.3.17 Pulse Compression Simulation

Transformation of pulse compression coefficients from a file to M1 samples

$\text{PC\_COEF\_FILE\_M1}_{m1} := 0$

$\text{PC\_COEF\_FILE\_M1}_{mT\_s+k} := \text{PC\_COEF\_FILE\_K}_k$

compression = 1   compression mode specified

Pulse compression coefficients selected

$$\text{PC\_COEF\_M1} := \begin{vmatrix} \text{PC\_COEF\_W\_M1} & \text{if compression} = 1 \\ \text{PC\_COEF\_FILE\_M1} & \text{if compression} = 2 \end{vmatrix} \quad \text{M1 samples}$$

$\text{PC\_COEF\_K}_k := \text{PC\_COEF\_M1}_{mT\_s+k}$ K samples

$\text{pc} := \text{Re}(\text{PC\_COEF\_K})$ real coefficients

Pulse compression simulation algorithm

$$\text{WF\_comp}(\text{WF\_unc}, \text{PC\_coef}) := \begin{vmatrix} \text{SP\_WF\_unc} \leftarrow \text{cfft}(\text{WF\_unc}) \\ \text{SP\_PC\_coef} \leftarrow \text{cfft}(\text{PC\_coef}) \\ \overrightarrow{\text{SP\_WF\_comp} \leftarrow \text{SP\_WF\_unc} \cdot \overline{\text{SP\_PC\_coef}}} \\ \text{WF\_compr} \leftarrow \text{icfft}(\text{SP\_WF\_comp}) \\ \text{for } \text{ind} \in 0 .. \text{m1\_0} - 1 \\ \qquad \begin{vmatrix} \text{WF0\_compr}_{\text{ind}} \leftarrow \text{WF\_compr}_{\text{m1\_0+ind}} \\ \text{WF0\_compr}_{\text{m1\_0+ind}} \leftarrow \text{WF\_compr}_{\text{ind}} \end{vmatrix} \\ \text{WF0\_compr} \end{vmatrix}$$

6.2.3.18 Compressed Waveform (M1 samples)

$\text{WF\_COMP} := \text{WF\_comp}(\text{WF\_UNC}, \text{PC\_COEF\_M1})$ compressed waveform

$\text{L} := \text{rows}(\text{WF\_COMP}) \qquad \text{L} = 408$ number of samples in the waveform

Magnitude of the compressed waveform, dB

$$\text{WF\_COMP\_db}_{m1} := 20 \cdot \log\left(\overrightarrow{\left| (\text{WF\_COMP}_{m1}) + 10^{-9} \right|} \right)$$

Normalized magnitude of the compressed waveform, dB

$$\text{WF\_COMP\_norm}_{m1} := \overrightarrow{\text{WF\_COMP\_db}_{m1} - \max(\text{WF\_COMP\_db})}$$

$\text{WF}_{m1} := \text{WF\_COMP\_norm}_{m1}$

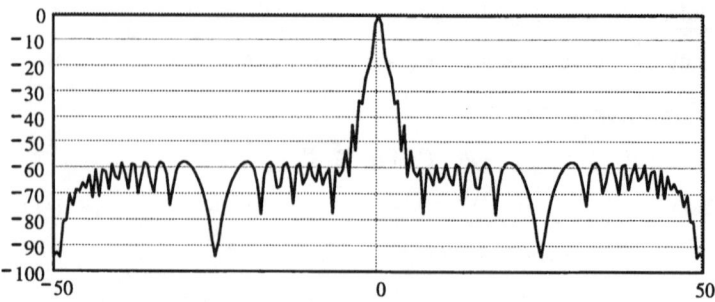

Figure 6.26 The magnitude of the compressed waveform (dB) vs. time in μs.

6.2.3.19 Compressed Waveform Spectrum (M1 Samples)

$SP\_WF\_COMP := cfft(WF\_COMP)$ spectrum of the compressed waveform

$SP0\_WF\_COMP_{r1} := SP\_WF\_COMP_{m1\_0+r1}$

$SP0\_WF\_COMP_{m1\_0+r1} := SP\_WF\_COMP_{r1}$

spectrum of the compressed waveform centered around zero

$SP0_{m1} := 20 \cdot log\left(\left|SP0\_WF\_COMP_{m1} + 10^{-10}\right|\right)$ dB, spectrum

$SP0\_MAX := max(SP0)$

$SP0N5 := SP0 - max(SP0)$

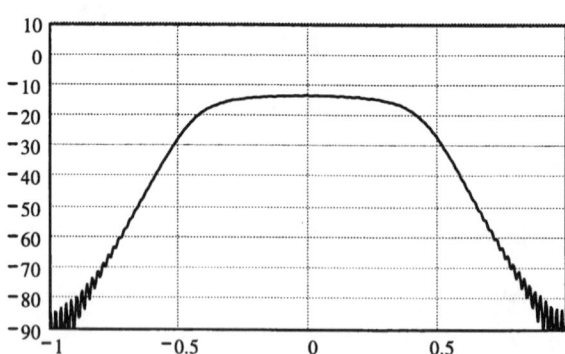

Figure 6.27 Spectrum of the compressed waveform (dB) vs. frequency (MHz).

6.2.4 OUTPUT DATA

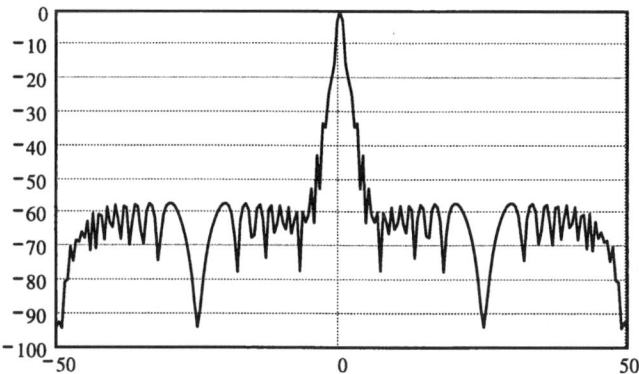

Figure 6.28 The magnitude of the compressed waveform (dB) vs. time in μs for input data specified.

6.2.5 REFERENCES

[1] Barton, D. K., and S. A. Leonov, (eds.), *Radar Technology Encyclopedia*, Norwood, MA: Artech House, 1997, pp. 315–318.

[2] Skolnik, M. I., (ed.), *Radar Handbook*, New York: McGraw-Hill, 1990, pp. 10.7–10.10.

6.2.6 EXAMPLES

Waveform before filtering and decimation   Spectrum

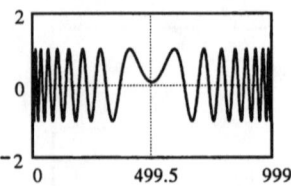 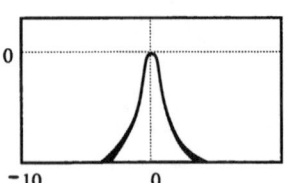

Waveform after filtering and decimation   Spectrum

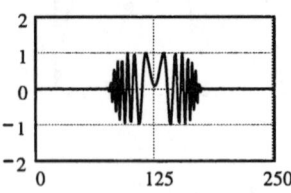 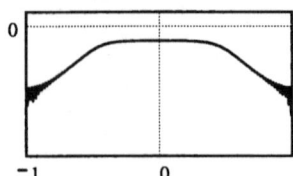

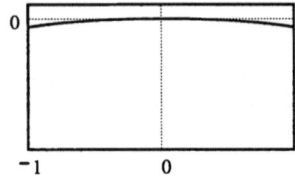

Weighting function

Pulse compression coefficients   Spectrum

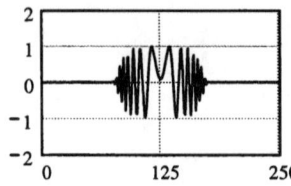 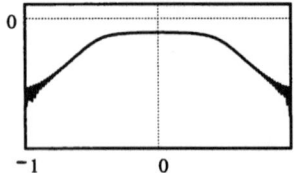

Compressed waveform

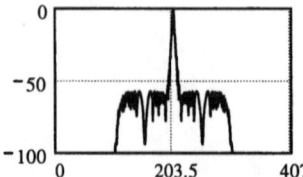

Spectrum

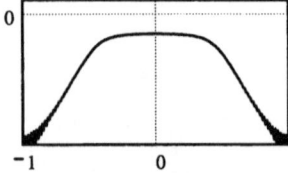

Figure 6.29 Pulse compression stages.

6.3 Frequency-Selective Filtering

6.3.1 DESCRIPTION

Simulates the time-domain and frequency-domain representation of the pulsed signal passing through the specified frequency-selective bandpass filter. The filter types covered by the model are as follows: (1) ideal rectangular; (2) Gaussian; (3) Chebyshev; and (4) Butterworth.

6.3.2 INPUT DATA

==

Parameter	*Dimension and Description*

==

Input Signal

Note: generate the waveform by a routine cited in Section 6.1 (Pulsed Waveform Generation) and enter the file here.

$S\_in := READPRN("wfmf5.prn")$ complex input signal

Bank of Filters

filter := 2

     1 -> ideal rectangular filter
     2 -> Gaussian filter
     3 -> Chebyshev filter
     4 -> Butterworth filter

$f_{0\_filter} := 5$ MHz, filter central frequency

Note: to match the central frequencies of the signal spectrum and filter frequency response, choose the filter central frequency equal to the signal carrier frequency in S_in.

BW := 1   MHz, bandwidth of the filter

------- *filters with poles (Chebyshev, Butterworth)* --------

$poles\_0 := 4$ number of poles in filter

par := -3   dB, peak-to-valley parameter for Chebyshev filter

Digital Sampling Data

$f\_s := 200$ MHz, digital sampling rate

Note: choose the sampling rate to be the same as that chosen in Section 6.1 when the input signal was generated.

==

6.3.3 MODEL

6.3.3.1 Constants

y_th := −3   dB, level to calculate spectrum width

$M := 2^{12}$ $M = 4.096 \times 10^3$ number of waveform samples for spectrum
calculations

6.3.3.2 Sampling Interval and Number of Samples

$\Delta t := \dfrac{1}{f\_s}$ $\Delta t = 5 \times 10^{-3}$ μs, sampling interval

$N := last(S\_in) + 1$ $N = 200$ number of samples in the input signal

$\tau := N \cdot \Delta t$ $\tau = 1$ μs, signal duration

$n := 0 .. N - 1$ $m := 0 .. M - 1$ cycles n and m

$n\_s := ceil\left(\dfrac{M - N}{2}\right)$ $n\_s = 1.948 \times 10^3$ the first nonzero sample
in the expanded signal (start)

$n\_0 := \dfrac{M}{2}$ $n\_0 = 2.048 \times 10^3$ the central sample

$n\_f := n\_s + N - 1$ $n\_f = 2.147 \times 10^3$ the last nonzero sample
in the expanded signal (finish)

$\Delta f := \dfrac{f\_s}{M}$ $\Delta f = 0.049$ MHz, frequency sampling interval

$f_m := (m - n\_0) \cdot \Delta f$ MHz, frequency counts

$f_{0\_filter} = 5$ MHz, the central frequency of the filter

$f_0 = -100$ MHz, frequency limits

$f_{M-1} = 99.951$

Note: the central filter frequency specified in the input data menu must be
within f_0–f_{M-1} range. To increase this range, increase sampling rate f_s.

6.3.3.3 Basic Definitions and Functions

$dim(a, b) := b - a + 1$ generic definition for the number of samples:
a - first sample, b - last sample

$POWER(V) := \left[\sum_m \left(|V_m|\right)^2\right] \cdot \dfrac{1}{M}$ [W], average power of the signal

$$\text{ENERGY(V)} := \left[\sum_m \left(|V_m| \right)^2 \right] \cdot \Delta t$$

[W*μs], energy of the signal

$$\text{filtering(SPC, H)} := \begin{vmatrix} \text{spec} \leftarrow \overrightarrow{\text{SPC} \cdot \text{H}} \\ \text{spec} \end{vmatrix}$$

filtering of the signal with the spectrum SPC by the filter with the frequency response H

MHz, frequency where the spectrum curve crosses the specified level upward

$$\text{f\_b(SPC)} := \begin{vmatrix} \text{for } m \in 1 .. M - 2 \\ \quad \text{break if } SPC_{m-1} \leq y\_th \text{ if } SPC_m > y\_th \\ df\_b \leftarrow \dfrac{SPC_m - y\_th}{SPC_m - SPC_{m-1}} \cdot \Delta f \\ f\_b \leftarrow f_m - df\_b \\ f\_b \end{vmatrix}$$

MHz, frequency where the spectrum curve crosses the specified level downward

$$\text{f\_e(SPC)} := \begin{vmatrix} \text{for } m \in 1 .. M - 1 \\ \quad \text{break if } SPC_{m-1} > y\_th \text{ if } SPC_m \leq y\_th \\ df\_e \leftarrow \dfrac{SPC_{m-1} - y\_th}{SPC_{m-1} - SPC_m} \cdot \Delta f \\ f\_e \leftarrow f_m + df\_e \\ f\_e \end{vmatrix}$$

Note: the spectrum width calculation works only for monotonous spectrum functions at levels higher than y_th dB. Be cautious when used for oscillating spectrum shapes with peak-to-valley spans more than y_th dB, such as those produced by Chebyshev filters.

6.3.3.4 The Signal at the Filter Input (N Samples)

$$V := \text{Re}(S\_in)$$

input signal voltage

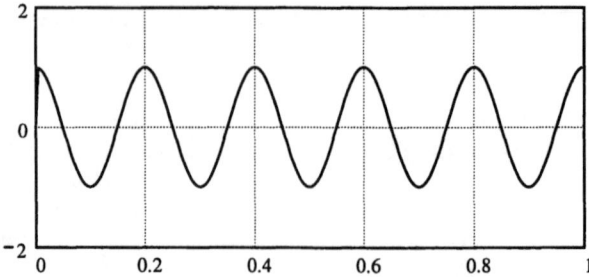

Figure 6.30 The signal voltage (V) vs. time (µs) at the filter input.

6.3.3.5 Expanded Input Signal for Better Transformation into Frequency Domain (M Samples)

$\dim(n\_s, n\_f) = 200$ number of samples

$ES\_in_m := 0$ sets all values to zero

$ES\_in_{n\_s+n} := S\_in_n$ sets nonzero values only for specified counts

$EV_m := Re(ES\_in_m)$ expanded signal voltage

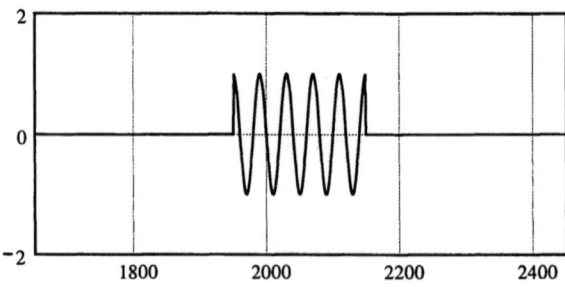

Figure 6.31 Expanded signal voltage (V) vs. number of samples at the filter input.

6.3.3.6 Spectrum of the Signal at the Filter Input (M Samples)

$SP\_S\_in := cfft(ES\_in)$ input signal spectrum

$p := 0 .. n\_0 - 1$ cycle p

$SP0\_S\_in_p := SP\_S\_in_{p+n\_0}$

    spectrum centered around zero ferquency

$SP0\_S\_in_{p+n\_0} := SP\_S\_in_p$

$$SP0\_in_m := 20 \cdot \log\left(\left|SP0\_S\_in_m + 10^{-10}\right|\right) \qquad \text{dB, spectrum}$$

$$SPN\_in := SP0\_in - \max(SP0\_in) \qquad \text{dB, normalized spectrum}$$

$$dSPN\_in := f\_e(SPN\_in) - f\_b(SPN\_in) \qquad \text{MHz, input signal spectrum width}$$

$$dSPN\_in = 0.937$$

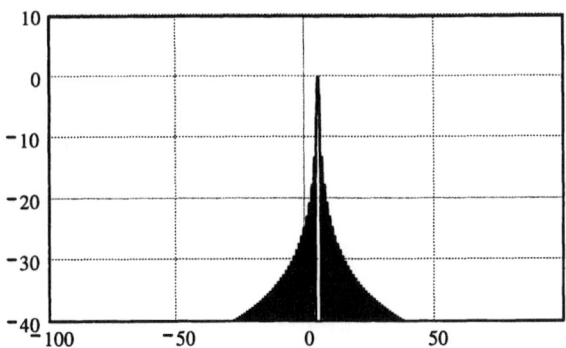

Figure 6.32 The spectrum (dB) of the signal at the filter input vs. frequency (MHz).

6.3.3.7 Filter Frequency Response

<u>A. Ideal Rectangular Filter</u>

filter frequency response

$$H\_FIL\_ideal_m := \begin{vmatrix} 1 & \text{if } \dfrac{-BW}{2} \le \left(f_m - f_{0\_filter}\right) \le \dfrac{BW}{2} \\ 0 & \text{otherwise} \end{vmatrix}$$

$$HF1_m := 20 \cdot \log\left(\left|H\_FIL\_ideal_m + 10^{-10}\right|\right) \qquad \text{dB, filter frequency response}$$

$$HF1db := HF1 - \max(HF1) \qquad \text{dB, normalized filter frequency response}$$

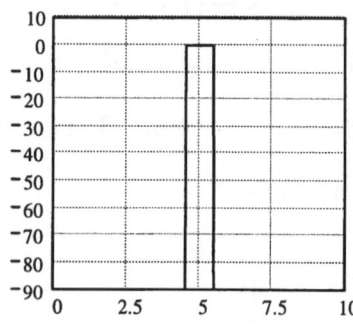

Figure 6.33 Filter frequency response (dB) vs. frequency (MHz) for ideal rectangular filter.

B. Gaussian Filter

$$x_m := 2 \cdot \left(f_m - f_{0\_filter} \right) \qquad \text{MHz, parameter}$$

$$\text{H\_FIL\_Gaussian}_m := \left| \begin{array}{l} 0 \quad \text{if} \quad \dfrac{3}{10} \cdot \left(\dfrac{x_m}{BW} \right)^2 \geq 300 \\[2em] \dfrac{1}{\sqrt{10^{\frac{3}{10} \left(\frac{x_m}{BW} \right)^2}}} \quad \text{otherwise} \end{array} \right.$$

filter frequency response

$$\text{HF2}_m := 20 \cdot \log\left(\left| \text{H\_FIL\_Gaussian}_m + 10^{-10} \right| \right) \quad \text{dB, filter frequency response}$$

$$\text{HF2db} := \text{HF2} - \max(\text{HF2}) \qquad \text{dB, normalized filter frequency response}$$

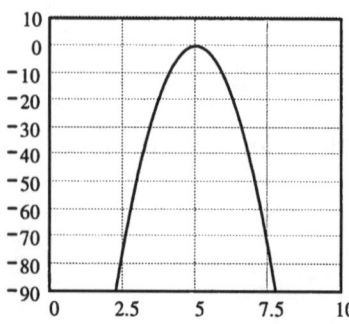

Figure 6.34 Filter frequency response (dB) vs. frequency (MHz) for Gaussian filter.

C. Chebyshev Filter

$$b_m := poles\_0 \cdot acosh\left(\frac{x_m}{BW}\right)$$ parameter

filter frequency response

$$H\_FIL\_Chebyshev_m := \frac{1}{\sqrt{1 + \left[\frac{1}{\left(10^{\frac{par}{20}}\right)^2} - 1\right] \cdot \left(\frac{e^{b_m} + e^{-b_m}}{2}\right)^2}}}$$

$$HF3_m := 20 \cdot log\left(\left|H\_FIL\_Chebyshev_m + 10^{-10}\right|\right) \quad dB, \text{ filter frequency response}$$

$$HF3db := HF3 - max(HF3) \qquad dB, \text{ normalized filter frequency response}$$

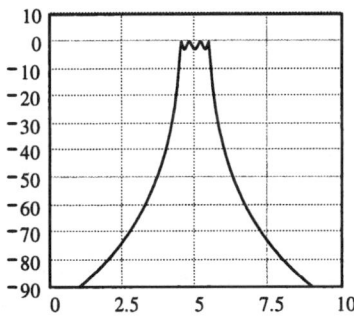

Figure 6.35 Filter frequency response (dB) vs. frequency (MHz) for Chebyshev filter.

D. Butterworth Filter

$$H\_FIL\_Butterworth_m := \frac{1}{\sqrt{1 + \left(\frac{x_m}{BW}\right)^{2 \cdot poles\_0}}}$$ filter frequency response

$$HF4_m := 20 \cdot log\left(\left|H\_FIL\_Butterworth_m + 10^{-10}\right|\right) \quad dB, \text{ filter frequency response}$$

$$HF4db := HF4 - max(HF4) \qquad dB, \text{ normalized filter frequency response}$$

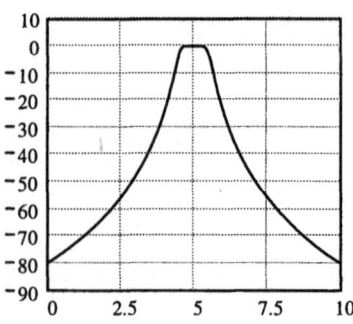

Figure 6.36 Filter frequency response (dB) vs. frequency (MHz) for Butterworth filter.

filter = 2   filter chosen

$$H\_FIL := \begin{vmatrix} H\_FIL\_ideal & if\ filter = 1 \\ H\_FIL\_Gaussian & if\ filter = 2 \\ H\_FIL\_Chebyshev & if\ filter = 3 \\ H\_FIL\_Butterworth & if\ filter = 4 \end{vmatrix}$$

filter frequency response

$$HF_m := 20 \cdot log\left(\left|H\_FIL_m + 10^{-10}\right|\right)$$ dB, filter frequency response

$$HFN := HF - max(HF)$$ dB, normalized filter frequency response

----

Writing the chosen filter response to a file

$$WRITEPRN("filter.prn") := H\_FIL$$

6.3.3.8 Spectrum of the Signal at the Filter Output (M Samples)

$$SP0\_S\_out := filtering(SP0\_S\_in, H\_FIL)$$ spectrum of the signal at the filter output

$$SP0\_out_m := 20 \cdot log\left(\left|SP0\_S\_out_m + 10^{-10}\right|\right)$$ dB, spectrum

$$SPN\_out := SP0\_out - max(SP0\_out)$$ dB, normalized spectrum

$$dSPN\_out := f\_e(SPN\_out) - f\_b(SPN\_out)$$

$$dSPN\_out = 0.719$$ MHz, output signal spectrum width

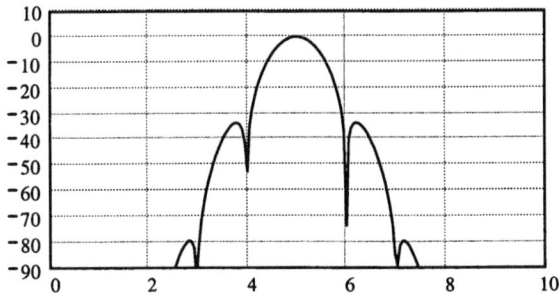

Figure 6.37 The spectrum (dB) of the signal at the filter output vs. frequency (MHz).

6.3.3.9 Expanded Signal at the Filter Output (M Samples)

Transformation of the spectrum for the *icfft* function

$$SP\_S\_out_p := SP0\_S\_out_{n\_0+p} \qquad SP\_S\_out_{n\_0+p} := SP0\_S\_out_p$$

$$ES\_out := icfft(SP\_S\_out) \qquad\qquad\qquad \text{expanded signal}$$

$$EV_m := Re(ES\_out_m) \qquad\qquad\qquad \text{expanded signal voltage}$$

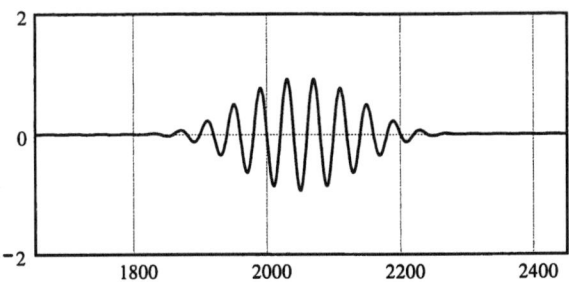

Figure 6.38 Expanded signal voltage (V) vs. number of samples at the filter output.

6.3.3.10 Energy Loss Due to Filtering

$$ENERGY(ES\_in) = 0.995 \qquad\qquad \text{energy of the signal at the filter input}$$

$$ENERGY(ES\_out) = 0.697 \qquad\qquad \text{energy of the signal at the filter output}$$

$$L := -\left(10 \cdot \log\left(\frac{ENERGY(SP0\_S\_out)}{ENERGY(SP0\_S\_in)}\right)\right) \quad L = 1.543 \quad \begin{array}{l}\text{dB, energy loss due to}\\ \text{filtering}\end{array}$$

Note: the loss L describes the loss of the signal energy due to filtering, and this is not the *signal-to-noise ratio (SNR) loss*. Since the noise is also filtered, actual SNR is typically improved after filtering because the noise energy is reduced more than the energy of the signal.

6.3.3.11 Input and Output Signals Representation (K Samples)

$C := 2 \qquad K := C \cdot 2 \cdot N - 1$ expands timescale

$k := 0 \ldots K$ cycle

input signal   output signal

$S\_in_k := ES\_in_{n\_0 - C \cdot N + k} \qquad S\_out_k := ES\_out_{n\_0 - C \cdot N + k}$

$V1 := Re(S\_in) \qquad\qquad V2 := Re(S\_out)$

$A_k := \left| S\_in_k \right| \qquad\qquad C_k := \left| S\_out_k \right|$

$B_k := - \left| S\_in_k \right| \qquad\qquad D_k := - \left| S\_out_k \right|$

6.3.4 REFERENCES

[1] Barton, D. K., and S. A. Leonov, (eds.), *Radar Technology Encyclopedia*, Norwood, MA: Artech House, 1997, p. 188.
[2] *Reference Data for Radio Engineers*, Howard W. Sams & Co., 1975, Ch. 8.

6.3.5 OUTPUT DATA

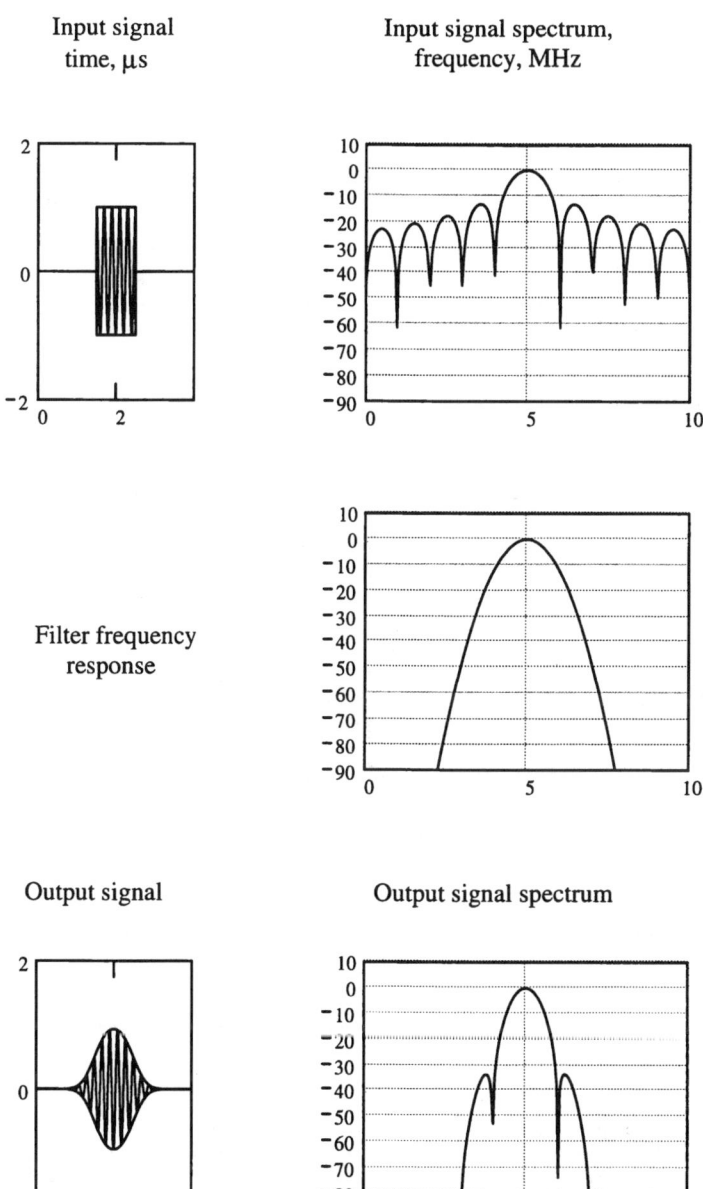

Figure 6.39 Frequency-selective filtering stages for input data specified.

6.3.6 EXAMPLES

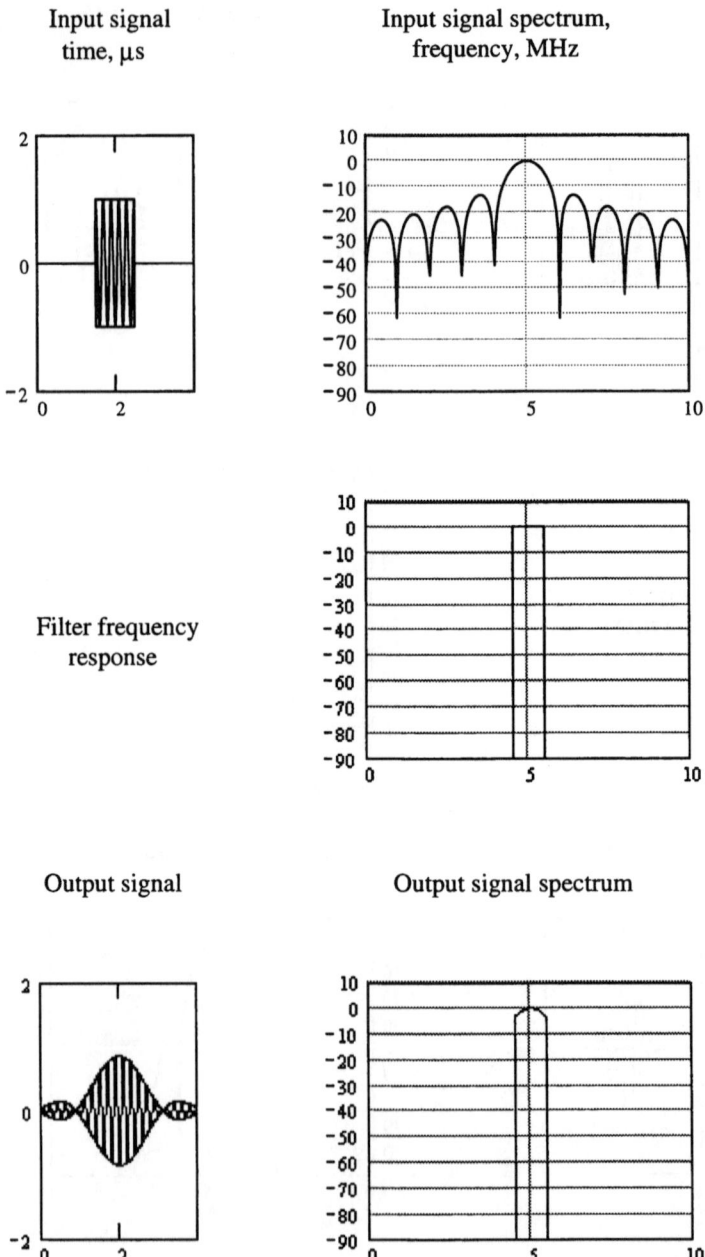

Figure 6.40 Frequency-selective filtering stages for ideal rectangular filter.

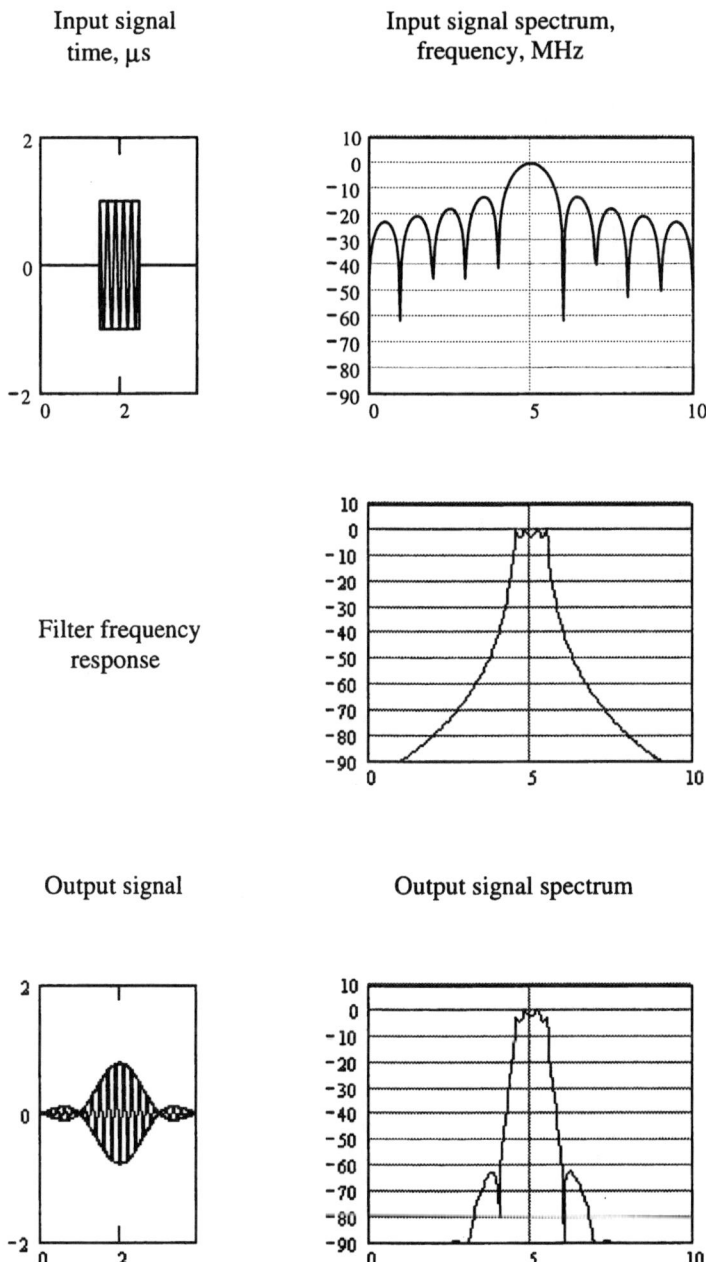

Figure 6.41 Frequency-selective filtering stages for Chebyshev filter.

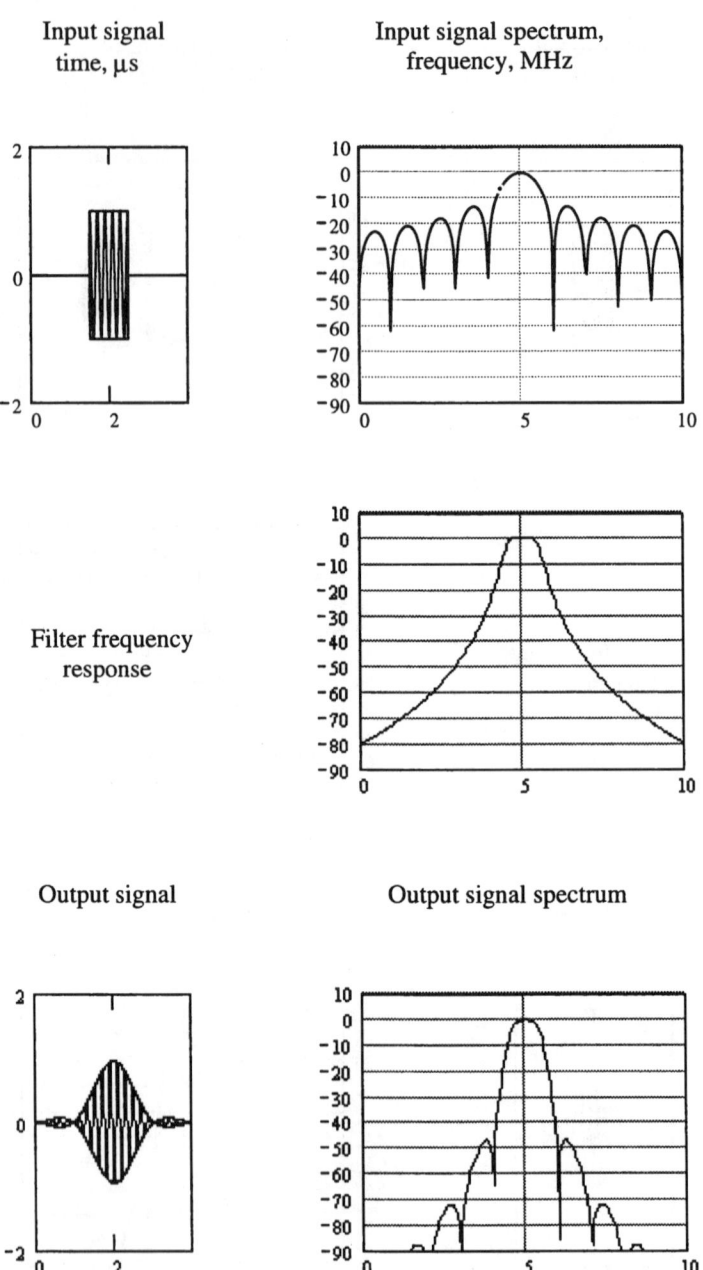

Figure 6.42 Frequency-selective filtering stages for Butterworth filter.

Chapter 7

Detection and Integration

7.1 Probability of Detection and False Alarm

7.1.1 DESCRIPTION

Calculates the probability of detection and false alarm for radar target detection. *Assumptions*: single-pulse detection (no integration) at the background of white Gaussian noise when the envelope of signal plus noise follows a Rician distribution.

7.1.2 INPUT DATA

===

Parameter   **Dimension and Description**

===

detection := 2
$$1 \rightarrow \text{noncoherent}$$
$$2 \rightarrow \text{coherent}$$

TH_db := 12
dB, detection threshold: a ratio of required threshold-to-noise power E_t^2/σ_N^2

SNR_db := 10
dB, signal-to-noise ratio

===

7.1.3 MODEL

7.1.3.1 Input Data Conversion

$\text{ratio10}(x) := 10^{0.1 \cdot x}$ $\text{ratio20}(x) := 10^{0.05 \cdot x}$ conversion equations

$\text{TH\_p} := \text{ratio10}(\text{TH\_db})$ $\text{TH\_p} = 15.849$ power detection threshold

$\text{TH\_v} := \text{ratio20}(\text{TH\_db})$ $\text{TH\_v} = 3.981$ voltage detection threshold

231

Note: the power detection threshold is the ratio of instantaneous peak power of a signal+noise sinusoid that would reach the threshold to mean noise power. The voltage detection threshold is the ratio of rms voltage of signal + noise sinusoid that would reach the threshold to rms noise power.

$$\text{SNR} := \text{ratio10(SNR\_db)} \qquad \text{SNR} = 10 \qquad\qquad \text{signal-to-noise ratio}$$

$$\text{erf(arg)} := \frac{2}{\sqrt{\pi}} \cdot \int_0^{\text{arg}} \exp(-t^2)\, dt \qquad\qquad \text{error function definition}$$

7.1.3.2 Single-Pulse Noncoherent Detection

$$P_{\text{fa\_non}} := \exp\left(-\frac{\text{TH\_p}}{2}\right) \qquad P_{\text{fa\_non}} = 3.618 \times 10^{-4} \qquad \text{false alarm probability}$$

$$n_{\text{f\_non}} := \frac{1}{P_{\text{fa\_non}}} \qquad n_{\text{f\_non}} = 2.764 \times 10^3 \qquad \text{false alarm number}$$

$$P_{\text{d\_non}} := 1 - \text{cnorm}\left(\sqrt{2 \cdot \ln(n_{\text{f\_non}})} - \sqrt{1 + 2 \cdot \text{SNR}}\right) \qquad \text{probability of detection}$$

$$P_{\text{d\_non}} = 0.726$$

7.1.3.3 Single-Pulse Coherent Detection

$$P1 := 0.5 \cdot (1 - \text{erf(TH\_v)}) \qquad P1 = 9.006 \times 10^{-9} \qquad\qquad \text{parameter}$$

$$P_{\text{fa\_coh}} := \begin{vmatrix} P1 & \text{if } P1 \geq 0 \\ 0 & \text{otherwise} \end{vmatrix} \qquad P_{\text{fa\_coh}} = 9.006 \times 10^{-9} \qquad \begin{array}{l}\text{false alarm}\\\text{probability}\end{array}$$

$$P_{\text{d\_coh}} := 1 - \text{cnorm}\left(\sqrt{2} \cdot \text{TH\_v} - \sqrt{2 \cdot \text{SNR}}\right) \qquad \text{probability of detection}$$

$$P_{\text{d\_coh}} = 0.123$$

7.1.3.4 Detection Specified

$$\text{detection} = 2 \qquad\qquad \text{detection type specified}$$

$$P_{fa} := \begin{vmatrix} P_{fa\_non} & \text{if detection} = 1 \\ P_{fa\_coh} & \text{if detection} = 2 \end{vmatrix} \qquad P_{fa} = 9.006 \times 10^{-9} \qquad \text{false alarm probability}$$

$$P_d := \begin{vmatrix} P_{d\_non} & \text{if detection} = 1 \\ P_{d\_coh} & \text{if detection} = 2 \end{vmatrix} \qquad P_d = 0.123 \qquad \text{detection probability}$$

7.1.4 OUTPUT DATA

Probability of detection $P_d = 0.123$ and false alarm $P_{fa} = 9.006 \times 10^{-9}$

for input data specified

7.1.5 REFERENCES

[1] Barton, D. K., and S. A. Leonov, (eds.), *Radar Technology Encyclopedia*, Norwood, MA: Artech House, 1997, p. 121.
[2] Barton, D. K., *Modern Radar System Analysis*, Norwood, MA: Artech House, 1989, pp. 65–68.

7.2 Detectability Factor

7.2.1 DESCRIPTION
Calculates detectability factor as the output of a signal processor for different fluctuation models. *Assumptions*: n-pulse detection at the background of white Gaussian noise when the envelope of signal plus noise follows a Rician distribution.

7.2.2 INPUT DATA

Parameter	Dimension and Description
$P_{fa} := 10^{-5}$	false alarm probability required
$P_d := 0.75$	detection probability required
$n := 10$	number of pulses integrated
Swerling := 1	$0 \rightarrow$ Swerling case 0 (nonfluctuating target) $1 \rightarrow$ Swerling case 1 fluctuation model $2 \rightarrow$ Swerling case 2 fluctuation model $3 \rightarrow$ Swerling case 3 fluctuation model $4 \rightarrow$ Swerling case 4 fluctuation model
$L_x := 1$	dB, signal processing losses
$M := 0.5$	dB, receiver mismatch loss
$L_p := 1.2$	dB, beamshape loss

7.2.3 MODEL

7.2.3.1 Basic Detectability Factor for Swerling 0 (Nonfluctuating) Target

$$D_0 := \left[\left[\sqrt{\ln\left(\frac{1}{P_{fa}}\right)} - \frac{1}{\sqrt{2}} \cdot \text{qnorm}\left[\left(1 - P_d\right), 0, 1 \right] \right]^2 - 0.5 \right]$$

$$D_{0\_db} := 10 \cdot \log(D_0) \qquad \text{dB, basic Swerling 0 detectability factor}$$

$$D_{0\_db} = 11.607$$

7.2.3.2 Integration Loss and Gain

$$D_c := \frac{1}{2} \cdot \left[\text{qnorm}\left[(1 - P_{fa}), 0, 1 \right] - \text{qnorm}\left[(1 - P_d), 0, 1 \right] \right]^2$$

$D_{c\_db} := 10 \cdot \log(D_c)$ $D_{c\_db} = 10.863$ dB, coherent detectability factor

$$L_i := \frac{1 + \sqrt{1 + \dfrac{9.2 \cdot n}{D_c}}}{1 + \sqrt{1 + \dfrac{9.2}{D_c}}}$$ integration loss

$L_{i\_db} := 10 \cdot \log(L_i)$ $L_{i\_db} = 2.273$ dB, integration loss

$G_{i\_db} := 10 \cdot \log(n)$ $G_{i\_db} = 10$ dB, integration gain

7.2.3.3 Fluctuation Loss

$$D_1 := \frac{\ln(P_{fa})}{\ln(P_d)} - 1$$ basic detectability factor for a single
pulse and Swerling 1 model

$D_{1\_db} := 10 \cdot \log(D_1)$ dB, basic Swerling 1 detectability factor

$D_{1\_db} = 15.913$

$$L_{f\_1} := \frac{D_1}{D_0}$$ $L_{f\_1} = 2.695$ fluctuation loss for Swerling case 1 target

$L_{f\_1\_db} := 10 \cdot \log(L_{f\_1}) \cdot (1 + 0.03 \cdot \log(n))$ dB, fluctuation loss for
Swerling case 1 target

$L_{f\_1\_db} = 4.435$

$\text{Swerling} = 1$ Swerling case specified

$$K(\text{Swerling}) := \begin{vmatrix} 10^{10} & \text{if Swerling} = 0 \\ 1 & \text{if Swerling} = 1 \\ 1 & \text{if Swerling} = 2 \\ 2 & \text{if Swerling} = 3 \\ 2 & \text{if Swerling} = 4 \end{vmatrix}$$

one-half of the number
of degrees of freedom of
a chi-square distribution
describing target fluctuation
pdf

$$n_e(Swerling) := \begin{vmatrix} 10^{10} & \text{if } Swerling = 0 \\ 1 & \text{if } Swerling = 1 \\ n & \text{if } Swerling = 2 \\ 1 & \text{if } Swerling = 3 \\ n & \text{if } Swerling = 4 \end{vmatrix}$$

number of independent target samples available for integration

$$L_{f\_db}(Swerling) := \frac{L_{f\_1\_db}}{K(Swerling) \cdot n_e(Swerling)}$$

dB, fluctuation loss

$$L_{f\_db}(Swerling) = 4.435$$

7.2.3.4 Detectability Factor at the Signal Processor Output

$$D_{e\_db} := D_{0\_db} + L_{f\_db}(Swerling) + L_{i\_db} - G_{i\_db}$$

$$D_{e\_db} = 8.315$$

dB, basic detectability factor

$$D_{x\_db} := D_{e\_db} + L_x + M + L_p$$

$$D_{x\_db} = 11.015$$

dB, effective detectability factor

Note: in a case where binary integration is used, different probabilities of false alarm and detection have to be used to calculate D_0 (P_{fa}, P_d at the input of binary integrator) and fluctuation loss (P_{fa}, P_d at the output of binary integrator). To relate probabilities at the input and output of a binary integrator, see Section 7.4.

7.2.4 OUTPUT DATA

Basic detectability factor $D_{e\_db} = 8.315$ dB   and effective detectability factor

$D_{x\_db} = 11.015$ dB   for input data specified.

7.2.5 REFERENCES

[1] Barton, D. K., and S. A. Leonov, (eds.), *Radar Technology Encyclopedia*, Norwood, MA: Artech House, 1997, p. 112.
[2] Barton, D. K., and W. F. Barton, *Modern Radar System Analysis Software*, Norwood, MA: Artech House, 1993, p. 131.

7.3 Minimum Detectable Signal

7.3.1 DESCRIPTION

Calculates minimum detectable signal (MDS) that is required for detection with a specified detectability factor. Sometimes MDS or sensitivity is defined as a signal level that is just equal to the noise level.

7.3.2 INPUT DATA

==

Parameter	Dimension and Description
$T_s := 500$	°K, system noise temperature
$B_n := 300 \cdot 10^6$	Hz, receiver bandwidth
$D_{x\_db} := 10$	dB, effective detectability factor

==

7.3.3 MODEL

7.3.3.1 Constants

$k := 1.38054 \cdot 10^{-23}$ 　　　　　　W*s/°K, Boltzmann's constant

7.3.3.2 MDS Calculation

$ratio(x) := 10^{0.1 \cdot x}$ 　　　　　　conversion equation

$D_x := ratio(D_{x\_db})$ 　　$D_x = 10$ 　　effective detectability factor

$MDS := k \cdot T_s \cdot B_n \cdot D_x$ 　　$MDS = 2.071 \times 10^{-11}$ 　W, minimum detectable signal

$MDS\_db := 10 \cdot \log(MDS)$ 　　　　　dB, minimum detectable signal

$MDS\_db = -106.8$

$MDS\_dbm := 10 \cdot \log\left(\dfrac{MDS}{10^{-3}}\right)$ 　　　　dBm, minimum detectable signal

$MDS\_dbm = -76.8$

7.3.4 OUTPUT DATA

Minimum detectable signal 　　$MDS - 2.071 \times 10^{-11}$ watt, or

$MDS\_db = -106.8$ 　dB or $MDS\_dbm = -76.8$ 　dBm for input data specified

7.3.5 REFERENCE

[1] Barton, D. K., and S. A. Leonov, (eds.), *Radar Technology Encyclopedia*, Norwood, MA: Artech House, 1997, pp. 113, 418.

7.4 M-out-of-N Integration

7.4.1 DESCRIPTION

Relates probabilities of detection and false alarms at the input and output of an M-out-of-N integrator.

7.4.2 INPUT DATA

Parameter	Dimension and Description
$N := 1$	number of successful detections required
$M := 2$	number of samples tested
conversion $:= 2$	$1 \to$ converts probability at the input to the output $2 \to$ converts probability at the output to the input
$P := 0.99$	probability of detection or false alarm at the input of the integrator (if conversion = 1) or at the output of the integrator (if conversion = 2)
$p0 := 0.5$	a guess value for input probability if conversion = 2

Note: for probabilities of detection a number $p0 = 0.5$ is a good first guess. For probabilities of false alarms make a guess about the value and use conversion = 1 to see if it gives a good match with output probability required.

7.4.3 MODEL

7.4.3.1 Output Probability as a Function of the Input Probability

$p := P$ input probability specified

$$P0(N,M,p) := \sum_{i=N}^{M} dbinom(i,M,p)$$ output probability

$P0(N,M,p) = 1$

7.4.3.2 Input Probability as a Function of the Output Probability

$P_{out} := P$ output probability specified

$$p1 := \text{root}\left(P_{out} - P0(N,M,p0),p0\right)$$ input probability

$$p1 = 0.899$$

7.4.3.3 Relationship Between Input and Output Probabilities

conversion = 2   conversion type specified

$$p := \begin{vmatrix} P & \text{if conversion} = 1 \\ p1 & \text{if conversion} = 2 \end{vmatrix}$$ $$p = 0.899$$ input probability

$$P := \begin{vmatrix} P0(N,M,p) & \text{if conversion} = 1 \\ P & \text{if conversion} = 2 \end{vmatrix}$$ $$P = 0.99$$ output probability

7.4.4 OUTPUT DATA

Probabilities at the input of integrator $p = 0.899$ and at the output of the integrator $P = 0.99$ for input data specified.

7.4.5 REFERENCES

[1] Barton, D. K., and S. A. Leonov, (eds.), *Radar Technology Encyclopedia*, Norwood, MA: Artech House, 1997, p. 220.
[2] Barton, D. K., *Modern Radar System Analysis*, Norwood, MA: Artech House, 1989, p. 73.

Chapter 8

Losses

8.1 Atmospheric Attenuation Loss

8.1.1 DESCRIPTION

Models atmospheric attenuation loss as the function of atmospheric conditions, radar frequency, target range and elevation (full-scale model, Blake), or frequency-dependent sea-level attenuation coefficients, target range and elevation (simplified model, Barton). *Assumptions*: for Blake's model attenuation is caused by absorption of oxygen molecules; a conventional exponential model of refractive index is used [2, p. 183]. Use Barton's model for frequencies higher than S band.

8.1.2 INPUT DATA

===

Parameter	*Dimension and Description*

===

$\theta := 1$

$R := 100$ nm, target range in vacuum

model := 1   1 → full-scale model (Blake's)
2 → simplified model (Barton's)

Blake's Model

$f := 1300$ MHz, radar frequency

$h_a := 70$

$K := 100$ number of points to calculate absorption integral

241

Barton's Model

$k_\alpha := 0.012$ attenuation coefficient
                                        (frequency-dependent, see Table 8.1 below)

Note: In table 8.1 r is the precipitation rate in mm/h. To find the attenuation coefficient at a specific radar frequency logarithmic interpolation should be performed.

Table 8.1 Two-Way Atmospheric Attenuation Coefficients

Band	Frequency (GHz)	k_α clear air	k_α/r rain	k_α/r snow
UHF	0.4	0.100	0	0
L	1.3	0.012	0.0003	0.0003
S	3.0	0.015	0.0013	0.0013
C	5.5	0.017	0.008	0.008
X	10	0.024	0.037	0.002
Ku	15	0.055	0.083	0.004
K	22	0.030	0.23	0.008
Ka	35	0.14	0.57	0.015
V	60	35	1.3	0.03
	90	0.80	2.0	0.06
	140	1.0	2.3	0.06
	240	15	2.2	0.08

===

8.1.3 MODEL

8.1.3.1 Constants

$N := 0.000313$ refractivity

$c_e := -0.1439$ 1/km, decay constant

$n0 := 1.000313$ refractive index

$r_{km} := 6370$ km, Earth radius

$r_m := 6370000$ m, Earth radius

$a_e := 8493333$ m, Earth effective radius

$p0 := 1013.25$ mbar, atmospheric pressure
                                                        constant

$\alpha1 := 5.2561222$

$\alpha2 := 0.034164794$ troposphere model constants

$\alpha3 := 11.388265$

$T0 := 300$ °K, standard temperature

$C := 2.0058$ the absorption coefficient constants

$z := \dfrac{\pi}{180}$ coefficient to transform degrees to radians

Oxygen Resonance Frequencies (GHz)

$\text{f\_N\_plus}_0 := 56.2648$	$\text{f\_N\_minus}_0 := 118.7505$	$Z_0 := 0$
$\text{f\_N\_plus}_1 := 56.2648$	$\text{f\_N\_minus}_1 := 118.7505$	$Z_1 := 1$
$\text{f\_N\_plus}_2 := 58.4466$	$\text{f\_N\_minus}_2 := 62.4862$	$Z_2 := 0$
$\text{f\_N\_plus}_3 := 58.4466$	$\text{f\_N\_minus}_3 := 62.4862$	$Z_3 := 1$
$\text{f\_N\_plus}_4 := 59.5910$	$\text{f\_N\_minus}_4 := 60.3061$	$Z_4 := 0$
$\text{f\_N\_plus}_5 := 59.5910$	$\text{f\_N\_minus}_5 := 60.3061$	$Z_5 := 1$
$\text{f\_N\_plus}_6 := 60.4348$	$\text{f\_N\_minus}_6 := 59.1642$	$Z_6 := 0$
$\text{f\_N\_plus}_7 := 60.4348$	$\text{f\_N\_minus}_7 := 59.1642$	$Z_7 := 1$
$\text{f\_N\_plus}_8 := 61.1506$	$\text{f\_N\_minus}_8 := 58.3239$	$Z_8 := 0$
$\text{f\_N\_plus}_9 := 61.1506$	$\text{f\_N\_minus}_9 := 58.3239$	$Z_9 := 1$
$\text{f\_N\_plus}_{10} := 61.8002$	$\text{f\_N\_minus}_{10} := 57.6125$	$Z_{10} := 0$
$\text{f\_N\_plus}_{11} := 61.8002$	$\text{f\_N\_minus}_{11} := 57.6125$	$Z_{11} := 1$
$\text{f\_N\_plus}_{12} := 62.4212$	$\text{f\_N\_minus}_{12} := 56.9682$	$Z_{12} := 0$
$\text{f\_N\_plus}_{13} := 62.4212$	$\text{f\_N\_minus}_{13} := 56.9682$	$Z_{13} := 1$
$\text{f\_N\_plus}_{14} := 62.9980$	$\text{f\_N\_minus}_{14} := 56.3634$	$Z_{14} := 0$
$\text{f\_N\_plus}_{15} := 62.9980$	$\text{f\_N\_minus}_{15} := 56.3634$	$Z_{15} := 1$
$\text{f\_N\_plus}_{16} := 63.5685$	$\text{f\_N\_minus}_{16} := 55.7839$	$Z_{16} := 0$
$\text{f\_N\_plus}_{17} := 63.5685$	$\text{f\_N\_minus}_{17} := 55.7839$	$Z_{17} := 1$
$\text{f\_N\_plus}_{18} := 64.1272$	$\text{f\_N\_minus}_{18} := 55.2214$	$Z_{18} := 0$
$\text{f\_N\_plus}_{19} := 64.1272$	$\text{f\_N\_minus}_{19} := 55.2214$	$Z_{19} := 1$
$\text{f\_N\_plus}_{20} := 64.6779$	$\text{f\_N\_minus}_{20} := 54.6728$	$Z_{20} := 0$

$f\_N\_plus_{21} := 64.6779$ $f\_N\_minus_{21} := 54.6728$ $Z_{21} := 1$

$f\_N\_plus_{22} := 65.2240$ $f\_N\_minus_{22} := 54.1294$ $Z_{22} := 0$

$f\_N\_plus_{23} := 65.2240$ $f\_N\_minus_{23} := 54.1294$ $Z_{23} := 1$

$f\_N\_plus_{24} := 65.7626$ $f\_N\_minus_{24} := 53.5960$ $Z_{24} := 0$

$f\_N\_plus_{25} := 65.7626$ $f\_N\_minus_{25} := 53.5960$ $Z_{25} := 1$

$f\_N\_plus_{26} := 66.2978$ $f\_N\_minus_{26} := 53.0695$ $Z_{26} := 0$

$f\_N\_plus_{27} := 66.2978$ $f\_N\_minus_{27} := 53.0695$ $Z_{27} := 1$

$f\_N\_plus_{28} := 66.8313$ $f\_N\_minus_{28} := 52.5458$ $Z_{28} := 0$

$f\_N\_plus_{29} := 66.8313$ $f\_N\_minus_{29} := 52.5458$ $Z_{29} := 1$

$f\_N\_plus_{30} := 67.3627$ $f\_N\_minus_{30} := 52.0259$ $Z_{30} := 0$

$f\_N\_plus_{31} := 67.3627$ $f\_N\_minus_{31} := 52.0259$ $Z_{31} := 1$

$f\_N\_plus_{32} := 67.8923$ $f\_N\_minus_{32} := 51.5091$ $Z_{32} := 0$

$f\_N\_plus_{33} := 67.8923$ $f\_N\_minus_{33} := 51.5091$ $Z_{33} := 1$

$f\_N\_plus_{34} := 68.4205$ $f\_N\_minus_{34} := 50.9949$ $Z_{34} := 0$

$f\_N\_plus_{35} := 68.4205$ $f\_N\_minus_{35} := 50.9949$ $Z_{35} := 1$

$f\_N\_plus_{36} := 68.9478$ $f\_N\_minus_{36} := 50.4830$ $Z_{36} := 0$

$f\_N\_plus_{37} := 68.9478$ $f\_N\_minus_{37} := 50.4830$ $Z_{37} := 1$

$f\_N\_plus_{38} := 69.4741$ $f\_N\_minus_{38} := 49.9730$ $Z_{38} := 0$

$f\_N\_plus_{39} := 69.4741$ $f\_N\_minus_{39} := 49.9730$ $Z_{39} := 1$

$f\_N\_plus_{40} := 70.0000$ $f\_N\_minus_{40} := 49.4648$ $Z_{40} := 0$

$f\_N\_plus_{41} := 70.0000$ $f\_N\_minus_{41} := 49.4648$ $Z_{41} := 1$

$f\_N\_plus_{42} := 70.5249$ $f\_N\_minus_{42} := 48.9582$ $Z_{42} := 0$

$f\_N\_plus_{43} := 70.5249$ $f\_N\_minus_{43} := 48.9582$ $Z_{43} := 1$

$f\_N\_plus_{44} := 71.0497$ $f\_N\_minus_{44} := 48.4530$ $Z_{44} := 0$

$f\_N\_plus_{45} := 71.0497$ $f\_N\_minus_{45} := 48.4530$ $Z_{45} := 1$

8.1.3.2 Full-Scale (Blake's) Model

$R_{km} := 1.852 \cdot R$ $R_{km} = 185.2$ km, target range

$R_m := 1852 \cdot R$ $R_m = 1.852 \times 10^5$ m, target range

$$h_m := R_m \cdot \sin(z \cdot \theta) + \frac{(R_m \cdot \cos(z \cdot \theta))^2}{2 \cdot a_e}$$ m, target height above antenna

$$h_m = 5.251 \times 10^3$$

$$h_{km} := \frac{h_m}{1000} \qquad h_{km} = 5.251$$ km, target height

$$\Delta h := \frac{h_{km}}{K} \qquad \Delta h = 0.053$$ km, target height increment

$$k := 0 .. K$$ cycle

$$h_k := \frac{h_a}{1000} + k \cdot \Delta h$$ km, current height

$$h_{g\_m_k} := \frac{r_m \cdot h_k \cdot 1000}{r_m + h_k \cdot 1000}$$ m, geopotential altitude

$$h_{g\_km_k} := \frac{h_{g\_m_k}}{1000} \qquad h_{g\_km_{100}} = 5.316$$ km, geopotential altitude

0K, temperature of the atmosphere

$$T_k := \begin{vmatrix} 288.16 - 0.0065 \cdot h_k \cdot 1000 & \text{if } h_{g\_km_k} \le 11 \\ 216.66 & \text{if } 11 < h_{g\_km_k} < 25 \\ 216.66 + 0.003 \cdot (h_k - 25) \cdot 1000 & \text{otherwise} \end{vmatrix}$$

Millibars, pressure of the atmosphere

$$p_k := \begin{vmatrix} p0 \cdot \left(\dfrac{T_k}{288.16}\right)^{\alpha 1} & \text{if } h_{g\_km_k} \le 11 \\ \dfrac{226.32}{T_k} \cdot c^{-\alpha 2 \cdot (h_k - 11) \cdot 1000} & \text{if } 11 < h_{g\_km_k} < 25 \\ 24.886 \cdot \left(\dfrac{216.66}{T_k}\right)^{\alpha 3} & \text{otherwise} \end{vmatrix}$$

$$f\_GHz := \frac{f}{1000} \qquad\qquad f\_GHz = 1.3 \qquad\qquad\qquad \text{GHz, radar frequency}$$

The line breadth constant parameter

$$g_k := \begin{vmatrix} 0.640 & \text{if } h_k \leq 8 \\ 0.640 + 0.04218 \cdot (h_k - 8) & \text{if } 8 < h_k \leq 25 \\ 1.357 & \text{otherwise} \end{vmatrix}$$

$$\Delta f_k := g_k \cdot \frac{P_k}{p0} \cdot \frac{T0}{T_k} \qquad\qquad\qquad \text{the line-breadth constant}$$

$$F0_k := \frac{\Delta f_k}{\left(\Delta f_k\right)^2 + f\_GHz^2} \qquad\qquad\qquad \text{the nonresonant contribution}$$

$$n := 1 .. 45 \qquad\qquad\qquad\qquad\qquad\qquad\qquad \text{cycle}$$

Parameters to calculate the absorption coefficient

$$\mu\_plus_n := \frac{n \cdot (2 \cdot n + 3)}{n + 1}$$

$$\mu\_minus_n := \frac{(n + 1) \cdot (2 \cdot n - 1)}{n}$$

$$\mu\_0_n := \frac{2 \cdot \left(n^2 + n + 1\right) \cdot (2 \cdot n + 1)}{n \cdot (n + 1)}$$

$$\Sigma1\_plus_{k,n} := \frac{\Delta f_k}{\left(f\_N\_plus_n - f\_GHz\right)^2 + \left(\Delta f_k\right)^2}$$

$$\Sigma2\_plus_{k,n} := \frac{\Delta f_k}{\left(f\_N\_plus_n + f\_GHz\right)^2 + \left(\Delta f_k\right)^2}$$

$$\Sigma 1\_minus_{k,n} := \frac{\Delta f_k}{\left(f\_N\_minus_n - f\_GHz\right)^2 + \left(\Delta f_k\right)^2}$$

$$\Sigma 2\_minus_{k,n} := \frac{\Delta f_k}{\left(f\_N\_minus_n + f\_GHz\right)^2 + \left(\Delta f_k\right)^2}$$

$$F\_N\_plus_{k,n} := \Sigma 1\_plus_{k,n} + \Sigma 2\_plus_{k,n}$$

$$F\_N\_minus_{k,n} := \Sigma 1\_minus_{k,n} + \Sigma 2\_minus_{k,n}$$

$$A1_{k,n} := \mu\_plus_n \cdot F\_N\_plus_{k,n}$$

$$A2_{k,n} := \mu\_minus_n \cdot F\_N\_minus_{k,n}$$

$$E_n := 2.06844 \cdot n \cdot (n + 1)$$

$$\Sigma_k := \sum_{n=1}^{45} Z_n \cdot \left[\left(A1_{k,n} + A2_{k,n} + \mu\_0_n \cdot F0_k\right) \cdot \exp\left(-\frac{E_n}{T_k}\right) \right]$$

$$\gamma_k := C \cdot p_k \cdot \left(T_k\right)^{-3} \cdot f\_GHz^2 \cdot \Sigma_k \qquad\qquad \text{dB/km, absorption coefficient}$$

$$m_k := 1 + N \cdot \exp\left(c_e \cdot h_k\right) \qquad\qquad \text{refractive index model}$$

Absorption loss integral calculation

$$S1 := \frac{\gamma_0}{\sqrt{1 - \left[\dfrac{n0 \cdot \cos(z \cdot \theta)}{m_0 \cdot \left(1 + \dfrac{h_0}{r_{km}}\right)}\right]^2}} \qquad\qquad S2 := \frac{\gamma_K}{\sqrt{1 - \left[\dfrac{n0 \cdot \cos(z \cdot \theta)}{m_K \cdot \left(1 + \dfrac{h_K}{r_{km}}\right)}\right]^2}}$$

$$S3 := \sum_{k=1}^{K-1} \frac{\gamma_k}{\sqrt{1 - \left[\dfrac{n0 \cdot \cos(z \cdot \theta)}{m_k \cdot \left(1 + \dfrac{h_k}{r_{km}}\right)}\right]^2}}$$

$L1_\alpha := 2 \cdot (S1 + S2 + 2 \cdot S3) \cdot \dfrac{\Delta h}{2}$ dB, two-way atmospheric attenuation loss

$L1_\alpha = 1.737$

8.1.3.3 Simplified (Barton's) Model

$\theta_{eff} := z \cdot \theta + \dfrac{2.5 \cdot 10^{-4}}{z \cdot \theta + 0.028}$ rad, effective elevation angle

$R_{eff} := \dfrac{3.0}{\sin(\theta_{eff})}$ km, effective range

$L2_\alpha := k_\alpha \cdot R_{eff} \cdot \left(1 - \exp\left(-\dfrac{R_{km}}{R_{eff}}\right)\right)$ dB, two-way atmospheric attenuation loss

$L2_\alpha = 1.188$

8.1.3.4 Atmospheric Attenuation Loss for the Model Specified

model = 1   model specified

$L_\alpha := \begin{vmatrix} L1_\alpha & \text{if model} = 1 \\ L2_\alpha & \text{otherwise} \end{vmatrix}$ dB, two-way atmospheric attenuation loss

8.1.4 OUTPUT DATA

Atmospheric loss $L_\alpha = 1.737$ dB for input data specified.

8.1.5 REFERENCES

[1] Barton, D. K., and S. A. Leonov, (eds.), *Radar Technology Encyclopedia*, Norwood, MA: Artech House, 1997, p. 247.
[2] Blake, L., *Radar Range Performance Analysis*, Lexington Books, 1980, p. 197.

8.2 Beamshape Loss

8.2.1 DESCRIPTION
Simulates the beamshape loss originated by covering the search sector with a rotating antenna that has a beam that differs from an ideal rectangular shape. *Assumption*: a Gaussian shape of antenna pattern.

8.2.2 INPUT DATA

==

Parameter	Dimension and Description

==

$\theta := 1$	deg, 3-dB beamwidth
$\omega_a := 20$	rpm, antenna rotation rate
$PRF := 1600$	Hz, pulse repetition frequency
$level := 2$	$1 \rightarrow$ loss at 3-dB level
	$2 \rightarrow$ loss at the level specified by n below
$n := 25$	number of pulses required for detection

==

8.2.3 MODEL

8.2.3.1 Time-on-Target Calculations

$$t_s := \frac{60}{\omega_a} \qquad t_s = 3 \qquad \text{s, a single scan time for rotating antenna}$$

$$t_{3db} := t_s \cdot \frac{\theta}{360} \qquad t_{3db} = 8.333 \times 10^{-3} \qquad \text{s, time-on-target at 3-dB beamwidth level}$$

$$n\_3db := t_{3db} \cdot PRF \qquad n\_3db = 13.333 \qquad \text{\# of pulses received at 3-dB level}$$

$$t_n := \frac{n}{PRF} \qquad t_n = 0.016 \qquad \text{s, time-on-target required to receive n pulses}$$

8.2.3.2 Antenna Pattern Simulation

$$T_s := -5 \cdot \frac{t_{3db}}{2} \qquad T_s = -0.021$$

s, pattern limits to be plotted

$$T_e := 5 \cdot \frac{t_{3db}}{2} \qquad T_e = 0.021$$

$N := 1000$ number of points to digitize antenna pattern

$$\Delta t := \frac{T_e - T_s}{N}$$ s, antenna pattern digitizing increment

$n := 0 .. N - 1$

$$f_n := \exp\left[-1.3863 \cdot \left(\frac{T_s + n \cdot \Delta t}{t_{3db}}\right)^2\right]$$ Gaussian antenna pattern

$level = 2$ level specified

$$t_0 := \begin{vmatrix} t_{3db} & \text{if } level = 1 \\ t_n & \text{otherwise} \end{vmatrix}$$ $t_0 = 0.016$ s, required time-on-target

Required time-on-target limits to plot

$$v_n := \begin{vmatrix} 1 & \text{if } -\left(\frac{t_0}{2}\right) \cdot 10^6 \leq \left[(T_s + n \cdot \Delta t) \cdot 10^6\right] \leq \left(-\frac{t_0}{2} + \Delta t\right) \cdot 10^6 \\ 0 & \text{if } \left(-\frac{t_0}{2} + \Delta t\right) \cdot 10^6 \leq \left[\left[T_s + (n + 0) \cdot \Delta t\right] \cdot 10^6\right] \leq \left(-\frac{t_0}{2} + 2 \cdot \Delta t\right) \cdot 10^6 \\ -1 & \text{otherwise} \end{vmatrix}$$

$$u_n := \begin{vmatrix} 1 & \text{if } \left(\frac{t_0}{2}\right) \cdot 10^6 \leq \left[(T_s + n \cdot \Delta t) \cdot 10^6\right] \leq \left(\frac{t_0}{2} + \Delta t\right) \cdot 10^6 \\ 0 & \text{if } \left(\frac{t_0}{2} + \Delta t\right) \cdot 10^6 \leq \left[\left[T_s + (n + 0) \cdot \Delta t\right] \cdot 10^6\right] \leq \left(\frac{t_0}{2} + 2 \cdot \Delta t\right) \cdot 10^6 \\ -1 & \text{otherwise} \end{vmatrix}$$

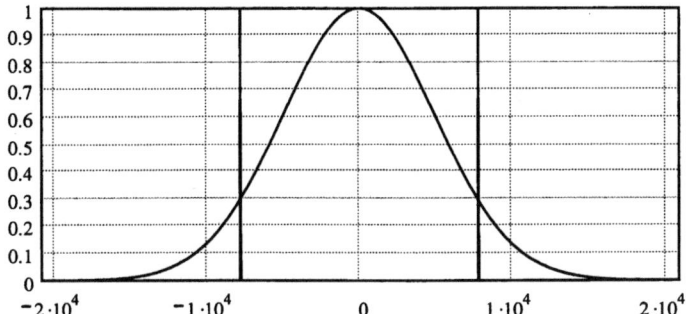

Figure 8.1 Normalized Gaussian antenna voltage pattern vs. time (µs) (required time-on-target boundaries are shown by vertical lines).

8.2.3.3 Beamshape Loss

$$f(t) := \exp\left[-1.3863 \cdot \left(\frac{t}{t_{3db}}\right)^2\right]$$ Gaussian voltage pattern as a function of time

$$S_f := \int_{-100 \cdot t_0}^{100 \cdot t_0} f(t)^4 \, dt$$ area covered by the actual power pattern (two-way propagation accounted for)

$$S_f = 6.272 \times 10^{-3}$$

$$S_r := 1 \cdot t_0 \qquad S_r = 0.016$$ area covered by the ideal rectangular pattern

$$L1_p := 10 \cdot \log\left(\frac{S_r}{S_f}\right) \qquad L1_p = 3.964$$ dB, beamshape loss

$$L_p := \begin{vmatrix} L1_p & \text{if } L1_p \geq 0 \\ 0 & \text{otherwise} \end{vmatrix} \qquad L_p = 3.964$$

8.2.4 OUTPUT DATA

Beamshape loss $L_p = 3.964$ dB for input data specified.

8.2.5 REFERENCE

[1] Barton, D. K., and S. A. Leonov, (eds.), *Radar Technology Encyclopedia*, Norwood, MA: Artech House, 1997, p. 248.

8.3 CFAR Loss

8.3.1 DESCRIPTION

Simulates constant-false-alarm rate (CFAR) circuit loss for a single-hit detection of the target with fluctuation models Swerling 0 (nonfluctuating), Swerling 1, Swerling 3 at the background of interference with Rayleigh (noise) and Weibull (clutter) distributions. *Assumption:* a circuit falls into the class of conventional cell-averaging CFARs; the estimate is the arithmetic mean of the reference cell's voltages; test samples are assumed statistically independent.

8.3.2 INPUT DATA

==

Parameter	*Dimension and Description*

==

$m := 12$	number of cells averaged
$P_{fa} := 10^{-4}$	probability of false alarm
$k0 := 0.35$	initial guess for parameter to solve equation with respect to false alarm probability

Note: if equation to determine parameter k does not converge try another k0 that satisfies:

$$\frac{\ln(P_{fa})}{m} := \ln(1 - k0) \qquad ∎$$

input := 1	$1 \rightarrow$ Pd specified $2 \rightarrow$ signal-to-noise ratio specified

if input = 1

$P_d := 0.75$

if input = 2

SNR_db := 20	dB, signal-to-noise ratio
distribution := 1	$1 \rightarrow$ Rayleigh distribution of interference $2 \rightarrow$ Weibull distribution of interference

if distribution = 2

$F_c := 0.5$	clutter pattern-propagation factor
$\theta := 3$	deg, antenna pattern beamwidth
$\tau := 10$	μs, processed pulsewidth
$R := 10$	km, range to a clutter cell

==

8.3.3 MODEL

8.3.3.1 Constants

$$z := \frac{\pi}{180}$$ coefficient to transform degrees to radians

$$c := 2.997925 \cdot 10^8$$ m/s, speed of light

8.3.3.2 Probability of Detection

$$k := \text{root}\left(\frac{\ln(P_{fa})}{m} - \ln(1 - k0), k0\right)$$ equation to determine parameter k based on probability of false alarm

$$k = 0.536$$

$$\frac{\ln(P_{fa})}{m} = -0.768 \quad \ln(1 - k) = -0.767$$ a check that both parts of equality are the same

$$SNR := 10^{\frac{SNR\_db}{10}}$$ signal-to-noise ratio specified

$$SNR = 100$$

$$P1_d := \left[\frac{(1 - k) \cdot (1 + SNR)}{1 + SNR \cdot (1 - k)}\right]^m$$ probability of detection

$$P1_d = 0.873$$

8.3.3.3 CFAR Loss for Rayleigh Interference

$$input = 1$$

$$P2_d := \begin{vmatrix} P_d & \text{if } input = 1 \\ P1_d & \text{if } input = 2 \end{vmatrix}$$ $P2_d = 0.75$ probability of detection

$$snr(m) := \frac{\left(\dfrac{P2_d}{P_{fa}}\right)^{\frac{1}{m}} - 1}{1 - P2_d^{\frac{1}{m}}}$$ $snr(m) = 46.58$ mean signal-to-noise ratio

$$\text{snr}_{\text{ref}} := \frac{\log\left(\dfrac{P_{fa}}{P2_d}\right)}{\log\left(P2_d\right)}$$

$$\text{snr}_{\text{ref}} = 31.016$$

signal-to-noise ratio when $m \to \infty$

$$L_{g\_R} := 10 \cdot \log\left(\frac{\text{snr}(m)}{\text{snr}_{\text{ref}}}\right)$$

$$L_{g\_R} = 1.766$$

dB, CFAR loss

8.3.3.4 Distribution Loss

$$\tau_s := \tau \cdot 10^{-6}$$

s, processed pulsewidth

$$a := 1 + \frac{28 \cdot \log\left(F_c\right)^2}{\log\left[\left(0.75 \cdot R \cdot 1000 \cdot z \cdot \theta\right) \cdot \left(\dfrac{\tau_s \cdot c}{2}\right)\right]}$$

$$a = 1.44$$

parameter

$$L_d := -6 \cdot \log\left(P_{fa}\right) \cdot \log(a)^{1.2}$$

$$L_d = 2.628$$

dB, distribution loss

8.3.3.5 CFAR Loss

distribution = 1

$$L_g := \begin{vmatrix} L_{g\_R} & \text{if distribution} = 1 \\ L_{g\_R} + L_d & \text{if distribution} = 2 \end{vmatrix}$$

$$L_g = 1.766$$

8.3.4 OUTPUT DATA

CFAR loss $L_g = 1.766$ dB for input data specified

8.3.5 REFERENCES

[1] Barton, D. K., and S. A. Leonov, (eds.), *Radar Technology Encyclopedia*, Norwood, MA: Artech House, 1997, p. 251.
[2] Barton, D. K., and W. F. Barton, *Modern Radar System Analysis Software*, Norwood, MA: Artech House, 1993, p. 164.
[3] Nitzberg, R., Analysis of the Arithmetic Mean CFAR Normalizer for Fluctuating Targets, *IEEE Trans. on Aerospace and Electronic Systems*, Vol. AES-14, No. 1, January 1978, pp. 44–47.

8.4 Fluctuation Loss

8.4.1 DESCRIPTION

Calculates fluctuation loss as a function of probabilities of detection and false alarms, number of pulses integrated and target fluctuation model.

8.4.2 INPUT DATA

==

Parameter	Dimension and Description

==

$P_{fa} := 10^{-4}$	false alarm probability required
$P_d := 0.75$	detection probability required
$n := 30$	number of pulses integrated
Swerling := 1	$0 \rightarrow$ Swerling case 0 (nonfluctuating target) $1 \rightarrow$ Swerling case 1 fluctuation model $2 \rightarrow$ Swerling case 2 fluctuation model $3 \rightarrow$ Swerling case 3 fluctuation model $4 \rightarrow$ Swerling case 4 fluctuation model

==

8.4.3 MODEL

8.4.3.1 Swerling 0 Basic Detectability Factor

$$D_0 := \left[\left[\sqrt{\ln\left(\frac{1}{P_{fa}}\right)} - \frac{1}{\sqrt{2}} \cdot qnorm\left[(1 - P_d), 0, 1\right] \right]^2 - 0.5 \right]$$

$D_{0\_db} := 10 \cdot \log(D_0)$ dB, basic Swerling 0 detectability factor

$D_{0\_db} = 10.731$

8.4.3.2 Fluctuation Loss

$$D_1 := \frac{\ln(P_{fa})}{\ln(P_d)} - 1$$ basic detectability factor for a single pulse and Swerling 1 model

$D_{1\_db} := 10 \cdot \log(D_1)$

$D_{1\_db} = 14.916$ dB, basic Swerling 1 detectability factor

$$L_{f\_1} := \frac{D_1}{D_0}$$ $L_{f\_1} = 2.621$ fluctuation loss for Swerling case 1 target

$$L_{f\_1\_db} := 10 \cdot \log\left(L_{f\_1}\right) \cdot (1 + 0.03 \cdot \log(n))$$

dB, fluctuation loss for
Swerling Case 1 target

Note: the factor $1 + 0.03\log(n)$ is an empirical adjustment to match the fluctuation loss to exact data.

$$L_{f\_1\_db} = 4.37$$

Swerling $= 1$ Swerling case specified

$$K(\text{Swerling}) := \begin{vmatrix} 10^{10} & \text{if Swerling} = 0 \\ 1 & \text{if Swerling} = 1 \\ 1 & \text{if Swerling} = 2 \\ 2 & \text{if Swerling} = 3 \\ 2 & \text{if Swerling} = 4 \end{vmatrix}$$

one-half of the number
of degrees of freedom of
a chi-square distribution
describing target fluctuation
pdf

$$n_e(\text{Swerling}) := \begin{vmatrix} 10^{10} & \text{if Swerling} = 0 \\ 1 & \text{if Swerling} = 1 \\ n & \text{if Swerling} = 2 \\ 1 & \text{if Swerling} = 3 \\ n & \text{if Swerling} = 4 \end{vmatrix}$$

number of independent target
samples available for integration

$$L_{f\_db}(\text{Swerling}) := \frac{L_{f\_1\_db}}{K(\text{Swerling}) \cdot n_e(\text{Swerling})}$$

dB, fluctuation loss

$$L_{f\_db}(\text{Swerling}) = 4.37$$

8.4.4 OUTPUT DATA

Fluctuation loss $L_{f\_db}(\text{Swerling}) = 4.37$ dB for input data specified.

8.4.5 REFERENCES

[1] Barton, D. K., and S. A. Leonov, (eds.), *Radar Technology Encyclopedia*, Norwood, MA: Artech House, 1997, p. 253.
[2] Barton, D. K., and W. F. Barton, *Modern Radar System Analysis Software*, Norwood, MA: Artech House, 1993, p. 84.

8.5 Integration Loss

8.5.1 DESCRIPTION

Calculates video integration loss (integration after envelope detection) relative to coherent integration of signal samples. *Assumptions*: detection at the background of white Gaussian noise when the envelope of signal plus noise follows a Rician distribution.

8.5.2 INPUT DATA

Parameter	Dimension and Description
$P_{fa} := 10^{-5}$	false alarm probability required
$P_d := 0.75$	detection probability required
$n := 10$	number of pulses integrated

8.5.3 MODEL

8.5.3.1 Coherent Detectability Factor

$$D_c := \frac{1}{2} \cdot \left[qnorm\left[(1 - P_{fa}), 0, 1 \right] - qnorm\left[(1 - P_d), 0, 1 \right] \right]^2$$

$$D_{c\_db} := 10 \cdot \log(D_c) \quad D_{c\_db} = 10.863 \qquad \text{dB, coherent detectability factor}$$

8.5.3.2 Integration Loss

$$L_i := \frac{1 + \sqrt{1 + \dfrac{9.2 \cdot n}{D_c}}}{1 + \sqrt{1 + \dfrac{9.2}{D_c}}} \qquad \text{integration loss}$$

$$L_{i\_db} := 10 \cdot \log(L_i) \qquad L_{i\_db} = 2.273 \qquad \text{dB, integration loss}$$

8.5.4 OUTPUT DATA

Integration loss $L_{i\_db} = 2.273$ dB for input data specified.

8.5.5 REFERENCES

[1] Barton, D. K., and S. A. Leonov, (eds.), *Radar Technology Encyclopedia*, Norwood, MA: Artech House, 1997, p. 254.

[2] Barton, D. K., *Modern Radar System Analysis*, Norwood, MA: Artech House, 1988, p. 71.

8.6 MTI Processing Loss

8.6.1 DESCRIPTION

Calculates moving target indicator (MTI) processing loss resulting from correlation of noise at the MTI filter output and loss of target at certain velocities caused by clutter rejection notches (blind speeds). *Assumption*: coherent detection at the background of white Gaussian noise when the envelope of signal plus noise follows a Rician distribution; I/Q processing technique is used; at least a single blind speed occurs within the range of possible target velocities.

8.6.2 INPUT DATA

Parameter	Dimension and Description
$N := 3$	number of pulses employed by MTI filter
$P_{fa} := 10^{-4}$	false alarm probability required
$P_d := 0.75$	detection probability required
$n := 10$	number of pulses integrated
Swerling := 2	$0 \rightarrow$ Swerling case 0 (nonfluctuating target) $1 \rightarrow$ Swerling case 1 fluctuation model $2 \rightarrow$ Swerling case 2 fluctuation model $3 \rightarrow$ Swerling case 3 fluctuation model $4 \rightarrow$ Swerling case 4 fluctuation model
canceler := 1	$1 \rightarrow$ a single canceler is used $2 \rightarrow$ a dual canceler is used $3 \rightarrow$ a triple canceler is used
stagger := 2	$1 \rightarrow$ PRF stagger is not used $2 \rightarrow$ PRF stagger is used

8.6.3 MODEL

8.6.3.1 Swerling 0 Basic Detectability Factor

$$\text{erf(arg)} := \frac{2}{\sqrt{\pi}} \cdot \int_0^{\text{arg}} \exp(-t^2)\, dt$$

error function definition

$$y := 2 \cdot P_d - 1$$

the initial error function argument

$$u0 := -0.5$$

a guess value to solve an error function equation with respect to its argument

$\text{erf}(u0) = -0.52$

error function for the guess value

$u := \text{root}(\text{erf}(u0) - y, u0)$ $u = 0.477$

accurate solution for error function argument

$z := -u$ $z = -0.477$

a sign change in error function argument

$$D_0 := \left[\left(\sqrt{\ln\left(\frac{1}{P_{fa}}\right)} - z \right)^2 - 0.5 \right]$$

basic detectability factor for a single pulse and Swerling 0 (nonfluctuating model)

$D_0 = 11.833$

$D_{0\_db} := 10 \cdot \log(D_0)$

dB, basic Swerling 0 detectability factor

$D_{0\_db} = 10.731$

8.6.3.2 Fluctuation Loss

$$a := \begin{vmatrix} \dfrac{2}{3} & \text{if } \text{canceler} = 1 \\[2mm] \dfrac{18}{35} & \text{if } \text{canceler} = 2 \\[2mm] \dfrac{20}{47} & \text{if } \text{canceler} = 3 \end{vmatrix}$$

$a = 0.667$

noise correlation correction coefficient

$$D_1 := \frac{\ln(P_{fa})}{\ln(P_d)} - 1$$

basic detectability factor for a single pulse and Swerling 1 model

$D_{1\_db} := 10 \cdot \log(D_1)$

dB, basic Swerling 1 detectability factor

$D_{1\_db} = 14.916$

$$L_{f\_1} := \frac{D_1}{D_0}$$ $L_{f\_1} = 2.621$

fluctuation loss for Swerling case 1 target

$10 \cdot \log(L_{f\_1}) = 4.185$

$L_{f\_1\_db} := 10 \cdot \log(L_{f\_1}) \cdot (1 + 0.03 \cdot \log(n))$

dB, fluctuation loss for Swerling case 1 target

$L_{f\_1\_db} = 4.31$

Swerling $= 2$ Swerling case specified

$$K(\text{Swerling}) := \begin{vmatrix} 10^{10} & \text{if Swerling} = 0 \\ 1 & \text{if Swerling} = 1 \\ 1 & \text{if Swerling} = 2 \\ 2 & \text{if Swerling} = 3 \\ 2 & \text{if Swerling} = 4 \end{vmatrix}$$

one-half of the number
of degrees of freedom of
a chi-square distribution
describing target fluctuation
pdf

$$n_e(\text{Swerling}) := \begin{vmatrix} 10^{10} & \text{if Swerling} = 0 \\ 1 & \text{if Swerling} = 1 \\ n & \text{if Swerling} = 2 \\ 1 & \text{if Swerling} = 3 \\ n & \text{if Swerling} = 4 \end{vmatrix}$$

number of independent target
samples available for integration

$$n1_e(\text{Swerling}) := \begin{vmatrix} 10^{10} & \text{if Swerling} = 0 \\ 1 & \text{if Swerling} = 1 \\ a{\cdot}n & \text{if Swerling} = 2 \\ 1 & \text{if Swerling} = 3 \\ a{\cdot}n & \text{if Swerling} = 4 \end{vmatrix}$$

number of independent target
samples available for integration
after correction

$$L_{f\_db}(\text{Swerling}) := \frac{L_{f\_1\_db}}{K(\text{Swerling}){\cdot}n_e(\text{Swerling})}$$

dB, fluctuation loss

$L_{f\_db}(\text{Swerling}) = 0.431$

$$L1_{f\_db}(\text{Swerling}) := \frac{L_{f\_1\_db}}{K(\text{Swerling}){\cdot}n1_e(\text{Swerling})}$$

dB, fluctuation loss
after correction

$L1_{f\_db}(\text{Swerling}) = 0.647$

$$L_f(\text{Swerling}) := 10^{\frac{L_{f\_db}(\text{Swerling})}{10}}$$

$L_f(\text{Swerling}) = 1.104$ fluctuation loss

$$L1_f(\text{Swerling}) := 10^{\frac{L1_{f\_db}(\text{Swerling})}{10}} \qquad L1_f(\text{Swerling}) = 1.161 \qquad \text{fluctuation loss after correction}$$

8.6.3.3 Loss Resulting from Correlation of Noise at the MTI Filter Output

$$D_{Sw} := D_0 \cdot L_f(\text{Swerling}) \qquad D_{Sw} = 13.068 \qquad \text{basic detectability factor}$$

$$D1_{Sw} := D_0 \cdot L1_f(\text{Swerling}) \qquad D1_{Sw} = 13.733 \qquad \text{basic detectability factor after correction}$$

$$L_{mti\_a} := \frac{D1_{Sw}}{D_{Sw}} \qquad L_{mti\_a} = 1.051 \qquad \text{noise correlation loss}$$

$$L_{mti\_a\_db} := 10 \cdot \log(L_{mti\_a})$$

$$L_{mti\_a\_db} = 0.216 \qquad \text{dB, noise correlation loss}$$

8.6.3.4 The Velocity Response Loss

$$\text{factor} := \begin{vmatrix} \left[\dfrac{0.05}{\left[(1 - P_d)^2\right]} \right]^{N-1} & \text{if } \text{stagger} = 1 \\[4mm] \dfrac{0.1}{1 - P_d} & \text{if } \text{stagger} = 2 \end{vmatrix}$$

$$L_{mti\_b} := 0.8 + \text{factor} \qquad L_{mti\_b} = 1.2 \qquad \text{velocity response loss}$$

$$L_{mti\_b\_db} := 10 \cdot \log(L_{mti\_b}) \qquad \text{dB, velocity response loss}$$

$$L_{mti\_b\_db} = 0.792$$

Note: when the first blind speed exceeds the maximum target velocity, the velocity response loss is 0 dB.

8.6.3.5 MTI Processing Loss

$$\text{canceler} = 1 \qquad \text{canceler specified}$$

$$\text{stagger} = 2 \qquad \text{stagger specified}$$

$$L_{mti\_db} := L_{mti\_a\_db} + L_{mti\_b\_db} \qquad \text{dB, MTI processing loss}$$

8.6.4 OUTPUT DATA

MTI processing loss $L_{mti\_db} = 1.007$ dB for input data specified.

8.6.5 REFERENCES

[1] Barton, D. K., and S. A. Leonov, (eds.), *Radar Technology Encyclopedia*, Norwood, MA: Artech House, 1997, p. 255.
[2] Barton, D. K., and W. F. Barton, *Modern Radar System Analysis Software*, Norwood, MA: Artech House, 1993, pp. 134, 135.

8.7 Propagation Loss

8.7.1 DESCRIPTION
Calculates propagation loss as a sum of free-space path loss and pattern-propagation loss.

8.7.2 INPUT DATA

Parameter	Dimension and Description
f := 3000	MHz, radar frequency
R := 100	m, propagation path
F := 1	pattern-propagation factor

8.7.3 MODEL

8.7.3.1 Constants

$c := 2.997925 \cdot 10^8$ $\qquad$ m/s, speed of light

8.7.3.2 Propagation Loss

$\lambda := \dfrac{c}{f \cdot 10^6}$ $\qquad \lambda = 0.1$ $\qquad$ m, wavelength

$L_{fs} := 10 \cdot \log\left[\left(4 \cdot \pi \cdot \dfrac{R}{\lambda}\right)^2\right]$ $\qquad L_{fs} = 81.99$ $\qquad$ dB, free-space path loss

$L_{prp} := L_{fs} - 20 \cdot \log(F)$ $\qquad L_{prp} = 81.99$ $\qquad$ dB, propagation loss

8.7.4 OUTPUT DATA
Propagation loss $\qquad L_{prp} = 81.99$ $\qquad$ dB for input data specified

8.7.5 REFERENCE
[1] Morchin, W., *Radar Engineers Sourcebook*, Norwood, MA: Artech House, 1993.

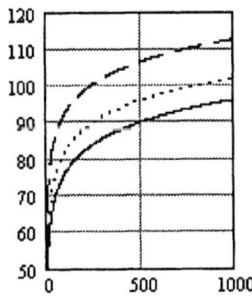

Figure 8.2 Free-space path loss (dB) vs. range (meters) for L band (solid line), S band (dotted line) and X band (dashed line).

8.8 Quantization Loss

8.8.1 DESCRIPTION

Calculates loss resulting from quantization of the receiver output data in analog-to-digital converter prior to digital processing.

8.8.2 INPUT DATA

===

Parameter	Dimension and Description
$m := 10$	number of bits in A/D converter

===

8.8.3 MODEL

$$L_{q\_db} := \frac{1.6}{m^2}$$
  dB, quantization loss

8.8.4 OUTPUT DATA

Quantization loss $L_{q\_db} = 0.016$ dB for input data specified.

8.8.5 REFERENCE

[1] Barton, D. K., and S. A. Leonov, (eds.), *Radar Technology Encyclopedia*, Norwood, MA: Artech House, 1997, p. 255.

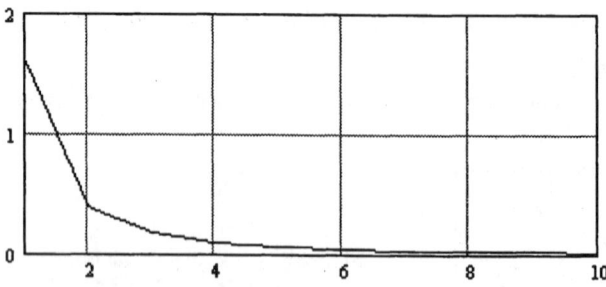

Figure 8.3 Quantization loss (dB) vs. number of bits in A/D converter.

8.9 Straddling Losses

8.9.1 DESCRIPTION

Calculates losses resulting from target position offset with respect to the system maximum response in angle, range and doppler frequency because of sampling in digital systems. Sometimes these losses are referred to as *sampling losses*.

8.9.2 INPUT DATA

==

Parameter	Dimension and Description

==

$P_d := 0.95$ — probability of detection

coordinate := 2

$1 \rightarrow$ angle
$2 \rightarrow$ range
$3 \rightarrow$ doppler

if coordinate = 1

$t_p := 5$ — µs, batch processing interval

$t_0 := 10$ — µs, integration time

Note: this loss results from noncoherent integration using batch processing in a scanning radar. For coherent integration beamshape loss has to be calculated rather than this loss.

if coordinate = 2

$\Delta t := 0.5$ — µs, sampling interval in time

$\tau := 1$ — µs, processed pulsewidth

if coordinate = 3

$\Delta f := 100$ — Hz, sampling interval in frequency

$B := 200$ — Hz, doppler filter bandwidth

==

8.9.3 MODEL

coordinate = 2 — coordinate specified

Equation parameters

$$a := \begin{vmatrix} t_p & \text{if} & \text{coordinate} = 1 \\ \Delta t & \text{if} & \text{coordinate} = 2 \\ \Delta f & \text{if} & \text{coordinate} = 3 \end{vmatrix} \qquad b := \begin{vmatrix} t_0 & \text{if} & \text{coordinate} = 1 \\ \tau & \text{if} & \text{coordinate} = 2 \\ B & \text{if} & \text{coordinate} = 3 \end{vmatrix}$$

$$a = 0.5 \qquad\qquad\qquad b = 1$$

$$u := \left(\frac{a}{b}\right)^2$$

$$L_{s\_db} := 1.25 \cdot P_d^{\frac{1}{3}} \cdot u$$

8.9.4 OUTPUT DATA

Straddling loss $L_{s\_db} = 0.307$ dB for input data specified.

8.9.5 REFERENCE

[1] Barton, D. K., and S. A. Leonov, (eds.), *Radar Technology Encyclopedia*, Norwood, MA: Artech House, 1997, pp. 245, 253, 256.

8.9.6 EXAMPLES

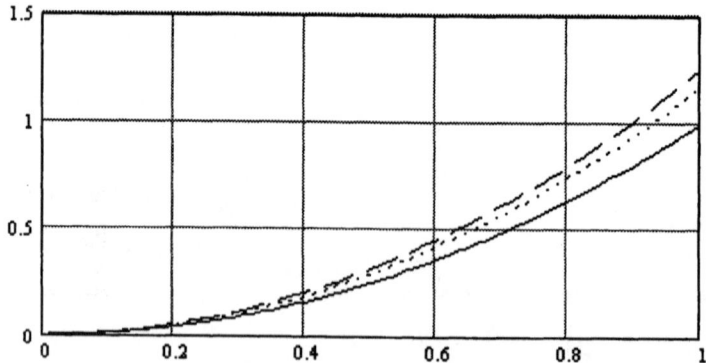

Figure 8.4 The sampling loss (dB) as a function of the ratio of the sampling interval to the duration of the resolution cell in the corresponding coordinate (Pd = 0.5 is a solid line, Pd = 0.8 is a dotted line, Pd = 0.99 is a dashed line).

8.10 Pulse Compression Weighting Loss

8.10.1 DESCRIPTION

Calculates pulse compression loss caused by use of a weighting function to bring the range sidelobes down. The weighting functions covered are Hamming window and Kaiser weighting.

8.10.2 INPUT DATA

Parameter	Dimension and Description
weight := 2	$0 \rightarrow$ no weighting applied $1 \rightarrow$ Hamming window $2 \rightarrow$ Kaiser weighting
N := 201	number of coefficients in the weighting function
Hamming window	
a0 := 0.53836	parameter #1
a1 := 1 − a0	parameter #2
Kaiser weighting	
α := 1	Kaiser parameter

8.10.3 MODEL

$$n := 0 .. N - 1 \qquad\qquad\qquad \text{cycle n}$$

$$n0 := \text{floor}\left(\frac{N}{2}\right) \qquad n0 = 100 \qquad \text{central coefficient number}$$

$$k_n := \frac{n - n0}{N} \qquad\qquad \text{normalized coefficients in Kaiser weighting}$$

$$I_0(z) := \frac{1}{2 \cdot \pi} \cdot \int_0^{2 \cdot \pi} \exp(z \cdot \cos(u)) \, du \qquad \text{Bessel function definition}$$

$$\text{weight} = 2 \qquad\qquad\qquad \text{weighting function specified}$$

$$c_n := \begin{vmatrix} 1 & \text{if } weight = 0 \\ a0 - a1 \cdot \cos\left(\dfrac{2 \cdot \pi \cdot n}{N}\right) & \text{if } weight = 1 \\ \dfrac{I_0\left[\pi \cdot \alpha \cdot \sqrt{1 - \left(\dfrac{k_n}{k_0}\right)^2}\right]}{I_0(\pi \cdot \alpha)} & \text{if } weight = 2 \end{vmatrix}$$

weighting coefficients

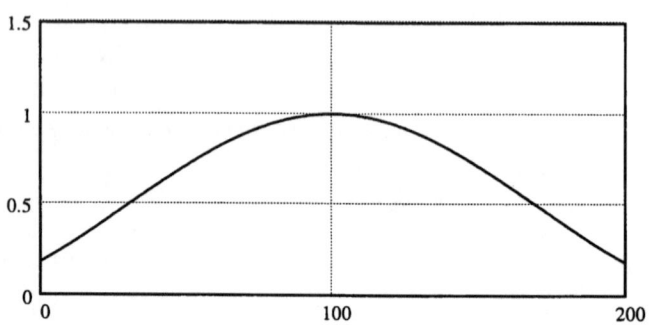

Figure 8.5 The weighting function vs. number of samples for weight = 2

$$L_w := \frac{N \cdot \left[\displaystyle\sum_n (c_n)^2\right]}{\left(\displaystyle\sum_n c_n\right)^2} \qquad L_w = 1.155$$

weighting loss

$$L_{w\_db} := 10 \cdot \log(L_w) \qquad L_{w\_db} = 0.624$$

dB, weighting loss

8.10.4 OUTPUT DATA

Pulse compression weighting loss $L_{w\_db} = 0.624$ dB for input data specified.

8.10.5 REFERENCE

[1] Barton, D. K., and S. A. Leonov, (eds.), *Radar Technology Encyclopedia*, Norwood, MA: Artech House, 1997, p. 258.

8.11 Clutter Distribution Loss

8.11.1 DESCRIPTION

Calculates loss that describes the increase in threshold level that results from peaks in the surface clutter, if false alarm probability is held constant with a two-parameter CFAR circuit or equivalent techniques and the clutter distribution differs from the Rayleigh distribution.

8.11.2 INPUT DATA

Parameter	Dimension and Description
distribution $:= 2$	$1 \rightarrow$ Rayleigh distribution $2 \rightarrow$ distribution other than Rayleigh
$P_{fa} := 10^{-4}$	false alarm probability
$R_{sc} := 10$	nm, surface clutter return range
$\tau_r := 0.5$	µs, received pulsewidth after processing
$\psi_3 := 1$	deg, azimuth beamwidth
$F_{sc} := 0.75$	pattern-propagation factor for the radar-clutter cell path

8.11.3 MODEL

8.11.3.1 Constants

$$c := 2.997925 \cdot 10^8 \qquad \text{m/s, speed of light}$$

$$z := \frac{\pi}{180} \qquad \text{coefficient to transform degrees to radians}$$

8.11.3.2 Weibull Clutter Spread Parameter

$$R_m := 1852 \cdot R_{sc} \qquad R_m = 1.852 \times 10^4 \qquad \text{m, clutter return range}$$

$$a := \begin{cases} 1 & \text{if distribution} = 1 \\[2ex] 1 + \dfrac{28 \cdot \left(\log\left(F_{sc}\right)\right)^2}{\log\left(\dfrac{0.75 \cdot R_m \cdot z \cdot \psi_3}{\tau_r \cdot 10^{-6} \cdot \dfrac{c}{2}}\right)} & \text{if distribution} = 2 \end{cases}$$

Weibull spread parameter

$$a = 1.857$$

8.11.3.3 Clutter Distribution Loss

$$L_{cd} := -6 \cdot \log\left(P_{fa}\right) \cdot (\log(a))^{1.2} \quad L_{cd} = 4.962 \qquad \text{dB, clutter distribution loss}$$

8.11.4 OUTPUT DATA

Clutter distribution loss $L_{cd} = 4.962$ dB for input data specified.

8.11.5 REFERENCES

[1] Barton, D. K., and S. A. Leonov, (eds.), *Radar Technology Encyclopedia*, Norwood, MA: Artech House, 1997, p. 250.
[2] Barton, D. K., and W. F. Barton, *Modern Radar System Analysis Software and User's Manual*, Norwood, MA: Artech House, 1993, pp. 163, 164.

Chapter 9

Maximum Detection Range

9.1 Maximum Detection Range for a Generic Pulsed Radar

9.1.1 DESCRIPTION

Calculates maximum detection range for a generic pulsed radar in an environment of thermal noise.

9.1.2 INPUT DATA

```
================================================================
```

Parameter   *Dimension and Description*

```
================================================================
```

Note: input data specification follow the typical inputs of the Blake chart. Thus sky temperature T_0, antenna loss L_a, transmission line loss in receiver channel L_r, and noise figure F_n are included in the system noise temperature T_s calculation; processing losses L_x (fluctuation loss, integration loss, CFAR loss, straddling losses, etc.) and processing gain G_x are included in the detectability factor D_x calculation while receiver mismatch loss M, transmission line loss in transmitter channel L_t, beamshape loss L_p, atmospheric attenuation loss L_α and other possible losses L_{oth} for the radar of interest (polarization loss, radome loss, etc.) are included directly in the radar range equation. Corresponding inputs for the maximum range calculation (T_s, L_α, L_p, F_n and others) should be calculated based on the corresponding models cited in Part 2.

Parameters to Calculate Free-Space Detection Range R_0

$P_t := 10000$ W, transmitter peak power

$\tau_t := 5$ μs, transmitted pulsewidth

$G_t := 40$ dB, transmit antenna gain

$G_r := 30$ dB, receive antenna gain

271

$\sigma := 5$ m^2, radar cross section of the target

$f := 2000$ MHz, radar frequency

$T_s := 650$ deg K, system noise temperature

$D_x := 10$ dB, detectability factor

$M := 1.0$ dB, receiver mismatch loss

$L_t := 2.0$ dB, transmission line loss in transmit channel

$L_p := 0.6$ dB, beamshape loss

$L_{oth} := 0$ dB, other losses

Parameters to Calculate Maximum Detection Range

Note: to calculate maximum detection range for atmospheric path in the presence of ground reflection effects follow the steps:

1) Maximum free-space detection range in vacuum R_0 has to be calculated using parameters cited above.

2) Pattern-propagation factor F and atmospheric loss L_α have to be calculated for the range R_0 using models cited in Sections 5.3 and 8.1, respectively.

3) Input parameters F and L_α found at step 2.

(See [2, p. 23] for a detailed description of how to go through the iteration process to modify a free-space range for the atmospheric path.)

$L_\alpha := 1.2$ dB, two-way atmospheric attenuation loss

$F := 1$ pattern-propagation factor

===

9.1.3 MODEL

9.1.3.1 Constants

$k := 1.38054 \cdot 10^{-23}$ W*s/°K, Boltzmann's constant

$c := 2.997925 \cdot 10^8$ m/s, speed of light

9.1.3.2 Transformation of Input Data

$\lambda := \dfrac{c}{f \cdot 10^6}$ $\lambda = 0.15$ m, wavelength

$ratio(x) := 10^{0.1 \cdot x}$ conversion equation

$G_{t\_dim} := ratio(G_t)$ $G_{t\_dim} = 1 \times 10^4$ transmit antenna gain

$$G_{r\_dim} := \text{ratio}(G_r) \qquad G_{r\_dim} = 1 \times 10^3 \qquad \text{receive antenna gain}$$

$$D_{x\_dim} := \text{ratio}(D_x) \qquad D_{x\_dim} = 10 \qquad \text{detectability factor}$$

$$M\_dim := \text{ratio}(M) \qquad M\_dim = 1.259 \qquad \text{receiver mismatch loss}$$

$$L_{t\_dim} := \text{ratio}(L_t) \qquad L_{t\_dim} = 1.585 \qquad \text{transmit loss}$$

$$L_{p\_dim} := \text{ratio}(L_p) \qquad L_{p\_dim} = 1.148 \qquad \text{beamshape loss}$$

$$L_{\alpha\_dim} := \text{ratio}(L_\alpha) \qquad L_{\alpha\_dim} = 1.318 \qquad \text{atmospheric loss}$$

$$L_{oth\_dim} := \text{ratio}(L_{oth}) \qquad L_{oth\_dim} = 1 \qquad \text{other losses}$$

9.1.3.3 Maximum Detection Range

$$\Pi_0 := \frac{P_t \cdot \tau_t \cdot 10^{-6} \cdot G_{t\_dim} \cdot G_{r\_dim} \cdot \lambda^2}{(4 \cdot \pi)^3 \cdot k \cdot T_s \cdot M\_dim \cdot L_{t\_dim} \cdot L_{p\_dim} \cdot L_{oth\_dim}} \qquad \begin{array}{l} \text{m}^2, \text{ free-space} \\ \text{radar potential} \end{array}$$

$$\Pi_0 = 2.754 \times 10^{20}$$

$$R_{0\_m} := \left(\frac{\Pi_0}{D_{x\_dim}} \cdot \sigma \right)^{\frac{1}{4}} \qquad \text{m, free-space maximum detection range in vacuum}$$

$$R_{0\_m} = 1.083 \times 10^5$$

Note: R_{0_m} is the free-space detection range in vacuum that has to be used to calculate pattern-propagation factor F and atmospheric loss L_α.

$$\Pi := \Pi_0 \cdot \frac{F^4}{L_{\alpha\_dim}} \qquad \Pi = 2.089 \times 10^{20} \qquad \text{m}^2, \text{ radar potential}$$

$$R_m := \left(\frac{\Pi}{D_{x\_dim}} \cdot \sigma \right)^{\frac{1}{4}} \qquad R_{0\_m} = 1.083 \times 10^5 \qquad \text{m, maximum detection range}$$

$$R_{km} := \frac{R_m}{1000} \qquad R_{km} = 101.096 \qquad \text{km, maximum detection range}$$

$$R_{nm} := \frac{R_m}{1852} \qquad R_{nm} = 54.587 \qquad \text{nm, maximum detection range}$$

9.1.4 OUTPUT DATA

Maximum detection range $R_{km} = 101.1$ km or $R_{nm} = 54.6$ nm

for input data specified

9.1.5 REFERENCES

[1] Barton, D. K., and S. A. Leonov, (eds.), *Radar Technology Encyclopedia*, Norwood, MA: Artech House, 1997, pp. 307, 378.

[2] Barton, D. K., *Modern Radar System Analysis*, Norwood, MA: Artech House, 1988, pp. 22, 23.

9.1.6 EXAMPLES

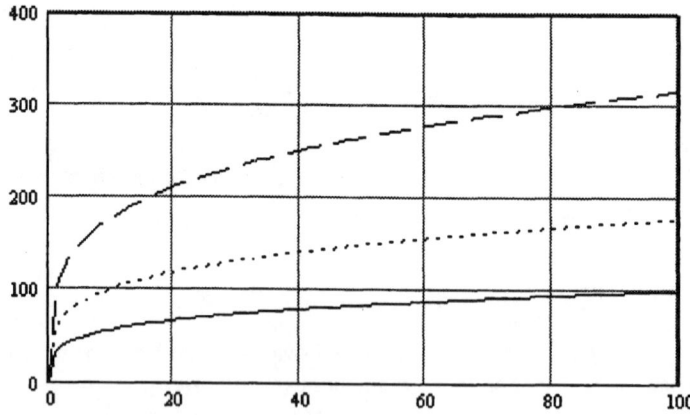

Figure 9.1 Maximum detection range (km) vs. radar potential (10^{20} m^2) for detectability factor 10 dB and target RCS 0.1 m^2 (solid line), 1 m^2 (dotted line) and 10 m^2 (dashed line).

9.2 Maximum Detection Range for a Generic Pulsed Radar in a Clutter Environment

9.2.1 DESCRIPTION
Calculates maximum detection range for a generic pulsed radar in an environment of surface and volume clutter. *Assumptions*: unambigous clutter range returns.

9.2.2 INPUT DATA

Parameter	Dimension and Description
Radar and Target Parameters	
$P_t := 10000$	W, transmitter peak power
$\tau_t := 5$	μs, transmitted pulsewidth
$\tau_r := 0.5$	μs, received pulsewidth after processing
$\psi_3 := 1$	deg, azimuth beamwidth
$\theta_3 := 3$	deg, elevation beamwidth
$G_{t\_t} := 40$	dB, transmit antenna gain in direction to the target
$G_{r\_t} := 30$	dB, receive antenna gain in direction to the target
$\sigma := 5$	m², radar cross section of the target
$f := 2000$	MHz, radar frequency
$T_s := 650$	°K, system noise temperature
$D_x := 20$	dB, detectability factor for noise environment
$M := 1.0$	dB, receiver mismatch loss
$L_{tx} := 2.0$	dB, transmission line loss in transmit channel
$L_p := 0.6$	dB, beamshape loss
$L_{oth} := 0$	dB, other losses

Note: follow the rules cited in Section 9.1 to calculate parameters F_t and L_{α_t}.

$F_t := 1$	pattern-propagation factor for the radar-target path
$L_{\alpha\_t} := 1.2$	dB, two-way atmospheric attenuation loss for the radar-target path

Clutter Environment

clutter := 1

$1 \to$ surface clutter environment

$2 \to$ volume clutter environment

Parameters for Surface Clutter Environment

$\sigma 0\_db := 1$

dB, clutter mean reflectivity

$G_{t\_sc} := 10$

dB, transmit antenna gain in direction to the clutter cell

$G_{r\_sc} := 10$

dB, receive antenna gain in direction to the clutter cell

$I_{sc} := 20$

dB, improvement factor in surface clutter environment

$L_{sc} := 3$

dB, surface clutter distribution loss

$L_{\alpha\_sc} := 0.3$

dB, two-way atmospheric attenuation loss for the radar-clutter cell path

$F_{sc} := 0.75$

pattern-propagation factor for the radar-clutter cell path

Note: the correspondence between clutter mean reflectivity and types of terrain or sea roughness is given in Section 11.3, Signal-to-Interference Ratio.

Parameters for Volume Clutter Environment

$G_{t\_vc} := 40$

dB, transmit antenna gain in direction to the clutter cell

$G_{r\_vc} := 30$

dB, receive antenna gain in direction to the clutter cell

$I_{vc} := 20$

dB, improvement factor in volume clutter environment

$ICR := 0$

dB, integrated cancellation ratio

Note: typically integrated cancellation ratio is equal to 0 dB for linear polarization and 15–20 dB for circular polarization. If circular polarization is chosen, a 3-dB loss should be included as L_{oth} in the Radar and Target Parameters section of the input menu.

$L_{\alpha\_vc} := 0.9$

dB, two-way atmospheric attenuation loss for the radar-clutter cell path

$F_{vc} := 1$

pattern-propagation factor for the radar-clutter cell path

Precipitation Parameters

$r := 4$

mm/hour, precipitation rate

$a := 200$

precipitation model parameter

$b := 1.6$

precipitation model parameter

$K := 0.93$

precipitation model parameter

Note: the table of values a, b, and K for different types of precipitation is given in Section 11.3, Signal-to-Interference Ratio.

===

9.2.3 MODEL

9.2.3.1 Constants

$k := 1.38054 \cdot 10^{-23}$ W*s/°K, Boltzmann's constant

$c := 2.997925 \cdot 10^{8}$ m/s, speed of light

$z := \dfrac{\pi}{180}$ coefficient to transform degrees to radians

9.2.3.2 Transformation of Input Data

$\lambda := \dfrac{c}{f \cdot 10^{6}} \qquad \lambda = 0.15$ m, wavelength

$ratio(x) := 10^{0.1 \cdot x}$ conversion equation

$G_{t\_t\_dim} := ratio(G_{t\_t}) \qquad G_{t\_t\_dim} = 1 \times 10^{4}$ transmit antenna gain in direction to the target

$G_{r\_t\_dim} := ratio(G_{r\_t}) \qquad G_{r\_t\_dim} = 1 \times 10^{3}$ receive antenna gain in direction to the target

$G_{t\_sc\_dim} := ratio(G_{t\_sc}) \qquad G_{t\_sc\_dim} = 10$ transmit antenna gain in direction to the surface clutter cell

$G_{r\_sc\_dim} := ratio(G_{r\_sc}) \qquad G_{r\_sc\_dim} = 10$ receive antenna gain in direction to the surface clutter cell

$G_{t\_vc\_dim} := ratio(G_{t\_vc}) \qquad G_{t\_vc\_dim} = 1 \times 10^{4}$ transmit antenna gain in direction to the volume clutter cell

$G_{r\_vc\_dim} := ratio(G_{r\_vc}) \qquad G_{r\_vc\_dim} = 1 \times 10^{3}$ receive antenna gain in direction to the volume clutter cell

$D_{x\_dim} := ratio(D_{x}) \qquad D_{x\_dim} = 100$ detectability factor

$M\_dim := ratio(M) \qquad M\_dim = 1.259$ receiver mismatch loss

$L_{tx\_dim} := ratio(L_{tx}) \qquad L_{tx\_dim} = 1.585$ transmit loss

$L_{p\_dim} := \text{ratio}(L_p)$ $L_{p\_dim} = 1.148$ beamshape loss

$L_{\alpha\_t\_dim} := \text{ratio}(L_{\alpha\_t})$ $L_{\alpha\_t\_dim} = 1.318$ atmospheric loss for radar-target path

$L_{\alpha\_sc\_dim} := \text{ratio}(L_{\alpha\_sc})$ $L_{\alpha\_sc\_dim} = 1.072$ atmospheric loss for radar-surface clutter path

$L_{\alpha\_vc\_dim} := \text{ratio}(L_{\alpha\_vc})$ $L_{\alpha\_vc\_dim} = 1.23$ atmospheric loss for radar-volume clutter path

$L_{oth\_dim} := \text{ratio}(L_{oth})$ $L_{oth\_dim} = 1$ other losses

$L_{c\_dim} := \text{ratio}(L_{sc})$ $L_{c\_dim} = 1.995$ clutter distribution loss

$I_{m\_dim} := \begin{vmatrix} \text{ratio}(I_{sc}) & \text{if clutter} = 1 \\ \text{ratio}(I_{vc}) & \text{if clutter} = 2 \end{vmatrix}$ $I_{m\_dim} = 100$ improvement factor

$\sigma 0 := \text{ratio}(\sigma 0\_db)$ $\sigma 0 = 1.259$ m²/m², surface clutter reflectivity

$\text{ICR\_dim} := \text{ratio}(\text{ICR})$ $\text{ICR\_dim} = 1$ integrated cancellation ratio

9.2.3.3 Maximum Detection Range in Noise Environment

$$\Pi_0 := \frac{P_t \cdot \tau_t \cdot 10^{-6} \cdot G_{t\_t\_dim} \cdot G_{r\_t\_dim} \cdot \lambda^2}{(4 \cdot \pi)^3 \cdot k \cdot T_s \cdot M\_dim \cdot L_{tx\_dim} \cdot L_{p\_dim} \cdot L_{oth\_dim}}$$ m², free-space radar potential

$$\Pi_0 = 2.754 \times 10^{20}$$

$$R_m := (\Pi_0 \cdot \sigma)^{\frac{1}{4}}$$ m, free-space maximum detection range in vacuum for $D_x = 0$ dB (MDS)

$$R_{km} := \frac{R_m}{1000}$$ $R_{km} = 192.634$ km, maximum detection range

$$R_{nm} := \frac{R_m}{1852}$$ $R_{nm} = 104.014$ nm, maximum detection range

$R_{max\_nm} := \text{ceil}(R_{nm})$ $R_{max\_nm} = 105$ nm, maximum possible MDS detection range

$R := 1 .. R_{max\_nm}$ cycle

9.2.3.4 Clutter + Noise Energy

$\sigma_{sc}(R) := \sigma 0$ $\qquad$ m2/m2, surface clutter reflectivity as a function of range

Antenna gains as a function of range
(assumed averaged over range in this example)

$G_{t\_t\_dim}(R) := G_{t\_t\_dim}$ $\qquad$ $G_{t\_sc\_dim}(R) := G_{t\_sc\_dim}$ $\qquad$ $G_{t\_vc\_dim}(R) := G_{t\_vc\_dim}$

$G_{r\_t\_dim}(R) := G_{r\_t\_dim}$ $\qquad$ $G_{r\_sc\_dim}(R) := G_{r\_sc\_dim}$ $\qquad$ $G_{r\_vc\_dim}(R) := G_{r\_vc\_dim}$

Pattern-propagation factor as a function of range
(assumed averaged over range in this example)

$F_t(R) := F_t$ $\qquad$ $F_{sc}(R) := F_{sc}$ $\qquad$ $F_{vc}(R) := F_{vc}$

Atmospheric loss as a function of range
(assumed averaged over range in this example)

$L_{\alpha\_t\_dim}(R) := L_{\alpha\_t\_dim}$ $\qquad$ $L_{\alpha\_sc\_dim}(R) := L_{\alpha\_sc\_dim}$ $\qquad$ $L_{\alpha\_vc\_dim}(R) := L_{\alpha\_vc\_dim}$

Note: for the sake of simplicity constant-range models within coverage volume are assumed for clutter mean reflectivity, antenna gains, pattern-propagation factor and atmospheric loss. A user can specifiy his own model of clutter reflectivity as a function of range if this is the case for a specific clutter environment. Antenna gains can be calculated as the function of elevation angle for each range increment if antenna pattern vs. elevation angle is known. Parameters $F_{sc}(R)$ and $L_{sc\_dim}(R)$ can be calculated as a matrix for each increment in R using models cited in Sections 5.3 and 8.1. These data can be input instead of constant-range values for more accurate calculation of maximum detection range in clutter environment.

W*s, range-independent factor for radar input surface clutter energy

$$E0_{sc\_in} := \frac{P_t \cdot \tau_t \cdot 10^{-6} \cdot \tau_r \cdot 10^{-6} \cdot \lambda^2 \cdot c \cdot z \cdot \psi_3 \cdot L_{c\_dim}}{(4 \cdot \pi)^3 \cdot 2 \cdot M\_dim \cdot L_{tx\_dim} \cdot T_{p\_dim} \cdot T_{oth\_dim}}$$

W*s, radar input surface clutter energy

$$E_{sc\_in}(R) := \frac{E0_{sc\_in} \cdot G_{t\_sc\_dim}(R) \cdot G_{r\_sc\_dim}(R) \cdot \sigma_{sc}(R) \cdot F_{sc}(R)^4}{L_{\alpha\_sc\_dim}(R) \cdot (R \cdot 1852)^3}$$

$$\eta_v := \frac{\dfrac{\pi^5}{\lambda^4} \cdot K \cdot a \cdot 10^{-18} \cdot r^b}{ICR\_dim} \qquad \eta_v = 1.036 \times 10^{-9} \qquad \text{m2/m3, volume precipitation reflectivity}$$

W*s, range-independent factor for radar input volume clutter energy

$$EO_{vc\_in} := \frac{P_t \cdot \tau_t \cdot 10^{-6} \cdot \lambda^2 \cdot \eta_v \cdot z \cdot \theta_3 \cdot \tau_r \cdot 10^{-6} \cdot c \cdot z \cdot \psi_3}{2 \cdot (4 \cdot \pi)^3 \cdot M\_dim \cdot L_{tx\_dim} \cdot L_{p\_dim}^2 \cdot L_{oth\_dim}}$$

W*s, radar input volume clutter energy

$$E_{vc\_in}(R) := \frac{EO_{vc\_in} \cdot G_{t\_vc\_dim}(R) \cdot G_{r\_vc\_dim}(R) \cdot F_{vc}(R)^4}{L_{\alpha\_vc\_dim}(R) \cdot (R \cdot 1852)^2}$$

clutter = 1   type of clutter specified

$$E_{c\_in}(R) := \begin{vmatrix} E_{sc\_in}(R) & \text{if } clutter = 1 \\ E_{vc\_in}(R) & \text{if } clutter = 2 \end{vmatrix}$$ W*s, radar input clutter energy

$$E_{c\_out}(R) := \frac{E_{c\_in}(R)}{I_{m\_dim}}$$ W*s, radar output clutter energy

$$E_{c\_out\_db}(R) := 10 \cdot \log\left(E_{c\_out}(R) + 10^{-100}\right)$$ dB, radar output clutter energy

$$E_n := k \cdot T_s \qquad\qquad E_n \cdot 10^{20} = 0.897$$ W*s, thermal noise energy

$$E_{n\_db} := 10 \cdot \log\left(E_n + 10^{-100}\right)$$ dB, thermal noise energy

$$E_{cn}(R) := E_n + E_{c\_out}(R)$$ W*s, combined clutter + noise energy

$$E_{cn\_db}(R) := 10 \cdot \log\left(E_{cn}(R) + 10^{-100}\right)$$ dB, combined clutter + noise energy

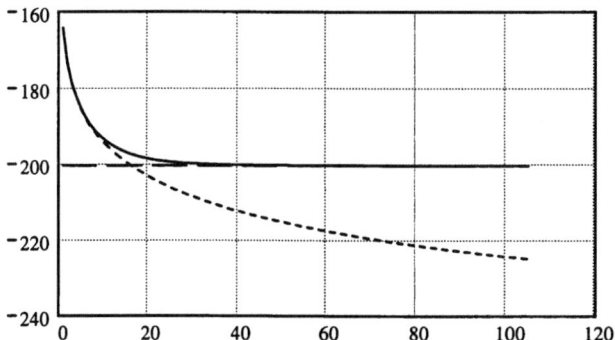

Figure 9.2 Clutter energy (dB, dotted line), noise energy (dB, dashed line), and clutter + noise energy (dB, solid line) vs. radar range (nm).

9.2.3.5 Signal Energy

W*s, range-independent factor for signal energy

$$E0_s := \frac{P_t \cdot \tau_t \cdot 10^{-6} \cdot \lambda^2 \cdot \sigma}{(4 \cdot \pi)^3 \cdot M\_dim \cdot L_{tx\_dim} \cdot L_{p\_dim} \cdot L_{oth\_dim}}$$

W*s, signal energy as a function of range

$$E_s(R) := \frac{E0_s \cdot G_{t\_t\_dim}(R) \cdot G_{r\_t\_dim}(R) \cdot F_t(R)^4}{L_{\alpha\_t\_dim}(R) \cdot (R \cdot 1852)^4}$$

$$E_{s\_db}(R) := 10 \cdot \log\left(E_s(R) + 10^{-100}\right) \qquad \text{dB, signal energy as a function of range}$$

9.2.3.6 Maximum Detection Range

$$E_{s\_cn\_db}(R) := 10 \cdot \log\left(10^{-100} + \frac{E_s(R)}{E_{cn}(R)}\right) \qquad \text{dB, signal-to-clutter+noise ratio}$$

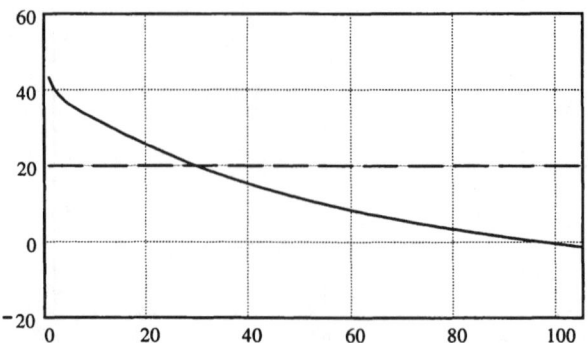

Figure 9.3 Detection threshold (dB, dotted line) and signal-to-clutter + noise energy (dB, solid line) vs. radar range (nm).

$$\text{DETECTION}_R := \begin{vmatrix} 1 & \text{if } E_{s\_cn\_db}(R) \geq D_x \\ 0 & \text{otherwise} \end{vmatrix} \qquad \text{detection criterion}$$

$$\text{DETECTION}_0 := 1$$

Maximum detection range in clutter, nm

$$R_{nm} := \begin{vmatrix} \text{for } R \in 1 .. R_{max\_nm} \\ \quad \begin{vmatrix} R_{nm} \leftarrow R - 1 & \text{if } \text{DETECTION}_R = 0 \\ \text{break if } \text{DETECTION}_{R-1} = 1 & \text{if } \text{DETECTION}_R = 0 \end{vmatrix} \\ R_{nm} \end{vmatrix}$$

$$R_{nm} = 29 \qquad\qquad\qquad \text{nm, maximum detection range}$$

$$R_{km} := R_{nm} \cdot 1.852 \qquad R_{km} = 53.7 \qquad\qquad \text{km, maximum detection range}$$

9.2.4 OUTPUT DATA

Maximum detection range $R_{km} = 53.7$ km or $R_{nm} = 29$ nm for input data specified.

9.2.5 REFERENCES

[1] Barton, D. K., and S. A. Leonov, (eds.), *Radar Technology Encyclopedia*, Norwood, MA: Artech House, 1997, p. 380.

[2] Barton, D. K., and W. F. Barton, *Modern Radar System Analysis Software and User's Manual*, Norwood, MA: Artech House, 1993, pp. 154–165.

9.3 Maximum Detection Range for a Generic Pulsed Radar in a Jamming Environment

9.3.1 DESCRIPTION

Calculates maximum detection range for a generic pulsed radar in an environment of organized interference (jamming). *Assumption*: interference is provided by a noise jammer at a fixed range separated from the target.

9.3.2 INPUT DATA

==

Parameter	*Dimension and Description*

==

Radar and Target Parameters

$P_t := 10 \cdot 10^3$	W, transmitter peak power
$\tau_t := 5$	μs, transmitted pulsewidth
$G_{t\_t} := 40$	dB, transmit antenna gain in direction to the target
$G_{r\_t} := 30$	dB, receive antenna gain in direction to the target
$\sigma := 5$	m2, radar cross section of the target
$f := 2000$	MHz, radar frequency
$T_s := 650$	°K, system noise temperature
$D_x := 10$	dB, detectability factor for noise environment
$M := 1.0$	dB, receiver mismatch loss
$L_{tx} := 1.2$	dB, transmission line loss in transmit channel
$L_p := 1.6$	dB, beamshape loss
$L_x := 8$	dB, processing loss
$G_x := 7$	dB, processing gain
$L_{oth} := 0$	dB, other losses

Note: follow the rules cited in Section 9.1 to calculate parameters F_t , L_{α_t} .

$L_{\alpha\_t} := 1.2$	dB, two-way atmospheric attenuation loss for the radar-target path
$F_t := 1$	pattern-propagation factor for the radar-target path
jamming := 2	1 → active jamming 2 → passive jamming (chaff)

Active Jamming Parameters

$P_{aj} := 10^6$ W, jammer peak power

$G_{aj} := 0$ dB, jammer antenna gain in direction to the radar

$G_{r\_aj} := 0$ dB, radar receive antenna gain in direction to the jammer

$B_{aj} := 10^3$ MHz, jammer noise bandwidth

$L_{\alpha\_aj} := 2.5$ dB, two-way atmospheric attenuation loss for the
                                      jammer-to-radar path

$F_{aj} := 1$ pattern-propagation factor for the jammer-to-radar path

Passive Jamming (Chaff) Parameters

$N := 1000$ total number of dipoles

$G_{t\_pj} := 40$ dB, transmit antenna gain in direction to chaff

$G_{r\_pj} := 30$ dB, receive antenna gain in direction to chaff

$L_{\alpha\_pj} := 1.5$ dB, two-way atmospheric attenuation loss for the
                                      chaff-to-radar path

$F_{pj} := 1$ pattern-propagation factor for the chaff-to-radar path

===

9.3.3 MODEL

9.3.3.1 Constants

$k := 1.38054 \cdot 10^{-23}$ W*s/°K, Boltzmann's constant

$c := 2.997925 \cdot 10^8$ m/s, speed of light

9.3.3.2 Total loss

$L_{\Sigma\_t} := M + L_{tx} + L_p + L_{oth}$ dB, total range-independent loss
                                                      for radar-target path
$L_{\Sigma\_t} = 3.8$

$L_{\Sigma\_j} := M + L_{tx} + L_p + L_x + L_{oth} - G_x$ dB, total range-independent loss
                                                                   for jammer (chaff)-radar path
$L_{\Sigma\_j} = 4.8$

Note: processing gain G_x and processing loss L_x (a sum of fluctuation loss, integration loss, CFAR loss, straddling losses, etc.) in target maximum detection range calculation in noise environment are typically included in detectability factor D_x.

9.3.3.3 Transformation of Input Data

$$\lambda := \frac{c}{f \cdot 10^6} \qquad \lambda = 0.15 \qquad\qquad\qquad \text{m, radar wavelength}$$

$$\text{ratio}(x) := 10^{0.1 \cdot x} \qquad\qquad\qquad\qquad\qquad \text{conversion equation}$$

$$G_{t\_t\_dim} := \text{ratio}\left(G_{t\_t}\right) \qquad G_{t\_t\_dim} = 1 \times 10^4 \qquad \begin{array}{r}\text{radar transmit antenna gain} \\ \text{in direction to the target}\end{array}$$

$$G_{r\_t\_dim} := \text{ratio}\left(G_{r\_t}\right) \qquad G_{r\_t\_dim} = 1 \times 10^3 \qquad \begin{array}{r}\text{radar receive antenna gain} \\ \text{in direction to the target}\end{array}$$

$$G_{aj\_dim} := \text{ratio}\left(G_{aj}\right) \qquad G_{aj\_dim} = 1 \qquad \begin{array}{r}\text{jammer antenna gain in direction} \\ \text{to the radar}\end{array}$$

$$G_{r\_aj\_dim} := \text{ratio}\left(G_{r\_aj}\right) \qquad G_{r\_aj\_dim} = 1 \qquad \begin{array}{r}\text{receive antenna gain in direction to} \\ \text{the jammer}\end{array}$$

$$G_{t\_pj\_dim} := \text{ratio}\left(G_{t\_pj}\right) \qquad G_{t\_pj\_dim} = 1 \times 10^4 \qquad \begin{array}{r}\text{transmit antenna gain in direction} \\ \text{to chaff}\end{array}$$

$$G_{r\_pj\_dim} := \text{ratio}\left(G_{r\_pj}\right) \qquad G_{r\_pj\_dim} = 1 \times 10^3 \qquad \begin{array}{r}\text{receive antenna gain in direction} \\ \text{to chaff}\end{array}$$

$$D_{x\_dim} := \text{ratio}\left(D_x\right) \qquad D_{x\_dim} = 10 \qquad\qquad \text{detectability factor}$$

$$L_{\Sigma\_t\_dim} := \text{ratio}\left(L_{\Sigma\_t}\right) \qquad L_{\Sigma\_t\_dim} = 2.399 \qquad \begin{array}{r}\text{total range-independent loss} \\ \text{for radar-target path}\end{array}$$

$$L_{\alpha\_t\_dim} := \text{ratio}\left(L_{\alpha\_t}\right) \qquad L_{\alpha\_t\_dim} = 1.318 \qquad \begin{array}{r}\text{atmospheric loss for radar-target} \\ \text{path}\end{array}$$

$$L_{\Sigma\_j\_dim} := \text{ratio}\left(L_{\Sigma\_j}\right) \qquad L_{\Sigma\_j\_dim} = 3.02 \qquad \begin{array}{r}\text{total range-independent loss} \\ \text{for jammer (chaff)-radar path}\end{array}$$

$$L_{\alpha\_aj\_dim} := \text{ratio}\left(L_{\alpha\_aj}\right) \qquad L_{\alpha\_aj\_dim} = 1.778 \qquad \begin{array}{r}\text{atmospheric loss for jammer-radar} \\ \text{path}\end{array}$$

$$L_{\alpha\_pj\_dim} := \text{ratio}\left(L_{\alpha\_pj}\right) \qquad L_{\alpha\_pj\_dim} = 1.413 \qquad \begin{array}{r}\text{atmospheric loss for chaff-radar} \\ \text{path}\end{array}$$

9.3.3.4 Maximum Detection Range in Noise Environment

$$\Pi_0 := \frac{P_t \cdot \tau_t \cdot 10^{-6} \cdot G_{t\_t\_dim} \cdot G_{r\_t\_dim} \cdot \lambda^2}{\left(4 \cdot \pi\right)^3 \cdot k \cdot T_s \cdot L_{\Sigma\_t\_dim}} \qquad\qquad \begin{array}{r}\text{m}^2\text{, free-space} \\ \text{radar potential}\end{array}$$

$$\Pi_0 = 2.63 \times 10^{20}$$

$$R_m := (\Pi_0 \cdot \sigma)^{\frac{1}{4}}$$ m, free-space maximum detection range in vacuum for $D_x = 0$ dB (MDS)

$$R_{km} := \frac{R_m}{1000}$$ $R_{km} = 190.429$ km, maximum detection range

$$R_{nm} := \frac{R_m}{1852}$$ $R_{nm} = 102.823$ nm, maximum detection range

$$R_{max\_nm} := ceil(R_{nm})$$ $R_{max\_nm} = 103$ nm, maximum possible MDS detection range

$$R := 1 .. R_{max\_nm}$$ cycle

Antenna gains as a function of range
(assumed averaged over range in this example)

$$G_{t\_t\_dim}(R) := G_{t\_t\_dim} \qquad G_{aj\_dim}(R) := G_{aj\_dim} \qquad G_{t\_pj\_dim}(R) := G_{t\_pj\_dim}$$

$$G_{r\_t\_dim}(R) := G_{r\_t\_dim} \qquad G_{r\_aj\_dim}(R) := G_{r\_aj\_dim} \qquad G_{r\_pj\_dim}(R) := G_{r\_pj\_dim}$$

Pattern-propagation factor as a function of range
(assumed averaged over range in this example)

$$F_t(R) := F_t \qquad F_{aj}(R) := F_{aj} \qquad F_{pj}(R) := F_{pj}$$

Atmospheric loss as a function of range
(assumed averaged over range in this example)

$$L_{\alpha\_t\_dim}(R) := L_{\alpha\_t\_dim} \qquad L_{\alpha\_aj\_dim}(R) := L_{\alpha\_aj\_dim} \qquad L_{\alpha\_pj\_dim}(R) := L_{\alpha\_pj\_dim}$$

Note: for the sake of simplicity constant-range models within coverage volume are assumed for antenna gains, pattern-propagation factor and atmospheric loss. Antenna gains can be calculated as the function of elevation angle for each range increment if antenna pattern vs. elevation angle is known. Parameters F(R) and L(R) can be calculated as a matrix for each increment in R using models cited in Sections 5.3 and 8.1. These data can be input instead of constant-range values for more accurate calculation of maximum detection range in jamming environment.

9.3.3.5 Active Jamming Energy

W*s, range-independent factor for jammer power spectral density

$$E0_{aj} := \frac{P_{aj} \cdot \lambda^2}{(4 \cdot \pi)^2 \cdot B_{aj} \cdot 10^6 \cdot L_{\Sigma\_j\_dim}}$$

W*s, jammer power spectral density as a function of range

$$E_{aj}(R) := E0_{aj} \cdot \frac{G_{aj\_dim}(R) \cdot G_{r\_aj\_dim}(R) \cdot F_{aj}(R)^2}{L_{\alpha\_aj\_dim}(R) \cdot (R \cdot 1852)^2}$$

9.3.3.6 Passive Jamming Energy

$$\sigma_d := 0.18 \cdot \lambda^2 \cdot N \qquad \sigma_d = 4.044 \qquad \text{m2, RCS of chaff dipoles}$$

W*s, range-inependent factor for chaff return power spectral density

$$E0_{pj} := \frac{P_t \cdot \tau_t \cdot 10^{-6} \cdot \lambda^2 \cdot \sigma_d}{(4 \cdot \pi)^3 \cdot L_{\Sigma\_j\_dim}}$$

W*s, chaff return power spectral density as a function of range

$$E_{pj}(R) := E0_{pj} \cdot \frac{\left(G_{t\_pj\_dim}(R) \cdot G_{r\_pj\_dim}(R) \cdot F_{pj}(R)^4 \right)}{\left[L_{\alpha\_pj\_dim}(R) \cdot (R \cdot 1852)^4 \right]}$$

jamming = 2   type of jamming specified

$$E_j(R) := \begin{vmatrix} E_{aj}(R) & \text{if } jamming = 1 \\ E_{pj}(R) & \text{if } jamming = 2 \end{vmatrix}$$ W*s, jamming energy

$$E_{j\_db}(R) := 10 \cdot \log\left(E_j(R) + 10^{-100} \right)$$ dB, jamming energy

$$E_n := k \cdot T_s \qquad E_n \cdot 10^{20} = 0.897$$ W*s, thermal noise energy

$$E_{n\_db} := 10 \cdot \log\left(E_n + 10^{-100} \right)$$ dB, thermal noise energy

$$E_{jn}(R) := E_n + E_j(R)$$ W*s, combined jamming + noise

$$E_{jn\_db}(R) := 10 \cdot \log\left(E_{jn}(R) + 10^{-100}\right) \qquad \text{dB, combined jamming + noise}$$

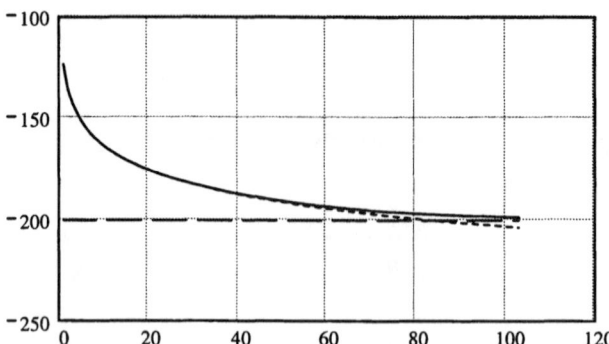

Figure 9.4 Jamming energy (dB, dotted line), noise energy (dB, dashed line), and jamming + noise energy (dB, solid) line vs. radar range (nm).

9.3.3.7 Signal Energy

W*s, range-independent factor for signal energy

$$E0_s := \frac{P_t \cdot \tau_t \cdot 10^{-6} \cdot \lambda^2 \cdot \sigma}{(4 \cdot \pi)^3 \cdot L_{\Sigma\_t\_dim}}$$

W*s, signal energy as a function of range

$$E_s(R) := \frac{E0_s \cdot G_{t\_t\_dim}(R) \cdot G_{r\_t\_dim}(R) \cdot F_t(R)^4}{L_{\alpha\_t\_dim}(R) \cdot (R \cdot 1852)^4}$$

$$E_{s\_db}(R) := 10 \cdot \log\left(E_s(R) + 10^{-100}\right) \quad \text{dB, signal energy as a function of range}$$

9.3.3.8 Maximum Detection Range

$$E_{s\_jn\_db}(R) := 10 \cdot \log\left(10^{-100} + \frac{E_s(R)}{E_{jn}(R)}\right) \qquad \text{dB, signal-to-jamming + noise ratio}$$

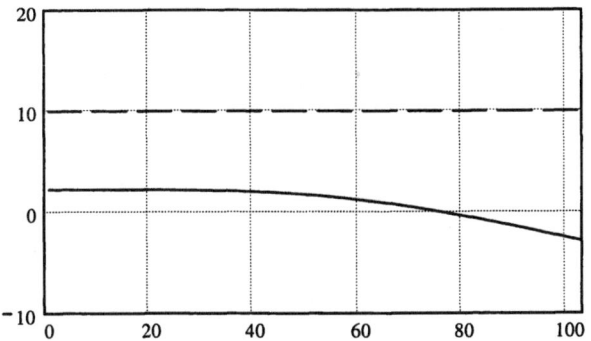

Figure 9.5 Detection threshold (dB, dotted line) and signal-to-jamming + noise energy (dB, solid line) vs. radar range (nm).

$$\text{DETECTION}_R := \begin{vmatrix} 1 & \text{if } E_{s\_jn\_db}(R) \geq D_x \\ 0 & \text{otherwise} \end{vmatrix} \qquad \text{detection criterion}$$

$$\text{DETECTION}_0 := 1$$

Maximum detection range in jamming, nm

$$R_{nm} := \begin{vmatrix} \text{for } R \in 1 .. R_{max\_nm} \\ \begin{vmatrix} R_{nm} \leftarrow R - 1 & \text{if } \text{DETECTION}_R = 0 \\ \text{break if } \text{DETECTION}_{R-1} = 1 & \text{if } \text{DETECTION}_R = 0 \end{vmatrix} \\ R_{nm} \end{vmatrix}$$

$$R_{nm} - 0 \qquad\qquad\qquad \text{nm, maximum detection range}$$

$$R_{km} := R_{nm} \cdot 1.852 \qquad R_{km} = 0 \qquad \text{km, maximum detection range}$$

9.3.4 OUTPUT DATA

Maximum detection range $\qquad R_{km} = 0 \quad$ km or $\quad R_{nm} - 0 \quad$ nm

for jamming = 2 and input data specified.

9.3.5 REFERENCE

[1] Barton, D. K., and S. A. Leonov, (eds.), *Radar Technology Encyclopedia*, Norwood, MA: Artech House, 1997, pp. 74, 381.

9.4 Maximum Detection Range for an Altimeter

9.4.1 DESCRIPTION

Calculates maximum detection range for an altimeter in a thermal noise environment.

9.4.2 INPUT DATA

==

Parameter	Dimension and Description

==

case $:= 2$	$1 \rightarrow$ beamwidth-limited case $2 \rightarrow$ pulsewidth-limited case
$\sigma 0 := 10^{-5}$	m²/m², the surface reflectivity
$\theta := 1$	deg, 3-dB beamwidth
$\tau_r := 0.5$	µs, received pulsewidth after processing
$P_t := 10000$	W, transmitter peak power
$\tau_t := 5$	µs, transmitted pulsewidth
$G_t := 40$	dB, transmit antenna gain
$G_r := 30$	dB, receive antenna gain
$f := 5000$	MHz, radar frequency
$T_s := 650$	°K, system noise temperature
$D_x := 10$	dB, detectability factor
$M := 1.0$	d B, receiver mismatch loss
$L_t := 1.2$	dB, transmission line loss in transmit channel
$L_p := 1.6$	dB, beamshape loss
$L_{oth} := 0$	dB, other losses

Note: follow the rules cited in Section 9.1 to calculate parameters F and L_α.

$L_\alpha := 1.2$	dB, two-way atmospheric attenuation loss
$F := 1$	pattern-propagation factor

==

9.4.3 MODEL
9.4.3.1 Constants

$k := 1.38054 \cdot 10^{-23}$	W*s/°K, Boltzmann's constant

$c := 2.997925 \cdot 10^8$ m/s, speed of light

$z := \dfrac{\pi}{180}$ coefficient to transform degrees to radians

9.4.3.2 Transformation of Input Data

$\lambda := \dfrac{c}{f \cdot 10^6}$ $\lambda = 0.06$ m, wavelength

$\text{ratio}(x) := 10^{0.1 \cdot x}$ conversion equation

$G_{t\_dim} := \text{ratio}(G_t)$ $G_{t\_dim} = 1 \times 10^4$ transmit antenna gain

$G_{r\_dim} := \text{ratio}(G_r)$ $G_{r\_dim} = 1 \times 10^3$ receive antenna gain

$D_{x\_dim} := \text{ratio}(D_x)$ $D_{x\_dim} = 10$ detectability factor

$M\_dim := \text{ratio}(M)$ $M\_dim = 1.259$ receiver mismatch loss

$L_{t\_dim} := \text{ratio}(L_t)$ $L_{t\_dim} = 1.318$ transmit loss

$L_{p\_dim} := \text{ratio}(L_p)$ $L_{p\_dim} = 1.445$ beamshape loss

$L_{\alpha\_dim} := \text{ratio}(L_\alpha)$ $L_{\alpha\_dim} = 1.318$ atmospheric loss

$L_{oth\_dim} := \text{ratio}(L_{oth})$ $L_{oth\_dim} = 1$ other losses

9.4.3.3 Maximum Detection Range

A. Beamwidth-Limited Case

$$R01_m := \left[\dfrac{P_t \cdot \tau_t \cdot 10^{-6} \cdot G_{t\_dim} \cdot G_{r\_dim} \cdot \lambda^2 \cdot \sigma0 \cdot (z \cdot \theta)^2}{(4 \cdot \pi)^3 \cdot k \cdot T_s \cdot D_{x\_dim} \cdot M\_dim \cdot L_{t\_dim} \cdot L_{p\_dim}^2 \cdot L_{oth\_dim}} \right]^{\frac{1}{2}}$$

$R01_m = 9.417 \times 10^4$ m, free-space maximum detection range in vacuum

$R1_m := R01_m \cdot \sqrt{\dfrac{F^4}{L_{\alpha\_dim}}}$ $R1_m = 8.202 \times 10^4$ m, maximum detection range

$R1_{km} := \dfrac{R1_m}{1000}$ $R1_{km} = 82.02$ km, maximum detection range

$$R1_{nm} := \frac{R1_m}{1852} \qquad R1_{nm} = 44.287 \qquad \text{nm, maximum detection range}$$

B. Pulsewidth-Limited Case

$$R02_m := \left[\frac{P_t \cdot \tau_t \cdot 10^{-6} \cdot G_{t\_dim} \cdot G_{r\_dim} \cdot \lambda^2 \cdot \pi \cdot \sigma 0 \cdot \tau_r \cdot 10^{-6} \cdot c}{(4 \cdot \pi)^3 \cdot k \cdot T_s \cdot D_{x\_dim} \cdot M\_dim \cdot L_{t\_dim} \cdot L_{p\_dim}^2 \cdot L_{oth\_dim}} \right]^{\frac{1}{3}}$$

$$R02_m = 2.393 \times 10^5 \qquad \text{m, free-space maximum detection range in vacuum}$$

$$R2_m := R02_m \cdot \left(\frac{F^4}{L_{\alpha\_dim}} \right)^{\frac{1}{3}} \qquad R2_m = 2.183 \times 10^5 \qquad \text{m, maximum detection range}$$

$$R2_{km} := \frac{R2_m}{1000} \qquad R2_{km} = 218.276 \qquad \text{km, maximum detection range}$$

$$R2_{nm} := \frac{R2_m}{1852} \qquad R2_{nm} = 117.86 \qquad \text{nm, maximum detection range}$$

C. Specified Case

case specified

$$\text{case} = 2$$

range in kilometers   range in nautical miles

$$R_{km} := \begin{vmatrix} R1_{km} & \text{if case} = 1 \\ R2_{km} & \text{if case} = 2 \end{vmatrix} \qquad R_{nm} := \begin{vmatrix} R1_{nm} & \text{if case} = 1 \\ R2_{nm} & \text{if case} = 2 \end{vmatrix}$$

9.4.4 OUTPUT DATA

Maximum detection range $R_{km} = 218.276$ km or $R_{nm} = 117.86$

for input data specified.

9.4.5 REFERENCE

[1] Barton, D. K., and S. A. Leonov, (eds.), *Radar Technology Encyclopedia*, Norwood, MA: Artech House, 1997, p. 379.

9.5 Maximum Detection Range for a Bistatic Radar

9.5.1 DESCRIPTION
Calculates maximum detection range for a bistatic pulsed radar in an environment of thermal noise.

9.5.2 INPUT DATA

===

Parameter	Dimension and Description

===

$P_t := 10000$	W, transmitter peak power
$\tau_t := 5$	μs, transmitted pulsewidth
$G_t := 40$	dB, transmit antenna gain
$G_r := 30$	dB, receive antenna gain
$\sigma_b := 3$	m^2, radar cross section of the target
$f := 5000$	MHz, radar frequency
$T_s := 650$	°K, system noise temperature
$D_x := 10$	dB, detectability factor
$M := 1.0$	dB, receiver mismatch loss
$L_t := 1.2$	dB, transmission line loss in transmit channel
$L_p := 1.6$	dB, beamshape loss
$L_{oth} := 0$	dB, other losses

Note: follow the rules cited in Section 9.1 to calculate parameters F_t, F_r and $L_{\alpha t}$, $L_{\alpha r}$.

$L_{\alpha t} := 1.2$	dB, atmospheric attenuation loss for transmit path
$L_{\alpha r} := 1.5$	dB, atmospheric attenuation loss for receive path
$F_t := 1$	pattern-propagation factor for transmit path
$F_r := 1$	pattern-propagation factor for receive path

===

9.5.3 MODEL

9.5.3.1 Constants

$k := 1.38054 \cdot 10^{-23}$ W*s/°K, Boltzmann's constant

$c := 2.997925 \cdot 10^{8}$ m/s, speed of light

9.5.3.2 Transformation of Input Data

$$\lambda := \frac{c}{f \cdot 10^{6}} \qquad \lambda = 0.06$$ m, wavelength

$ratio(x) := 10^{0.1 \cdot x}$ conversion equation

$G_{t\_dim} := ratio(G_t)$ $G_{t\_dim} = 1 \times 10^{4}$ transmit antenna gain

$G_{r\_dim} := ratio(G_r)$ $G_{r\_dim} = 1 \times 10^{3}$ receive antenna gain

$D_{x\_dim} := ratio(D_x)$ $D_{x\_dim} = 10$ detectability factor

$M\_dim := ratio(M)$ $M\_dim = 1.259$ receiver mismatch loss

$L_{t\_dim} := ratio(L_t)$ $L_{t\_dim} = 1.318$ transmit loss

$L_{p\_dim} := ratio(L_p)$ $L_{p\_dim} = 1.445$ beamshape loss

$L_{\alpha t\_dim} := ratio(L_{\alpha t})$ $L_{\alpha t\_dim} = 1.318$ atmospheric loss

$L_{\alpha r\_dim} := ratio(L_{\alpha r})$ $L_{\alpha r\_dim} = 1.413$

$L_{oth\_dim} := ratio(L_{oth})$ $L_{oth\_dim} = 1$ other losses

9.5.3.3 Maximum Detection Range

$$RO_m := \left[\frac{P_t \cdot \tau_t \cdot 10^{-6} \cdot G_{t\_dim} \cdot G_{r\_dim} \cdot \lambda^2 \cdot \sigma_b}{(4 \cdot \pi)^3 \cdot k \cdot T_s \cdot D_{x\_dim} \cdot M\_dim \cdot L_{t\_dim} \cdot L_{p\_dim} \cdot L_{oth\_dim}} \right]^{\frac{1}{4}}$$

$RO_m = 5.961 \times 10^4$ 　　　　　m, free-space maximum detection range in vacuum

$$R_m := RO_m \cdot \left[\frac{\left(F_t^2 \cdot F_r^2 \right)}{\left(L_{\alpha t\_dim} \cdot L_{\alpha r\_dim} \right)} \right]^{\frac{1}{4}}$$ 　　　　　m, maximum detection range

$R_m = 5.103 \times 10^4$

$$R_{km} := \frac{R_m}{1000} \qquad R_{km} = 51.027 \qquad \text{km, maximum detection range}$$

$$R_{nm} := \frac{R_m}{1852} \qquad R_{nm} = 27.552 \qquad \text{nm, maximum detection range}$$

Note: maximum detection range is the geometric mean of ranges from transmitter to target and receiver to target.

9.5.4 OUTPUT DATA

Maximum detection range 　　$R_{km} = 51.027$ 　km or 　$R_{nm} = 27.552$ 　nm

for input data specified.

9.5.5 REFERENCE

[1] Barton, D. K., and S. A. Leonov, (eds.), *Radar Technology Encyclopedia*, Norwood, MA: Artech House, 1997, pp. 368, 379.

9.6 Maximum Detection Range for a Laser Radar

9.6.1 DESCRIPTION
Calculates maximum detection range for a coherent laser radar in an environment of a thermal noise.

9.6.2 INPUT DATA

==

Parameter	Dimension and Description

==

$P_t := 10000$	W, transmitter laser power
$f := 10^{14}$	Hz, transmission frequency
$B := 10^{12}$	Hz, electronic bandwidth of the receiver
$\rho := 0.8$	average reflectivity of the target
$A_t := 2$	m2, projected area of the illuminated part of the target
$A_r := 2$	m2, receiving aperture area
$\Omega_t := 2$	minutes, the solid angle of the transmitted beam
$\Omega_r := 3$	minutes, the solid angle of the reflected beam
$\eta_t := 0.9$	transmission coefficient for the transmitted beam
$\eta_r := 0.9$	transmission coefficient for the received beam
$\eta_d := 0.9$	detector quantum efficiency
$D_x := 10$	dB, detectability factor

==

9.6.3 MODEL

9.6.3.1 Constants

$h := 6.626 \cdot 10^{-34}$	W*s*s, Planck's constant
$z := \dfrac{\pi}{180}$	coefficient to transform degrees to radians

9.6.3.2 Maximum Detection Range

$$D_{x\_dim} := 10^{\frac{D_x}{10}} \qquad D_{x\_dim} = 10 \qquad\qquad \text{detectability factor}$$

$$R_m := \left[\frac{P_t \cdot \rho \cdot A_t \cdot A_r \cdot \eta_t \cdot \eta_r \cdot \eta_d}{\left(\dfrac{\Omega_t}{60} \cdot z\right)\left(\dfrac{\Omega_r}{60} \cdot z\right) \cdot h \cdot f \cdot B \cdot D_{x\_dim}} \right]^{\frac{1}{4}} \qquad \text{m, maximum detection range}$$

$$R_m = 1.623 \times 10^4$$

$$R_{km} := \frac{R_m}{1000} \qquad R_{km} = 16.228 \qquad\qquad \text{km, maximum detection range}$$

$$R_{nm} := \frac{R_m}{1852} \qquad R_{nm} = 8.762 \qquad\qquad \text{nm, maximum detection range}$$

9.6.4 OUTPUT DATA

Maximum detection range $R_{km} = 16.228$ km or $R_{nm} = 8.762$ nm
for input data specified.

9.6.5 REFERENCE

[1] Barton, D. K., and S. A. Leonov, (eds.), *Radar Technology Encyclopedia*, Norwood, MA: Artech House, 1997, p. 382.

9.7 Maximum Detection Range for a Meteorological Radar

9.7.1 DESCRIPTION

Calculates maximum detection range for a meteorological pulsed radar in an environment of thermal noise. *Assumption*: precipitation reflectivity meets the conditions of Rayleigh scattering region: $\lambda \gg 2\pi D/2$, where λ is a wavelength and D is a diameter of a scattering particle.

9.7.2 INPUT DATA

===

Parameter   *Dimension and Description*

===

Radar Parameters

$P_t := 10000$	W, transmitter peak power
$\tau_t := 5$	μs, transmitted pulsewidth
$\tau_r := 0.5$	μs, received pulsewidth after processing
$G_t := 40$	dB, transmit antenna gain
$G_r := 30$	dB, receive antenna gain
$\theta_3 := 3$	deg, 3-dB beamwidth in elevation plane
$\psi_3 := 1$	deg, 3-dB beamwidth in azimuth plane
$f := 5000$	MHz, radar frequency
$T_s := 650$	°K, system noise temperature
$D_x := 10$	dB, detectability factor
$M := 1.0$	dB, receiver mismatch loss
$L_t := 1.2$	dB, transmission line loss in transmit channel
$L_p := 1.6$	dB, beamshape loss
$L_{oth} := 0$	dB, other losses

Note: follow the rules cited in Section 9.1 to calculate parameters F and L_α.

$L_\alpha := 1.2$	dB, two-way atmospheric attenuation loss
$F := 1$	pattern-propagation factor

Precipitation Parameters

$r := 4$ mm/hour, precipitation rate

$a := 200$ precipitation model parameter

$b := 1.6$ precipitation model parameter

$K := 0.93$ precipitation model parameter

Note: the table of values a, b, K for different types of precipitation is given in Section 11.3, Signal-to-Interference Ratio.

==

9.7.3 MODEL

9.7.3.1 Constants

$k := 1.38054 \cdot 10^{-23}$ W*s/°K, Boltzmann's constant

$c := 2.997925 \cdot 10^{8}$ m/s, speed of light

$z := \dfrac{\pi}{180}$ coefficient to transform degrees to radians

9.7.3.2 Transformation of Input Data

$\lambda := \dfrac{c}{f \cdot 10^{6}}$ $\lambda = 0.06$ m, wavelength

$\text{ratio}(x) := 10^{0.1 \cdot x}$ conversion equation

$G_{t\_dim} := \text{ratio}(G_t)$ $G_{t\_dim} = 1 \times 10^{4}$ transmit antenna gain

$G_{r\_dim} := \text{ratio}(G_r)$ $G_{r\_dim} = 1 \times 10^{3}$ receive antenna gain

$D_{x\_dim} := \text{ratio}(D_x)$ $D_{x\_dim} = 10$ detectability factor

$M\_dim := \text{ratio}(M)$ $M\_dim = 1.259$ receiver mismatch loss

$L_{t\_dim} := \text{ratio}(L_t)$ $L_{t\_dim} = 1.318$ transmit loss

$L_{p\_dim} := \text{ratio}(L_p)$ $L_{p\_dim} = 1.445$ beamshape loss

$$L_{\alpha\_dim} := ratio(L_\alpha) \qquad L_{\alpha\_dim} = 1.318 \qquad\qquad \text{atmospheric loss}$$

$$L_{oth\_dim} := ratio(L_{oth}) \qquad L_{oth\_dim} = 1 \qquad\qquad \text{other losses}$$

9.7.3.3 Reflectivity of Precipitation

$$\eta_v := \frac{\pi^5}{\lambda^4} \cdot K \cdot a \cdot 10^{-18} \cdot r^b \qquad \eta_v = 4.047 \times 10^{-8} \qquad \text{m}^2/\text{m}^3, \text{ volume precipitation}$$
$$\text{reflectivity for matched polarization}$$

9.7.3.4 Maximum Detection Range

$$RO_m := \left[\frac{P_t \cdot \tau_t \cdot 10^{-6} \cdot G_{t\_dim} \cdot G_{r\_dim} \cdot \lambda^2 \cdot \eta_v \cdot z \cdot \theta_3 \cdot \tau_r \cdot 10^{-6} \cdot c \cdot z \cdot \psi_3}{2 \cdot (4 \cdot \pi)^3 \cdot k \cdot T_s \cdot D_{x\_dim} \cdot M\_dim \cdot L_{t\_dim} \cdot L_{p\_dim}^2 \cdot L_{oth\_dim}} \right]^{\frac{1}{2}}$$

$$RO_m = 8.983 \times 10^4 \qquad\qquad \text{m, free-space maximum detection range in vacuum}$$

$$R_m := RO_m \cdot \sqrt{\frac{F^4}{L_{\alpha\_dim}}} \qquad R_m = 7.824 \times 10^4 \qquad\qquad \text{m, maximum detection range}$$

$$R_{km} := \frac{R_m}{1000} \qquad R_{km} = 78.241 \qquad\qquad \text{km, maximum detection range}$$

$$R_{nm} := \frac{R_m}{1852} \qquad R_{nm} = 42.247 \qquad\qquad \text{nm, maximum detection range}$$

9.7.4 OUTPUT DATA

Maximum detection range $\qquad R_{km} = 78.241 \quad$ km or $\quad R_{nm} = 42.247 \quad$ nm

for input data specified.

9.7.5 REFERENCE

[1] Barton, D. K., and S. A. Leonov, (eds.), *Radar Technology Encyclopedia*, Norwood, MA: Artech House, 1997, p. 382.

9.8 Maximum Detection Range for a Search Radar

9.8.1 DESCRIPTION

Calculates maximum detection range for a search radar in an environment of thermal noise. *Assumption*: the radar scans uniformly over a solid angle Ψ in a frame time t_s.

9.8.2 INPUT DATA

==

Parameter	*Dimension and Description*

==

$P_{av} := 1000$ W, transmitter average power

Note: for continuous-wave radar average power is equal to peak power: for pulsed radar the average power is equal to the product of the peak power and the radar duty factor.

$t_s := 5$ s, frame time

$A_r := 5$ m2, effective aperture of the receive antenna

$\Psi := 90$ deg2, solid search angle

$\sigma := 5$ m2, radar cross section of the target

$T_s := 650$ °K, system noise temperature

$D_x := 10$ dB, detectability factor

$L_s := 20$ dB, total search losses

==

9.8.3 MODEL

9.8.3.1 Constants

$k := 1.38054 \cdot 10^{-23}$ W*s/°K, Boltzmann's constant

$c := 2.997925 \cdot 10^{8}$ m/s, speed of light

$z := \dfrac{\pi}{180}$ coefficient to transform degrees to radians

9.8.3.2 Transformation of Input Data

$\text{ratio}(x) := 10^{0.1 \cdot x}$ conversion equation

$D_{x\_dim} := \text{ratio}(D_x)$ $D_{x\_dim} = 10$ detectability factor

$L_{s\_dim} := \text{ratio}(L_s)$ $L_{s\_dim} = 100$ total search loss

9.8.3.3 Maximum Detection Range

$$R_m := \left(\frac{P_{av} \cdot A_r \cdot t_s \cdot \sigma}{4 \cdot \pi \cdot k \cdot T_s \cdot D_{x\_dim} \cdot z^2 \cdot \Psi \cdot L_{s\_dim}} \right)^{\frac{1}{4}}$$ m, maximum detection range

$R_m = 4.484 \times 10^5$

$R_{km} := \dfrac{R_m}{1000}$ $R_{km} = 448.42$ km, maximum detection range

$R_{nm} := \dfrac{R_m}{1852}$ $R_{nm} = 242.128$ nm, maximum detection range

9.8.4 OUTPUT DATA

Maximum detection range $R_{km} = 448.42$ km or $R_{nm} = 242.128$ nm
for input data specified.

9.8.5 REFERENCE

[1] Barton, D. K., and S. A. Leonov, (eds.), *Radar Technology Encyclopedia*, Norwood, MA: Artech House, 1997, pp. 250, 384.

9.9 System Operating Range for a Secondary Radar

9.9.1 DESCRIPTION
Calculates system operating range for a secondary radar in an environment of thermal noise.

9.9.2 INPUT DATA

==

Parameter   *Dimension and Description*

==

Interrogator (Radar)

$P_i := 10000$

W, peak power

$\tau_i := 5$

μs, signal duration

$G_i := 10$

dB, antenna gain

$f_i := 900$

MHz, frequency

$T_i := 650$

°K, system noise temperature

$D_i := 15$

dB, detectability factor

$L_i := 20$

dB, total system loss

Note: follow the rules cited in Section 9.1 to calculate parameters F_i and $L_{\alpha i}$.

$L_{\alpha i} := 1.2$

dB, one-way atmospheric attenuation loss from interrogator to transponder

$F_i := 1$

pattern-propagation factor

Transponder (Beacon)

$P_t := 50$

W, peak power

$\tau_t := 5$

μs, signal duration

$G_t := 10$

dB, antenna gain

$f_t := 900$

MHz, frequency

$T_t := 650$

°K, system noise temperature

$D_t := 40$

dB, detectability factor

$L_t := 20$

dB, total system loss

Note: follow the rules cited in Section 9.1 to calculate parameters F_t and $L_{\alpha t}$.

$L_{\alpha t} := 1.2$ dB, one-way atmospheric attenuation loss from transponder to interrogator

$F_t := 1$ pattern-propagation factor

===

9.9.3 MODEL

9.9.3.1 Constants

$k := 1.38054 \cdot 10^{-23}$ W*s/°K, Boltzmann's constant

$c := 2.997925 \cdot 10^{8}$ m/s, speed of light

9.9.3.2 Transformation of Input Data

$$\lambda_i := \frac{c}{f_i \cdot 10^6} \qquad \lambda_i = 0.333$$ m, interrogator wavelength

$$\lambda_t := \frac{c}{f_t \cdot 10^6} \qquad \lambda_t = 0.333$$ m, transponder wavelength

$ratio(x) := 10^{0.1 \cdot x}$ conversion equation

$G_{i\_dim} := ratio(G_i) \qquad G_{i\_dim} = 10$ interrogator antenna gain

$G_{t\_dim} := ratio(G_i) \qquad G_{t\_dim} = 10$ transponder antenna gain

$D_{i\_dim} := ratio(D_i) \qquad D_{i\_dim} = 31.623$ interrogator detectability factor

$D_{t\_dim} := ratio(D_t) \qquad D_{t\_dim} = 1 \times 10^4$ transponder detectability factor

$L_{i\_dim} := ratio(L_i) \qquad L_{i\_dim} = 100$ interrogator total system loss

$L_{t\_dim} := ratio(L_t) \qquad L_{t\_dim} = 100$ transponder total system loss

$$L_{\alpha i\_dim} := \text{ratio}(L_{\alpha i}) \qquad L_{\alpha i\_dim} = 1.318 \qquad \text{interrogator atmospheric loss}$$

$$L_{\alpha t\_dim} := \text{ratio}(L_{\alpha t}) \qquad L_{\alpha t\_dim} = 1.318 \qquad \text{transponder atmospheric loss}$$

9.9.3.3 Interrogator Maximum Detection Range

$$R0\_i_m := \left[\frac{P_i \cdot \tau_i \cdot 10^{-6} \cdot G_{i\_dim} \cdot G_{t\_dim} \cdot \lambda_i^2}{(4 \cdot \pi)^2 \cdot k \cdot T_i \cdot D_{t\_dim} \cdot L_{i\_dim}} \right]^{\frac{1}{2}}$$

$$R0\_i_m = 6.257 \times 10^5 \qquad \text{m, free-space maximum detection range in vacuum}$$

$$R\_i_m := R0\_i_m \cdot \sqrt{\frac{F_i^2}{L_{\alpha i\_dim}}} \qquad R\_i_m = 5.45 \times 10^5 \quad \text{m, maximum detection range}$$

$$R\_i_{km} := \frac{R\_i_m}{1000} \qquad R\_i_{km} = 544.97 \qquad \text{km, maximum detection range}$$

$$R\_i_{nm} := \frac{R\_i_m}{1852} \qquad R\_i_{nm} = 294.26 \qquad \text{nm, maximum detection range}$$

9.9.3.4 Transponder Maximum Detection Range

$$R0\_t_m := \left[\frac{P_t \cdot \tau_t \cdot 10^{-6} \cdot G_{i\_dim} \cdot G_{t\_dim} \cdot \lambda_t^2}{(4 \cdot \pi)^2 \cdot k \cdot T_t \cdot D_{i\_dim} \cdot L_{t\_dim}} \right]^{\frac{1}{2}}$$

$$R0\_t_m = 7.868 \times 10^5 \qquad \text{m, free-space maximum detection range in vacuum}$$

$$R\_t_m := R0\_t_m \cdot \sqrt{\frac{F_t^2}{L_{\alpha t\_dim}}} \qquad R\_t_m = 6.853 \times 10^5 \quad \text{m, maximum detection range}$$

$$R\_t_{km} := \frac{R\_t_m}{1000} \qquad R\_t_{km} = 685.263 \qquad \text{km, maximum detection range}$$

$$R\_t_{nm} := \frac{R\_t_m}{1852} \qquad R\_t_{nm} = 370.012 \qquad \text{nm, maximum detection range}$$

9.9.3.5 System Operating Range

$$R_m := \begin{vmatrix} R\_i_m & \text{if } R\_i_m \leq R\_t_m \\ R\_t_m & \text{otherwise} \end{vmatrix} \qquad \text{m, system operating range}$$

$$R_m = 5.45 \times 10^5$$

$$R_{km} := \frac{R_m}{1000} \qquad R_{km} = 544.97 \qquad \text{km, system operating range}$$

$$R_{nm} := \frac{R_m}{1852} \qquad R_{nm} = 294.26 \qquad \text{nm, system operating range}$$

9.9.4 OUTPUT DATA

System operating range $R_{km} = 544.97$ km or $R_{nm} = 294.26$ nm
for input data specified.

9.9.5 REFERENCE

[1] Barton, D. K., and S. A. Leonov, (eds.), *Radar Technology Encyclopedia*, Norwood, MA: Artech House, 1997, p. 384.

9.10 Maximum Detection Range for a Synthetic-Aperture Radar

9.10.1 DESCRIPTION

Calculates maximum detection range for a synthetic-aperture radar in an environment of thermal noise.

9.10.2 INPUT DATA

===

Parameter	Dimension and Description

===

$P_{av} := 1000$	W, transmitter average power
$G := 40$	dB, antenna gain
$f := 3000$	MHz, radar frequency
$\sigma 0 := 10^{-4}$	m²/m², surface reflectivity
$\Delta_R := 1$	m, range resolution cell width
$V_p := 300$	knots, radar platform velocity
$\alpha := 45$	deg, beam angle from broadside
$T_s := 650$	°K, system noise temperature
$D_x := 10$	dB, detectability factor
$L_0 := 20$	dB, total loss

===

9.10.3 MODEL

9.10.3.1 Constants

$k := 1.38054 \cdot 10^{-23}$	W*s/°K, Boltzmann's constant
$c := 2.997925 \cdot 10^8$	m/s, speed of light
$z := \dfrac{\pi}{180}$	coefficient to transform degrees to radians

9.10.3.2 Transformation of Input Data

$$V_{m\_s} := V_p \cdot \frac{1852}{3600} \qquad V_{m\_s} = 154.333 \qquad \text{m/s, radar platform velocity}$$

$$\lambda := \frac{c}{f \cdot 10^6} \qquad \lambda = 0.1 \qquad\qquad\qquad \text{m, wavelength}$$

$$\text{ratio}(x) := 10^{0.1 \cdot x} \qquad\qquad\qquad\qquad\qquad \text{conversion equation}$$

$$G_{dim} := \text{ratio}(G) \qquad G_{dim} = 1 \times 10^4 \qquad\qquad \text{antenna gain}$$

$$D_{x\_dim} := \text{ratio}(D_x) \qquad D_{x\_dim} = 10 \qquad\qquad \text{detectability factor}$$

$$L_{0\_dim} := \text{ratio}(L_0) \qquad L_{0\_dim} = 100 \qquad\qquad \text{total search loss}$$

9.10.3.3 Maximum Detection Range

m, maximum detection range

$$R_m := \left[\frac{P_{av} \cdot G_{dim}{}^2 \cdot \lambda^3 \cdot \sigma 0 \cdot \Delta_R}{(4 \cdot \pi)^3 \cdot k \cdot T_s \cdot D_{x\_dim} \cdot L_{0\_dim} \cdot 2 \cdot V_{m\_s} \cdot \sin(z \cdot \alpha)} \right]^{\frac{1}{3}}$$

$$R_m = 1.369 \times 10^5$$

$$R_{km} := \frac{R_m}{1000} \qquad R_{km} = 136.934 \qquad \text{km, maximum detection range}$$

$$R_{nm} := \frac{R_m}{1852} \qquad R_{nm} = 73.938 \qquad \text{nm, maximum detection range}$$

9.10.4 OUTPUT DATA

Maximum detection range $\qquad R_{km} = 136.934$ km or $\quad R_{nm} = 73.938$ nm
for input data specified.

9.10.5 REFERENCE

[1] Barton, D. K., and S. A. Leonov, (eds.), *Radar Technology Encyclopedia*, Norwood, MA: Artech House, 1997, p. 384.

9.11 Maximum Detection Range for a Tracking Radar

9.11.1 DESCRIPTION
Calculates maximum range when a nonaccelerating target can be tracked with specified accuracy in an environment of thermal noise.

9.11.2 INPUT DATA

===

Parameter   *Dimension and Description*

===

Basic Radar and Target Parameters

$P_t := 10000$ W, transmitter peak power

$\tau_t := 5$ μs, transmitted pulsewidth

$G_t := 40$ dB, transmit antenna gain

$G_r := 30$ dB, receive antenna gain

$\sigma := 5$ m^2, radar cross section of the target

$f := 5000$ MHz, radar frequency

$T_s := 650$ °K, system noise temperature

$M := 1.0$ dB, receiver mismatch loss

$L_t := 1.2$ dB, transmission line loss in transmit channel

$L_{oth} := 0$ dB, other losses

Note: follow the rules cited in Section 9.1 to calculate parameters F and L_α.

$L_\alpha := 1.2$ dB, two-way atmospheric attenuation loss

$F := 1$ pattern-propagation factor

Tracking Loop Parameters

error := 3   %, allowable rms noise error in tracking (a ratio of root-mean-square tracking error in the specified coordinate to the resolution cell width in this coordinate)

$\beta_{n0} := 2$ Hz, design tracking loop bandwidth for strong signals

$k_x := 1.4$ normalized tracking error slope

$t_p := 2000$ μs, processing interval

Note: processing interval is the pulse repetition interval for noncoherent systems or for coherent systems is the interval over which the signal-to-noise ratio equals SNR required to maintain specified accuracy of tracking.

===

9.11.3 MODEL

9.11.3.1 Constants

$$k := 1.38054 \cdot 10^{-23}$$

W*s/°K, Boltzmann's constant

$$c := 2.997925 \cdot 10^{8}$$

m/s, speed of light

9.11.3.2 Transformation of Input Data

$$\lambda := \frac{c}{f \cdot 10^{6}} \qquad \lambda = 0.06$$

m, wavelength

$$\text{ratio}(x) := 10^{0.1 \cdot x}$$

conversion equation

$$G_{t\_dim} := \text{ratio}(G_t) \qquad G_{t\_dim} = 1 \times 10^{4}$$

transmit antenna gain

$$G_{r\_dim} := \text{ratio}(G_r) \qquad G_{r\_dim} = 1 \times 10^{3}$$

receive antenna gain

$$M\_dim := \text{ratio}(M) \qquad M\_dim = 1.259$$

receiver mismatch loss

$$L_{t\_dim} := \text{ratio}(L_t) \qquad L_{t\_dim} = 1.318$$

transmit loss

$$L_{\alpha\_dim} := \text{ratio}(L_\alpha) \qquad L_{\alpha\_dim} = 1.318$$

atmospheric loss

$$L_{oth\_dim} := \text{ratio}(L_{oth}) \qquad L_{oth\_dim} = 1$$

other losses

9.11.3.3 Detectability Factor

$$ERR := error \cdot 0.01 \qquad ERR = 0.03$$

allowable tracking error

$$SNR := \frac{\beta_{n0} \cdot t_p \cdot 10^{-6}}{k_x^{2} \cdot ERR^{2}} - 1 \qquad SNR = 1.268$$

signal-to-noise ratio required to maintain specified accuracy

Note: only the thermal noise error is considered here, and in a generic case tracking may not be possible due to other errors when the calculated signal-to-noise ratio is below unity (SNR < 0). If SNR < 0, the allowable noise error will not be reached even in the absence of signal, and tracking range is equal to zero [2].

$$D_{x\_dim} := \begin{vmatrix} SNR & \text{if } SNR \geq 0 \\ 10^{100} & \text{otherwise} \end{vmatrix} \qquad D_{x\_dim} = 1.268 \qquad \text{detectability factor}$$

9.11.3.4 Maximum Tracking Range

m, range-independent factor for maximum tracking range

$$R0_m := \left[\frac{P_t \cdot \tau_t \cdot 10^{-6} \cdot G_{t\_dim} \cdot G_{r\_dim} \cdot \lambda^2 \cdot \sigma}{(4 \cdot \pi)^3 \cdot k \cdot T_s \cdot D_{x\_dim} \cdot M\_dim \cdot L_{t\_dim} \cdot L_{oth\_dim}} \right]^{\frac{1}{4}}$$

$$R_m := R0_m \cdot \left(\frac{F^4}{L_{\alpha\_dim}} \right)^{\frac{1}{4}} \qquad R_m = 1.161 \times 10^5 \quad \text{m, maximum tracking range}$$

$$R_{km} := \frac{R_m}{1000} \qquad R_{km} = 116.15 \qquad \text{km, maximum tracking range}$$

$$R_{nm} := \frac{R_m}{1852} \qquad R_{nm} = 62.716 \qquad \text{nm, maximum tracking range}$$

9.11.4 OUTPUT DATA

Maximum tracking range $R_{km} = 116.15$ km or $R_{nm} = 62.716$ nm

for allowable noise error and input data specified.

9.11.5 REFERENCES

[1] Barton, D. K., and S. A. Leonov, (eds.), *Radar Technology Encyclopedia*, Norwood, MA: Artech House, 1997, p. 378.
[2] Barton, D. K., *Modern Radar System Analysis*, Norwood, MA: Artech House, 1988, p. 468.

Chapter 10

Measurement Errors

10.1 Angle Measurement Error

10.1.1 DESCRIPTION

Calculates rms angle (azimuth or elevation) measurement error for a tracking pulsed radar as a function of target range in an environment of thermal noise. *Assumption*: monopulse technique is used for angle tracking.

10.1.2 INPUT DATA

==

Parameter	Dimension and Description

==

Radar Parameters

$P_t := 10000$	W, transmitter peak power
$\tau_t := 5$	μs, transmitted pulsewidth
$\Theta_3 := 3$	deg, 3-dB beamwidth for a specified angle (azimuth or elevation)
$G_t := 40$	dB, transmit antenna gain
$G_r := 30$	dB, receive antenna gain
$f := 5000$	MHz, radar frequency
$T_s := 650$	°K, system noise temperature
$M := 1.0$	dB, receiver mismatch loss
$L_t := 1.2$	dB, transmission line loss in transmit channel
$L_{oth} := 0$	dB, other losses

Note: parameters F and L_α have to be calculated using models cited in Sections 5.3 and 8.1 for target range R specified in the Target Parameters section of this menu.

$L_\alpha := 1.2$ dB, two-way atmospheric attenuation loss

$F := 1$ pattern-propagation factor

$k_\Theta := 1.4$ normalized tracking error slope

$n_i := 50$ number of samples integrated by a tracking loop for high signal-to-noise ratio

$t_p := 1000$ μs, processing interval (coherent processing interval or pulse repetition interval for noncoherent radar

$\sigma_0 := 0.1$ deg, instrumental error

Target Parameters

$R := 80$ nm, target range

$\sigma_t := 5$ m2, radar cross section of the target

$a_{t\_\Theta} := 10$ m/s2, target acceleration component in specified angle

$L_h := 10$ m, effective target extent in height

===

10.1.3 MODEL

10.1.3.1 Constants

$k := 1.38054 \cdot 10^{-23}$ W*s/°K, Boltzmann's constant

$c := 2.997925 \cdot 10^8$ m/s, speed of light

$z := \dfrac{\pi}{180}$ coefficient to transform degrees to radians

10.1.3.2 Transformation of Input Data

$\lambda := \dfrac{c}{f \cdot 10^6}$ $\lambda = 0.06$ m, wavelength

$\text{ratio}(x) := 10^{0.1 \cdot x}$ conversion equation

$G_{t\_dim} := \text{ratio}(G_t)$ $G_{t\_dim} = 1 \times 10^4$ transmit antenna gain

$G_{r\_dim} := \text{ratio}(G_r)$ $G_{r\_dim} = 1 \times 10^3$ receive antenna gain

$M\_dim := \text{ratio}(M)$ $M\_dim = 1.259$ receiver mismatch loss

$L_{t\_dim} := \text{ratio}(L_t)$ $L_{t\_dim} = 1.318$ transmit loss

$L_{\alpha\_dim} := \text{ratio}(L_\alpha)$ $L_{\alpha\_dim} = 1.318$ atmospheric loss

$L_{oth\_dim} := \text{ratio}(L_{oth})$ $L_{oth\_dim} = 1$ other losses

10.1.3.3 Noise Error

$$\Pi := \frac{P_t \cdot \tau_t \cdot 10^{-6} \cdot G_{t\_dim} \cdot G_{r\_dim} \cdot \lambda^2 \cdot F^4}{(4 \cdot \pi)^3 \cdot k \cdot T_s \cdot M\_dim \cdot L_{t\_dim} \cdot L_{\alpha\_dim} \cdot L_{oth\_dim}}$$ m2, radar potential

$$\Pi = 4.614 \times 10^{19}$$

$$SNR(R) := \frac{\Pi \cdot \sigma_t}{(R \cdot 1852)^4}$$ signal-to-noise ratio

$$\sigma_n(R) := \Theta_3 \cdot \frac{1}{k_\Theta \cdot \sqrt{2 \cdot n_i \cdot (1 + SNR(R))}}$$

$$\sigma_n(R) = 0.176$$ deg, rms angle measurement noise error

10.1.3.4 Instrumental, Dynamic Lag and Target Glint Error

$$\sigma_0 = 0.1$$ deg, instrumental error

$$\sigma_1 := \frac{\dfrac{a_{t\_\Theta}}{1852 \cdot R} \cdot \left[2.5 \cdot \left(\dfrac{1}{2 \cdot n_i \cdot t_p \cdot 10^{-6}} \right)^2 \cdot \dfrac{SNR(R)^4}{(1 + SNR(R))^4} \right]^{-1}}{z}$$

$$\sigma_1 = 1.408 \times 10^{-3} \qquad\qquad\qquad\qquad \text{deg, dynamic lag error}$$

$$\sigma_{d\_cr} := \frac{\Theta_3}{2} \qquad\qquad \sigma_{d\_cr} = 1.5 \qquad \text{deg, critical dynamic lag error for which the track is lost}$$

$$\sigma_2 := \frac{L_h \cdot (3 \cdot 1852 \cdot R)^{-1}}{z} \qquad\qquad \sigma_2 = 1.289 \times 10^{-3} \quad \text{deg, target glint error}$$

10.1.3.5 Total Angular Error

$$\sigma_\Theta := \left| \sqrt{\sum_{k=0}^{2} \left(\sigma_k\right)^2 + \sigma_n(R)^2} \quad \text{if } \sigma_1 \leq \sigma_{d\_cr} \qquad\qquad \text{deg, total angle error} \right.$$

$$\left| 180 \quad \text{otherwise} \qquad\qquad\qquad\qquad \text{a break-lock condition} \right.$$

Note: in case the error has to be corrected for propagation effects or platform-dependent errors, additional components of angle measurement error can be calculated using models cited in Sections 10.4 and 10.5.

10.1.4 OUTPUT DATA

Angle measurement error $\sigma_\Theta = 0.203$ degrees for input data specified.

10.1.5 REFERENCES

[1] Barton, D. K., and S. A. Leonov, (eds.), *Radar Technology Encyclopedia*, Norwood, MA: Artech House, 1997, pp. 164, 385.

[2] Barton, D. K., and W. F. Barton, *Modern Radar System Analysis Software*, Norwood, MA: Artech House, 1993, pp. 174–180.

[3] Barton, D. K., and H. R. Ward, *Handbook of Radar Measurement*, Norwood, MA: Artech House, 1984, pp. 208–228.

[4] Barton, D. K., *Modern Radar System Analysis*, Norwood, MA: Artech House, 1988, pp. 533–548.

[5] Barton, D. K., *Radar System Analysis*, Norwood, MA: Artech House, 1985, pp. 263–316.

[6] Leonov, A. I., and K. I. Fomichev, *Monopulse Radar*, Norwood, MA: Artech House, 1986, pp. 131–168.

10.1.6 EXAMPLES

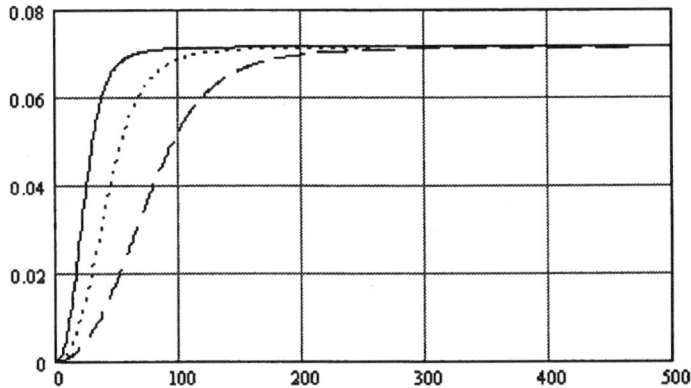

Figure 10.1 Angle measurement noise error normalized to 3-dB beamwidth vs. range (nm) for monopulse tracking radar when 50 samples are integrated: RCS = 1 m^2, radar potential $\Pi = 10^{19}$ m^2 (solid line), $\Pi = 10^{20}$ m^2 (dotted line), $\Pi = 10^{21}$ m^2 (dashed line).

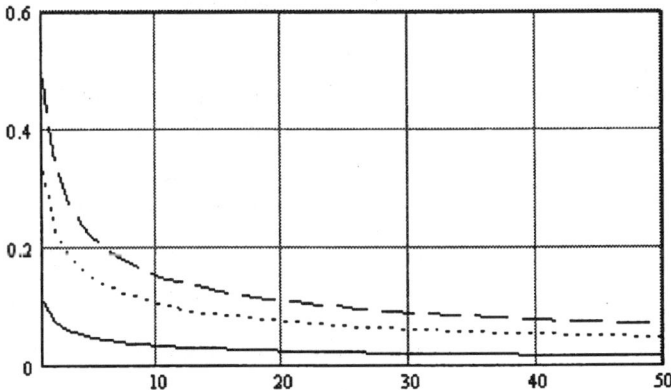

Figure 10.2 Angle measurement noise error normalized to 3-dB beamwidth vs. number of samples integrated for monopulse tracking radar with radar potential $\Pi = 10^{20}$ m^2: RCS = 1 m^2, range R = 25 nm (solid line), R = 50 nm (dotted line), R = 100 nm (dashed line).

10.2 Range Measurement Error

10.2.1 DESCRIPTION

Calculates rms range measurement error for a tracking pulsed radar as a function of target range in an environment of thermal noise.

10.2.2 INPUT DATA

===

Parameter	Dimension and Description

===

Radar Parameters

$P_t := 10000$	W, transmitter peak power
$\tau_t := 5$	μs, transmitted pulsewidth
$\tau_r := 0.5$	μs, received pulsewidth after processing
$G_t := 40$	dB, transmit antenna gain
$G_r := 30$	dB, receive antenna gain
$f := 5000$	MHz, radar frequency
$T_s := 650$	°K, system noise temperature
$M := 1.0$	dB, receiver mismatch loss
$L_t := 1.2$	dB, transmission line loss in transmit channel
$L_{oth} := 0$	dB, other losses

Note: parameters F and L_α have to be calculated using models cited in Sections 5.3 and 8.1 for target range R specified in the Target Parameters section of this menu.

$L_\alpha := 1.2$	dB, two-way atmospheric attenuation loss
$F := 1$	pattern-propagation factor
$k_R := 1.0$	normalized tracking error slope
$n_i := 10$	number of samples integrated by a tracking loop
$t_p := 1000$	μs, processing interval (coherent processing interval or pulse repetition interval for noncoherent radar)
$\sigma_0 := 1$	m, instrumental error

Target Parameters

$R := 80$	nm, target range
$\sigma_t := 5$	m2, radar cross section of the target

$a_{t\_R} := 10$ m/s2, radial component of the target
                                              acceleration in range

$L_R := 10$ m, effective target radial extent

==

10.2.3 MODEL

10.2.3.1 Constants

$k := 1.38054 \cdot 10^{-23}$ W*s/°K, Boltzmann's constant

$c := 2.997925 \cdot 10^{8}$ m/s, speed of light

10.2.3.2 Transformation of Input Data

$\lambda := \dfrac{c}{f \cdot 10^{6}}$ $\lambda = 0.06$ m, wavelength

$ratio(x) := 10^{0.1 \cdot x}$ conversion equation

$G_{t\_dim} := ratio(G_t)$ $G_{t\_dim} = 1 \times 10^{4}$ transmit antenna gain

$G_{r\_dim} := ratio(G_r)$ $G_{r\_dim} = 1 \times 10^{3}$ receive antenna gain

$M\_dim := ratio(M)$ $M\_dim = 1.259$ receiver mismatch loss

$L_{t\_dim} := ratio(L_t)$ $L_{t\_dim} = 1.318$ transmit loss

$L_{\alpha\_dim} := ratio(L_\alpha)$ $L_{\alpha\_dim} = 1.318$ atmospheric loss

$L_{oth\_dim} := ratio(L_{oth})$ $L_{oth\_dim} = 1$ other losses

10.2.3.3 Noise Error

$$\Pi := \dfrac{P_t \cdot \tau_t \cdot 10^{-6} \cdot G_{t\_dim} \cdot G_{r\_dim} \cdot \lambda^2 \cdot F^4}{(4 \cdot \pi)^3 \cdot k \cdot T_s \cdot M\_dim \cdot L_{t\_dim} \cdot L_{\alpha\_dim} \cdot L_{oth\_dim}}$$ m2, radar potential

$\Pi = 4.614 \times 10^{19}$

$SNR(R) := \dfrac{\Pi \cdot \sigma_t}{(R \cdot 1852)^4}$ signal-to-noise ratio

$$\sigma_n(R) := \frac{\tau_r \cdot 10^{-6} \cdot c}{2} \cdot \frac{1}{k_R \cdot \sqrt{2 \cdot n_i \cdot (1 + SNR(R))}}$$

$\sigma_n(R) = 13.781$ m, rms range measurement noise error

10.2.3.4 Instrumental, Dynamic Lag and Target Glint Error

$\sigma_0 = 1$ m, instrumental error

$$\sigma_1 := a_{t\_R} \cdot \left[2.5 \cdot \left(\frac{1}{2 \cdot n_i \cdot t_p \cdot 10^{-6}} \right)^2 \cdot \frac{SNR(R)^4}{(1 + SNR(R))^4} \right]^{-1}$$

$\sigma_1 = 0.146$ m, dynamic lag error

$$\sigma_{d\_cr} := \frac{\tau_r \cdot 10^{-6} \cdot c}{4}$$ $\sigma_{d\_cr} = 37.474$ m, critical dynamic lag error for which the track is lost

$$\sigma_2 := \frac{L_R}{3}$$ $\sigma_2 = 3.333$ m, target glint error

10.2.3.5 Total Range Error

$$\sigma_R := \left| \begin{array}{l} \sqrt{\displaystyle\sum_{k=0}^{2} (\sigma_k)^2 + \sigma_n(R)^2} \quad \text{if } \sigma_1 \le \sigma_{d\_cr} \\ \infty \quad \text{otherwise} \end{array} \right.$$

       m, total range error

       a break-lock condition

Note: in case the error has to be corrected for propagation effects or platform-dependent errors, additional components of range measurement error can be simulated using models cited in Sections 10.4 and 10.5.

10.2.4 OUTPUT DATA

Range measurement error $\sigma_R = 14.215$ meters for input data specified.

10.2.5 REFERENCES

[1] Barton, D. K., and S. A. Leonov, (eds.), *Radar Technology Encyclopedia*, Norwood, MA: Artech House, 1997, pp. 171, 385.

[2] Barton, D. K., and W. F. Barton, *Modern Radar System Analysis Software*, Norwood, MA: Artech House, 1993, pp. 180, 181.

10.2.6 EXAMPLES

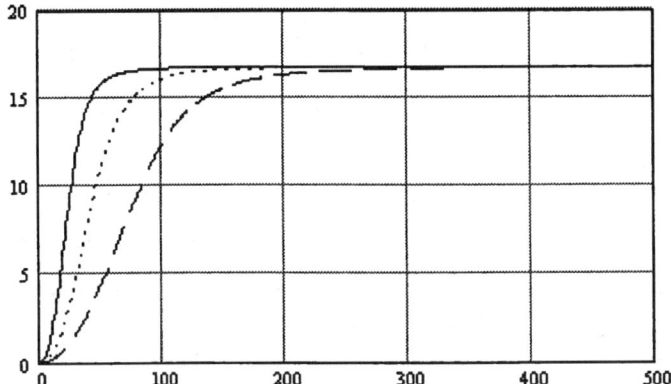

Figure 10.3 Range measurement noise error (m) vs. range (nm) for monopulse tracking radar when 10 samples are integrated: receive pulsewidth = 0.5 μs, RCS = 1 m², radar potential $\Pi = 10^{19}$ m² (solid line), $\Pi = 10^{20}$ m² dotted line), $\Pi = 10^{21}$ m² (dashed line).

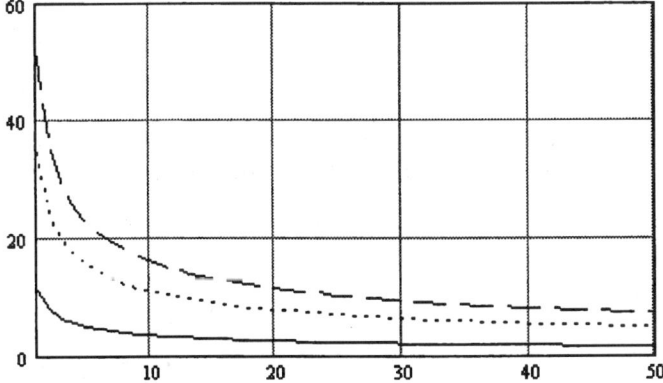

Figure 10.4 Range measurement noise error (m) vs. number of samples integrated for monopulse tracking radar with radar potential $\Pi = 10^{20}$ m²: RCS = 1 m², range R = 25 nm (solid line), R = 50 nm (dotted line), R = 100 nm (dashed line).

10.3 Velocity Measurement Error

10.3.1 DESCRIPTION

Calculates radial target velocity measurement error for a tracking pulsed radar as a function of target range in an environment of thermal noise. *Assumption*: radial target velocity is measured based on doppler frequency shift, no modulated scatterers producing false spectrum lines imitating target velocities are present.

10.3.2 INPUT DATA

===

Parameter	Dimension and Description

===

	Radar Parameters
$P_t := 10000$	W, transmitter peak power
$\tau_t := 5$	µs, transmitted pulsewidth
$\tau_{ef} := 50$	µs, effective signal duration after processing
$G_t := 40$	dB, transmit antenna gain
$G_r := 30$	dB, receive antenna gain
$f := 5000$	MHz, radar frequency
$T_s := 650$	°K, system noise temperature
$M := 1.0$	dB, receiver mismatch loss
$L_t := 1.2$	dB, transmission line loss in transmit channel
$L_{oth} := 0$	dB, other losses

Note: parameters F and L_α have to be calculated using models cited in Sections 5.3 and 8.1 for target range R specified in Target Parameters section of this menu.

$L_\alpha := 1.2$	dB, two-way atmospheric attenuation loss
$F := 1$	pattern-propagation factor
$k_V := 1.0$	normalized tracking error slope
$n_i := 10$	number of samples integrated by a tracking loop
$\sigma_0 := 10$	m/s, instrumental error
	Target Parameters
$R := 80$	nm, target range
$\sigma_t := 5$	m2, radar cross section of the target
$a_{t\_r} := 10$	m/s2, radial component of the target acceleration

===

10.3.3 MODEL

10.3.3.1 Constants

$k := 1.38054 \cdot 10^{-23}$ W*s/°K, Boltzmann's constant

$c := 2.997925 \cdot 10^{8}$ m/s, speed of light

10.3.3.2 Transformation of Input Data

$\lambda := \dfrac{c}{f \cdot 10^{6}}$ $\lambda = 0.06$ m, wavelength

$\text{ratio}(x) := 10^{0.1 \cdot x}$ conversion equation

$G_{t\_dim} := \text{ratio}(G_t)$ $G_{t\_dim} = 1 \times 10^{4}$ transmit antenna gain

$G_{r\_dim} := \text{ratio}(G_r)$ $G_{r\_dim} = 1 \times 10^{3}$ receive antenna gain

$M\_dim := \text{ratio}(M)$ $M\_dim = 1.259$ receiver mismatch loss

$L_{t\_dim} := \text{ratio}(L_t)$ $L_{t\_dim} = 1.318$ transmit loss

$L_{\alpha\_dim} := \text{ratio}(L_{\alpha})$ $L_{\alpha\_dim} = 1.318$ atmospheric loss

$L_{oth\_dim} := \text{ratio}(L_{oth})$ $L_{oth\_dim} = 1$ other losses

10.3.3.3 Noise Error

$$\Pi := \frac{P_t \cdot \tau_t \cdot 10^{-6} \cdot G_{t\_dim} \cdot G_{r\_dim} \cdot \lambda^{2} \cdot F^{4}}{(4 \cdot \pi)^{3} \cdot k \cdot T_s \cdot M\_dim \cdot L_{t\_dim} \cdot L_{\alpha\_dim} \cdot L_{oth\_dim}}$$ m², radar potential

$\Pi = 4.614 \times 10^{19}$

$$\text{SNR}(R) := \frac{\Pi \cdot \sigma_t}{(R \cdot 1852)^{4}}$$ signal-to-noise ratio

$$\sigma_n(R) := \frac{\lambda}{2 \cdot \tau_{ef} \cdot 10^{-6}} \cdot \frac{1}{k_V \cdot \sqrt{2 \cdot n_i \cdot (1 + \text{SNR}(R))}}$$

$\sigma_n(R) = 110.252$ m/s, rms velocity measurement noise error

10.3.3.4 Instrumental and Dynamic Errors

$\sigma_0 = 10$
                                                          m/s, instrumental error

$$\varepsilon(t) := a_{t\_r}\left[2.5\cdot\left(\frac{1}{2\cdot n_i\cdot t}\right)^2\cdot\frac{SNR(R)^4}{(1 + SNR(R))^4}\right]^{-1}$$ range dynamic lag error function

$$\varepsilon(t) := 3.2\cdot a_{t\_r}\cdot n_i^2\cdot\frac{t}{SNR(R)^4}\cdot(1 + SNR(R))^4$$ derivative of the range dynamic lag error function

$\sigma_1 := \varepsilon\left(\tau_{ef}\cdot 10^{-6}\right)$ $\sigma_1 = 14.562$ m/s, velocity dynamic lag error

$\sigma_1 = 14.562$ m, dynamic lag error

$$\sigma_{d\_cr} := \frac{\lambda}{4\cdot\tau_{ef}\cdot 10^{-6}}   \qquad \sigma_{d\_cr} = 299.793$$ m/s, critical dynamic lag error for which the track is lost

10.3.3.5 Total Velocity Error

$$\sigma_V := \left|\sqrt{\sum_{k=0}^{1}\left(\sigma_k\right)^2 + \sigma_n(R)^2}\right. \quad \text{if } \sigma_1 \le \sigma_{d\_cr}$$ m, total range error

∞ otherwise   a break-lock condition

Note: in case the error has to be corrected for propagation effects or platform-dependent errors, additional components of range measurement error can be simulated using models cited in Sections 10.4 and 10.5.

10.3.4 OUTPUT DATA

Velocity measurement error $\sigma_V = 111.7$ m/s for input data specified.

10.3.5 REFERENCES

[1] Barton, D. K., and S. A. Leonov, (eds.), *Radar Technology Encyclopedia*, Norwood, MA: Artech House, 1997, pp. 165, 385.
[2] Barton, D. K., *Modern Radar System Analysis*, Norwood, MA: Artech House, 1988, p. 551.

10.3.6 EXAMPLES

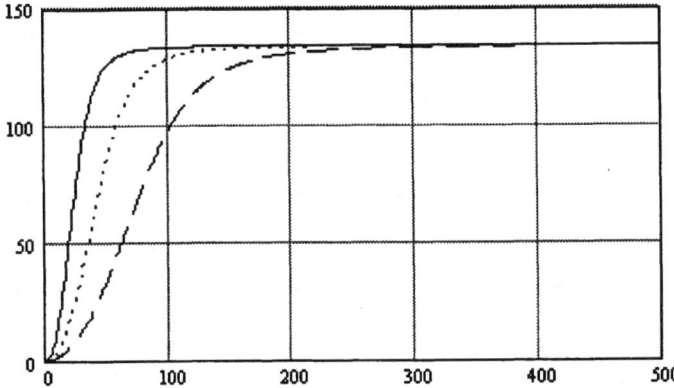

Figure 10.5 Velocity measurement noise error (m/s) vs. range (nm) for monopulse tracking radar when 10 samples are integrated: effective signal duration = 50 μs, RCS = 1 m^2, radar potential $\Pi = 10^{19}$ m^2 (solid line), $\Pi = 10^{20}$ m^2 (dotted line), $\Pi = 10^{21}$ m^2 (dashed line).

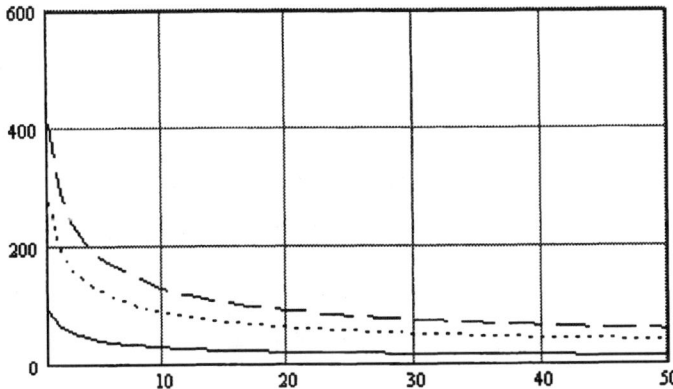

Figure 10.6 Velocity measurement noise error (m/s) vs. number of samples integrated for monopulse tracking radar with radar potential $\Pi = 10^{20}$ m^2: RCS = 1 m^2, range R = 25 nm (solid line), R = 50 nm (dotted line), R = 100 nm (dashed line).

10.4 Propagation Errors

10.4.1 DESCRIPTION

Calculates additional errors in radar measurements caused by propagation effects. The types of errors covered are: *multipath error* in measurement of elevation, *atmospheric refraction error* in measurement of elevation and range. *Assumptions*: monopulse technique is used for angle tracking, specular reflection is a predominant effect causing elevation multipath errors.

10.4.2 INPUT DATA

===

Parameter	Dimension and Description

===

Target and Radar Parameters

$R := 80$	nm, target range
$h_t := 10000$	m, a target height above surface
$h_a := 25$	m, radar antenna height above surface

Multipath Propagation Effects

Surface Parameters

$\rho_0 := 0.9$	Fresnel reflection coefficient
$\rho_s := 0.8$	specular scattering coefficient
$\rho_v := 0.7$	vegetation coefficient
$\beta := 2$	deg, rms surface slope

Elevation Plane

$\theta_3 := 2$	deg, half-power (3-dB) elevation beamwidth
$\theta_{max} := 1$	deg, elevation angle of beam axis
$k_{me} := 1.4$	normalized tracking error slope in elevation channel

Atmospheric Refraction Effects

$N_s := 313$	refractivity at the radar site
fluctuations := 1	1 → low fluctuations in refractive index 2 → average fluctuations in refractive index 3 → high fluctuations in refractive index

===

10.4.3 MODEL

10.4.3.1 Constants

$$z := \frac{\pi}{180}$$

coefficient to transform degrees to radians

$$a_e := 8493333$$

m, effective Earth radius with 4/3 approximation

10.4.3.2 Monopulse Antenna Pattern Modeling

Monopulse voltage sum pattern as the function of angle from the beam axis normalized to the beamwidth.

$$f_\Sigma(u) := \exp\left(-1.3866 \cdot u^2\right)$$

Monopulse voltage difference pattern as the function of angle from the beam axis normalized to the beamwidth.

$$f_\Delta(u) := 2.354 \cdot u \cdot f_\Sigma(u)$$

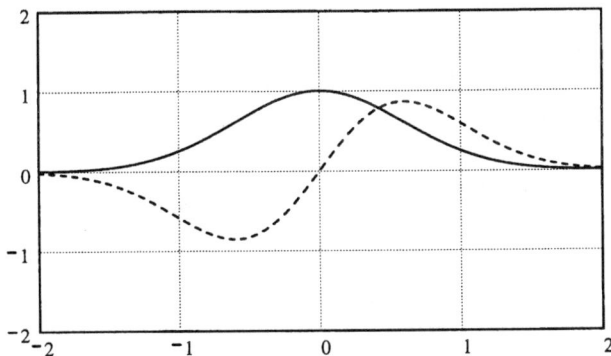

Figure 10.7 Normalized monopulse antenna sum and difference voltage patterns vs. angle from the beam axis normalized to the beamwidth.

10.4.3.3 Elevation Multipath Error

Parameters to calculate the target height above the plane tangent

$$R_t := 1852 \cdot R \qquad\qquad R_t = 1.482 \times 10^5 \qquad\qquad \text{m, target range}$$

$$p := \sqrt{\frac{4 \cdot a_e \cdot (h_a + h_t) + R_t^{\,2}}{3}} \qquad\qquad p = 3.476 \times 10^5$$

$$\Phi := a\cos\left[\frac{2 \cdot a_e \cdot (h_t - h_a) \cdot R_t}{p^3}\right] \qquad\qquad \Phi = 0.93$$

$$d_1 := \frac{R_t}{2} - p \cdot \cos\left(\frac{\Phi + \pi}{3}\right) \qquad\qquad d_1 = 423.615$$

$$H_1 := h_a - \frac{d_1^{\,2}}{2 \cdot a_e} \qquad\qquad H_1 = 24.989$$

$$H_2 := \frac{(R_t - d_1) \cdot H_1}{d_1} \qquad\qquad H_2 = 8.715 \times 10^3 \qquad\qquad \text{m, target height above}$$
$$\text{the plane tangent}$$

$$\theta_t := \frac{1}{z} \cdot a\sin\left(\frac{H_2 - H_1}{R_t}\right) \qquad \theta_t = 3.363 \qquad\qquad \text{deg, target elevation angle}$$

$$\psi_0 := \frac{1}{z} \cdot a\sin\left(\frac{H_2 + H_1}{R_t}\right) \qquad \psi_0 = 3.382 \qquad\qquad \text{deg, grazing angle for specular}$$
$$\text{reflection}$$

$$\theta_d := -\theta_{max} + \theta_t \qquad \theta_d = 2.363 \qquad\qquad \text{deg, off-axis angle of target}$$

$$f_\Sigma := f_\Sigma\left(\frac{\theta_d}{\theta_3}\right) \qquad f_\Sigma = 0.144 \qquad\qquad \text{monopulse voltage sum pattern response}$$

$$F_\Sigma := f_\Sigma \qquad F_\Sigma = 0.144 \qquad\qquad \text{sum channel pattern-propagation factor}$$

$$\theta_r := -\theta_{max} - \psi_0 \qquad \theta_r = -4.382 \qquad\qquad \text{deg, angle of reflected return}$$

$$f_\Delta := f_\Delta\left(\frac{\theta_r}{\theta_3}\right) \qquad f_\Delta = -6.634 \times 10^{-3}$$

monopulse difference-channel pattern
factor for specular reflection

$$F_\Delta := f_\Delta \cdot \rho_0 \cdot \rho_s \cdot \rho_v$$

difference channel pattern-propagation factor

$$F_\Delta = -3.343 \times 10^{-3}$$

$$\sigma_{e\_s} := \frac{\theta_3}{\sqrt{2} \cdot k_{me}} \cdot \sqrt{\frac{F_\Delta^2}{F_\Sigma^2}}$$

deg, specular reflection component
of elevation multipath error

$$\sigma_{e\_s} = 0.023$$

$$\sigma_{e\_s\_max} := \frac{\theta_3}{2} \qquad \sigma_{e\_s\_max} = 1$$

deg, maximum value for specular
reflection component of elevation
multipath error

$$\sigma_{e\_m} := \begin{vmatrix} \sigma_{e\_s} & \text{if } \sigma_{e\_s} \le \sigma_{e\_s\_max} \\ \sigma_{e\_s\_max} & \text{otherwise} \end{vmatrix}$$

deg, elevation multipath error

$$\sigma_{e\_m} = 0.023$$

10.4.3.5 Elevation Error Due to Atmospheric Refraction

$$\theta_{eff} := z \cdot \theta_t + \frac{2.5 \cdot 10^{-4}}{z \cdot \theta_t + 0.028} \qquad \theta_{eff} = 0.062$$

rad, effective target elevation

$$C_1 := \begin{vmatrix} 1 \cdot 10^{-6} & \text{if fluctuations} = 1 \\ 2 \cdot 10^{-6} & \text{if fluctuations} = 2 \\ 4 \cdot 10^{-6} & \text{if fluctuations} = 3 \end{vmatrix}$$

rad/(km)$^{1/2}$, refractive index
fluctuations constant

$$C_1 = 1 \times 10^{-6}$$

$$\sigma_0 := \frac{C_1}{z} \cdot \sqrt{\frac{3}{\sin(\theta_{eff})}} \qquad \sigma_0 = 4.001 \times 10^{-4}$$

deg, random component of refractive error due to refractive index fluctuations

$$b := \begin{vmatrix} 10^{-7} \cdot \cot(\theta_t) \cdot (14.5 + 4.5 \cdot \log(\theta_t)) & \text{if } \theta_t \le 0.1 \\ 10^{-6} \cdot \cot(\theta_t) & \text{otherwise} \end{vmatrix}$$

rad/N_s, parameter

$$a := \begin{vmatrix} -0.15 \cdot 10^{-3} \cdot \cot(\theta_t) \cdot (1 + 0.3 \cdot \log(\theta_t)) & \text{if } \theta_t < 0.02 \\ 5 \cdot 10^{-5} \cdot \cot(\theta_t) \cdot (1 + 1.4 \cdot \log(\theta_t)) & \text{if } 0.02 \le \theta_t < 0.2 \\ 0 & \text{if } \theta_t \ge 0.2 \end{vmatrix}$$

rad/N_s, parameter

$$\Delta\theta := \begin{vmatrix} 0.019 & \text{if } 0 \le \theta_t < 0.001 \\ (b \cdot N_s + a) \cdot \left(\dfrac{h_t}{11000 + h_t} \right) & \text{otherwise} \end{vmatrix}$$

rad, refractive bias error

$$\Delta\theta = 6.636 \times 10^{-4}$$

$$\sigma_1 := \frac{\Delta\theta}{20 \cdot z} \cdot \left[1 + \frac{10^{-3}}{(0.01 + \theta_t)^2} \right]$$

deg, rms refractive bias error corrected based on reflectivity profile measurements

$$\sigma_1 = 1.901 \times 10^{-3}$$

$$\sigma_{e\_r} := \sqrt{\sum_{k=0}^{1} (\sigma_k)^2} \qquad \sigma_{e\_r} = 1.943 \times 10^{-3}$$

deg, rms error of elevation measurement

10.4.3.6 Range Error Due to Atmospheric Refraction

$$\sigma_0 := 0.1$$

m, random component of refraction error

$$p := \frac{h_t - h_a + 3.3 \cdot 10^{-4} \cdot \left(h_t - h_a\right)^2}{15000 + h_t - h_a + 3.3 \cdot 10^{-4} \cdot \left(h_t - h_a\right)^2}$$ range refraction parameter

$p = 0.741$

$\Delta R := 0.0072 \cdot N_s \cdot \csc\left(\theta_t\right) \cdot p$ m, the range refraction bias error

$\Delta R = -7.615$

$\sigma_1 := \dfrac{\Delta R}{20}$ $\sigma_1 = -0.381$ m, corrected error based on local surface reflectivity measurements

$\sigma_{R\_r} := \sqrt{\displaystyle\sum_{k=0}^{1} \left(\sigma_k\right)^2}$ $\sigma_{R\_r} = 0.394$ m, rms error of range measurement

10.4.4 OUTPUT DATA

Elevation multipath error: $\sigma_{e\_m} = 0.023$ deg,

Elevation refraction error: $\sigma_{e\_r} = 1.943 \times 10^{-3}$ deg,

Range refraction error: $\sigma_{R\_r} = 0.394$ m

for input data specified.

10.4.5 REFERENCES

[1] Barton, D. K., and S. A. Leonov, (eds.), *Radar Technology Encyclopedia*, Norwood, MA: Artech House, 1997, p. 170.

[2] Barton, D. K., and W. F. Barton, *Modern Radar System Analysis Software*, Norwood, MA: Artech House, 1993, pp. 174–181.

[3] Barton, D. K., *Modern Radar System Analysis*, Norwood, MA: Artech House, 1988, pp. 512–530.

[4] Barton, D. K., and H. R. Ward, *Handbook of Radar Measurement*, Norwood, MA: Artech House, 1984, pp. 129–160.

[5] Kolosov, M. A., N. A. Armand, and O. I. Yakovlev, *Radio Waves Propagation in Space Communications*, (in Russian), Moscow: Svyaz, 1969, pp. 17–30.

10.5 Platform-Dependent Errors

10.5.1 DESCRIPTION
Calculates the angle and range measurement errors caused by the errors in knowledge of position and orientation of a radar platform.

10.5.2 INPUT DATA

Parameter	Dimension and Description
$R := 0.1$	nm, target range
$\psi := 45$	deg, azimuth of the target
$\theta := 45$	deg, elevation of the target
$\Delta\alpha_1 := 3$	ang.min, the error of platform yaw angle measurement
$\Delta\alpha_2 := 3$	ang.min, the error of platform pitch angle measurement
$\Delta\alpha_3 := 3$	ang.min, the error of platform roll angle measurement
$\Delta x_1 := 30$	m, the error of the platform center of gravity position measurement along the X1 axis
$\Delta x_2 := 30$	m, the error of the platform center of gravity position measurement along the X2 axis
$\Delta x_3 := 30$	m, the error of the platform center of gravity position measurement along the X3 axis

Note: for definition of platform position and orientation errors see Figure 10.8.

10.5.3. MODEL

10.5.3.1 Constants

$$z := \frac{\pi}{180}$$ coefficient to transform degrees to radians

10.5.3.2 Angular Measurement Errors Caused by Platform Orientation Errors

$\Theta := 90 - \theta$ deg, an angle complement to the target elevation angle

$\Theta = 45$

Coordinate Transformation Matrix Elements

$$c_1 := \cos\left(z \cdot \frac{\Delta\alpha_1}{60}\right) \qquad s_1 := \sin\left(z \cdot \frac{\Delta\alpha_1}{60}\right) \qquad c_2 := \cos\left(z \cdot \frac{\Delta\alpha_2}{60}\right)$$

$$s_2 := \sin\left(z \cdot \frac{\Delta\alpha_2}{60}\right) \qquad c_3 := \cos\left(z \cdot \frac{\Delta\alpha_3}{60}\right) \qquad s_3 := \sin\left(z \cdot \frac{\Delta\alpha_3}{60}\right)$$

$$A_0 := \cos(z \cdot \psi) \cdot \sin(z \cdot \Theta) \cdot c_1 \cdot c_2 \qquad A_1 := \sin(z \cdot \psi) \cdot \sin(z \cdot \Theta) \cdot s_1 \cdot c_2$$

$$A_2 := \cos(z \cdot \Theta) \cdot s_2 \qquad B_0 := -\cos(z \cdot \psi) \cdot \sin(z \cdot \Theta) \cdot (s_1 \cdot c_3 + c_1 \cdot s_2 \cdot s_3)$$

$$B_1 := -\sin(z \cdot \psi) \cdot \sin(z \cdot \Theta) \cdot (s_1 \cdot s_2 \cdot s_3 - c_1 \cdot c_3) \qquad B_2 := \cos(z \cdot \Theta) \cdot c_2 \cdot s_3$$

$$C_0 := \cos(z \cdot \psi) \cdot \sin(z \cdot \Theta) \cdot (s_1 \cdot s_3 - c_1 \cdot s_2 \cdot c_3)$$

$$C_1 := -\sin(z \cdot \psi) \cdot \sin(z \cdot \Theta) \cdot (c_1 \cdot s_3 + s_1 \cdot s_2 \cdot c_3) \qquad C_2 := \cos(z \cdot \Theta) \cdot c_2 \cdot c_3$$

$$k := 0 .. 2 \qquad\qquad\qquad\qquad\qquad\qquad\qquad \text{cycle}$$

$$U := \frac{\displaystyle\sum_k A_k}{\sqrt{\left(\displaystyle\sum_k A_k\right)^2 + \left(\displaystyle\sum_k B_k\right)^2}} \qquad V := \frac{\displaystyle\sum_k C_k}{\sqrt{\left(\displaystyle\sum_k A_k\right)^2 + \left(\displaystyle\sum_k B_k\right)^2 + \left(\displaystyle\sum_k C_k\right)^2}}$$

$$\Delta\psi1 := \left| \begin{array}{l} \dfrac{\text{acos}(U) - z \cdot \psi}{z} \cdot 60 \quad \text{if } 0 \leq z \cdot \psi \leq \pi \\[2mm] \dfrac{2 \cdot \pi - \text{acos}(U) - z \cdot \psi}{z} \cdot 60 \quad \text{otherwise} \end{array} \right.$$

$$\Delta\psi1 = -2.998 \qquad\qquad \text{ang. min., the error of azimuth measurement}$$

$$\Delta\theta1 := -\left(\frac{\text{acos}(V) - z \cdot \Theta}{z} \cdot 60\right)$$

$$\Delta\theta1 = -4.243 \qquad\qquad \text{ang. min., the error of elevation measurement}$$

10.5.3.3 Angular and Range Measurement Errors Caused by
   Platform Position Errors

$$R_m := R \cdot 1852 \qquad\qquad R_m = 185.2 \qquad\qquad\qquad \text{m, target range}$$

Transformation of the Platform Position Error Components to Polar Coordinates

$$\eta_1 := \sqrt{\Delta x_1^2 + \Delta x_2^2 + \Delta x_3^2} \quad \eta_1 = 51.962 \qquad \text{m, position error vector modulus}$$

$$\eta_2 := \text{atan}\left(\frac{\Delta x_2}{\Delta x_1 + 10^{-10}} \right) \qquad\qquad \text{rad, azimuth of position error vector}$$

$$\frac{\eta_2}{z} = 45 \qquad\qquad\qquad\qquad \text{deg, azimuth of position error vector}$$

$$\eta_3 := \text{atan}\left(\frac{\sqrt{\Delta x_1^2 + \Delta x_2^2}}{\Delta x_3 + 10^{-10}} \right) \qquad \begin{array}{l} \text{rad, angle of position error vector complement to} \\ \text{elevation of this vector } (\pi/2 \text{ - elevation}) \end{array}$$

$$\frac{\eta_3}{z} = 54.736 \qquad\qquad \begin{array}{l} \text{deg, angle of position error vector complement to} \\ \text{elevation of this vector } (\pi/2 \text{ - elevation}) \end{array}$$

Coordinate Transformation Matrix Elements

$$D_0 := \cos(z \cdot \psi) \cdot \sin(z \cdot \Theta) - \frac{\eta_1}{R_m} \cdot \cos(\eta_2) \cdot \sin(\eta_3)$$

$$D_1 := \sin(z \cdot \psi) \cdot \sin(z \cdot \Theta) - \frac{\eta_1}{R_m} \cdot \sin(\eta_2) \cdot \sin(\eta_3)$$

$$D_2 := \cos(z \cdot \Theta) - \frac{\eta_1}{R_m} \cdot \cos(\eta_3)$$

$$U := \frac{D_0}{\sqrt{\displaystyle\sum_{k=0}^{1} (D_k)^2}} \qquad\qquad V := \frac{D_2}{\sqrt{\displaystyle\sum_{k=0}^{2} (D_k)^2}}$$

$$\Delta\psi2 := \begin{vmatrix} \dfrac{\mathrm{acos}(U) - z\cdot\psi}{z}\cdot 60 & \text{if } 0 \le z\cdot\psi \le \pi \\[3mm] \dfrac{2\cdot\pi - \mathrm{acos}(U) - z\cdot\psi}{z}\cdot 60 & \text{otherwise} \end{vmatrix}$$

$\Delta\psi2 = 2.745 \times 10^{-9}$ ang. min., the error of azimuth measurement

$$\Delta\theta2 := -\left(\dfrac{\mathrm{acos}(V) - z\cdot\Theta}{z}\cdot 60 \right)$$

$\Delta\theta2 = 225.124$ ang. min., the error of elevation measurement

$$a_R := \sin(z\cdot\Theta)\cdot\cos\big(z\cdot\psi - \eta_2\big)\cdot\sin(\eta_3) + \cos(z\cdot\Theta)\cdot\cos(\eta_3)$$

$$\delta_R := \sqrt{1 - 2\cdot\dfrac{\eta_1}{R_m}\cdot a_R + \left(\dfrac{\eta_1}{R_m}\right)^2}$$

$\Delta R := R\cdot\big(\delta_R - 1\big)$ $\Delta R = -0.027$ m, range measurement error

10.5.4 OUTPUT DATA

The errors of azimuth and elevation angles measurements caused by the errors in platform orientation knowledge are:

$\Delta\psi1 = -2.998$ $\Delta\theta1 = -4.243$ angular minutes

The errors of azimuth, elevation and range measurements caused by the errors in platform position knowledge are:

$\Delta\psi2 = 2.745 \times 10^{-9}$ $\theta2 = 225.124$ angular minutes, $\Delta R = -0.027$ m

for input data specified.

Note: the actual value of a coordinate C (where C is ψ, θ or R), the measured value C_m and measurement error ΔC are related as $C = C_m + \Delta C$. If the error has a minus sign it means that actual coordinate is less than a measured value. The models may lead to discontinuous functions when azimuth ψ and elevation θ are close to 0, π or 2π, so exercise caution when calculating errors for those angles.

10.5.5 REFERENCES

[1] Barton, D. K., and S. A. Leonov, (eds.), *Radar Technology Encyclopedia*, Norwood, MA: Artech House, 1997, p. 169.
[2] Leonov, S. A., et al., *Radar Test*, (in Russian), Moscow: Radio i Svyaz, 1990, Ch. 8.

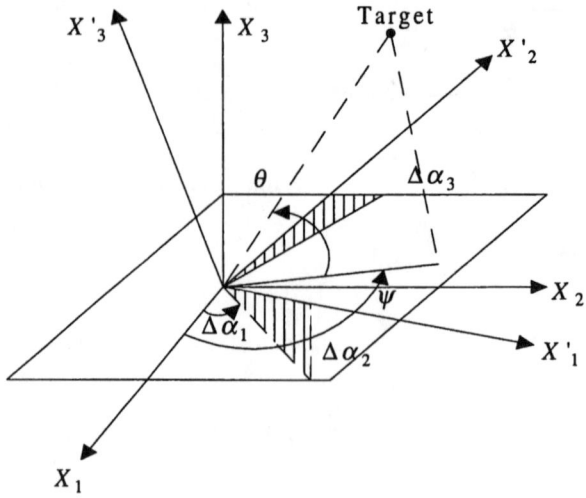

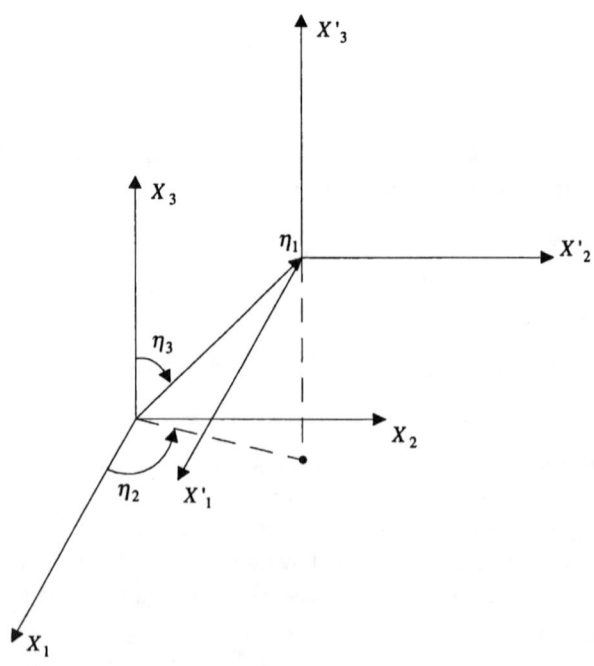

Figure 10.8 Parameters of radar platform position errors.

10.5.6 EXAMPLES

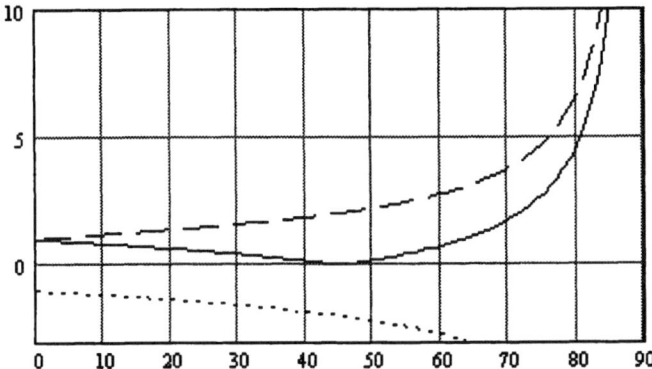

Figure 10.9 Azimuth measurement error (angular minutes) vs. target elevation angle (degrees) caused by platform orientation errors $\Delta\alpha_1 = \Delta\alpha_2 = \Delta\alpha_3 = 1$ ang. min. (solid line $\psi = 0°$, dotted line $\psi = 90°$, dashed line $\psi = 180°$).

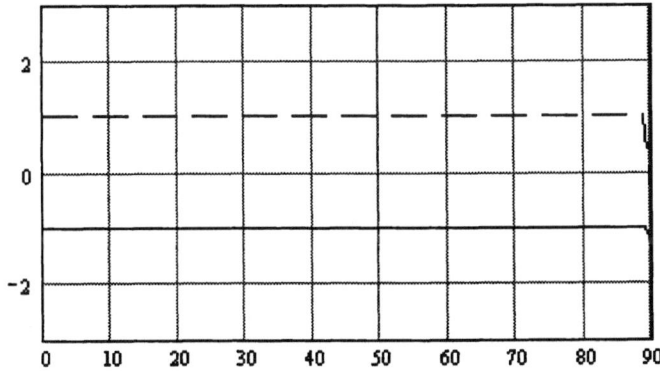

Figure 10.10 Elevation measurement error (angular minutes) vs. target elevation angle (degrees) caused by platform orientation errors $\Delta\alpha_1 = \Delta\alpha_2 = \Delta\alpha_3 = 1$ ang. min. (solid line $\psi = 0°$, dashed line $\psi = 180°$).

Chapter 11

Noise and Interference Immunity

11.1 Noise and Interference Simulation

11.1.1 DESCRIPTION

Simulates independent samples of noise or interference with specified mean value, rms value and probability distribution function (the following common functions are covered: Gaussian, Rayleigh, Ricean, exponential, log-normal).

11.1.2 INPUT DATA

```
================================================================
```

Parameter	*Dimension and Description*

```
================================================================
```

$t_c := 10^{-2}$ s, correlation time

$T := 1$ s, interference sample length

distribution := 1 $1 \rightarrow$ Gaussian probability distribution function
     $2 \rightarrow$ Rayleigh probability distribution function
     $3 \rightarrow$ Ricean probability distribution function
     $4 \rightarrow$ exponential probability distribution function
     $5 \rightarrow$ log-normal probability distribution function

Gaussian pdf

$VG := 1$ V, mean magnitude

$\sigma G := 3$ V, rms magnitude

Note: the simulated pdf is $f(x) := \dfrac{1}{\sigma G \cdot \sqrt{2 \cdot \pi}} \cdot \exp\left[-\dfrac{(x - VG)^2}{2 \cdot \sigma G^2} \right]$

Describes random thermal noise or radar returns from many scattering centers (without a predominant one) with random phase. Typical for land clutter (relatively flat land, deserts), sea clutter (quiet sea observed with low resolution radar at high grazing angles) and chaff jamming.

Rayleigh pdf

$\sigma R := 1$
V, rms magnitude

Note: the simulated pdf is $\quad f(x) := \dfrac{x}{\sigma R^2} \cdot \exp\left(-\dfrac{x^2}{2 \cdot \sigma R^2}\right) \; x >= 0$

Describes envelope of the random thermal noise with Gaussian distribution or phase-independent radar returns from a large number of small scatterers. Typically, underestimates the range of values that may be obtained from real clutter.

Ricean pdf

$\qquad\qquad\qquad\qquad\qquad\qquad\qquad\qquad\qquad V$, rms magnitude

$\sigma RC := 2$

$\alpha := 1 \qquad\qquad\qquad\qquad\qquad\qquad\qquad\qquad\qquad V$, Ricean parameter

Note: the simulated pdf is $\quad f(x) := \dfrac{x}{\sigma RC^2} \cdot \exp\left(-\dfrac{x^2 + \alpha^2}{2 \cdot \sigma RC^2}\right) \cdot I_0\left(\dfrac{\alpha \cdot x}{\sigma RC^2}\right) \; x >= 0$

where $\quad I_0(u) := \dfrac{1}{2 \cdot \pi} \cdot \displaystyle\int_0^{2 \cdot \pi} e^{-j \cdot (j \cdot u) \cdot \cos(t)} \, dt$

Describes radar returns from a constant amplitude dominant scatterer and large number of small scatterers (for example, urban clutter that is a combination of distributed clutter with Rayleigh statistics and discrete clutter—buildings or other point structures).

Exponential pdf

$VE := 1 \qquad\qquad\qquad\qquad\qquad\qquad\qquad\qquad\qquad W$, mean power

Note: the simulated pdf is $\quad f(x) := \dfrac{1}{VE} \cdot \exp\left(-\dfrac{x}{VE}\right) \quad , x >= 0$

Typically used to describe distribution of power returns from many scattering centers with random phase (for example, power returns by precipitation clutter or RCS of extended target when return signal has Gaussian statistics).

Log-Normal pdf

VLN := 1

V, mean magnitude
(specify VLN = 10^{-10} if VLN = 0)

mM := 5

mean-to-median ratio

Note: the simulated pdf is

$$f(x) := \frac{1}{\sigma LN \cdot \sqrt{2 \cdot \pi}} \cdot \exp\left[-\frac{(\ln(x) - VLN)^2}{2 \cdot \sigma LN^2} \right]^{\blacksquare}$$

where $\sigma LN := VLN \cdot \sqrt{mM^2 - 1}$. Describes radar returns from a collection of scatterers of different sizes including small number of large scatterers (the logarithms of radar return are normally distributed). Used to describe land clutter (cultivated land, forests, jungle, mountainous areas), sea clutter (rough sea observed with high-resolution radar) and returns from flocks of birds and swarms of insects). Typically, overestimates the range of values that may be obtained from real clutter.

===

11.1.3 MODEL

$\Delta t := t_c$

s, sampling interval

$N := floor\left(\dfrac{T}{\Delta t}\right)$ $N = 100$

number of points in interference sample

$n := 0 .. N - 1$

cycle

11.1.3.1 Gaussian Distribution

$k := 0 .. 1$

cycle

$u_{n,k} := rnd(1)$

random variables with uniform distribution at (0,1)

$V1_n := VG + \sigma G \cdot \sqrt{-2 \cdot \ln(u_{n,0} + 10^{-10})} \cdot \cos(2 \cdot \pi \cdot u_{n,1})$

V, Gaussian process

11.1.3.2 Rayleigh Distribution

$z_n := rnd(1)$

random variable with uniform distribution at (0,1)

$$V2_n := \sigma R \cdot \sqrt{-2 \cdot \ln\left(z_n + 10^{-10}\right)}$$

V, Rayleigh process

11.1.3.3 Ricean Distribution

$k := 0 .. 1$ cycle

$u_{n,k} := rnd(1)$ random variables with uniform distribution at (0,1)

$$V3_n := \left| \begin{array}{l} p \leftarrow 2 \cdot \alpha \cdot \sigma RC \cdot \sqrt{-2 \cdot \ln\left(u_{n,0} + 10^{-10}\right)} \cdot \cos\left(2 \cdot \pi \cdot u_{n,1}\right) \\[2mm] mag_n \leftarrow \sqrt{\alpha^2 - 2 \cdot \sigma RC^2 \cdot \ln\left(u_{n,0} + 10^{-10}\right)} - p \end{array} \right.$$

V, Ricean process

11.1.3.4 Exponential Distribution

$z_n := rnd(1)$ random variable with uniform distribution at (0,1)

$$V4_n := VE \cdot -\ln\left(z_n\right)$$

11.1.3.5 Log-Normal Distribution

$k := 0 .. 1$ cycle

$u_{n,k} := rnd(1)$ random variables with uniform distribution at (0,1)

$$V5_n := \left| \begin{array}{l} p \leftarrow \sqrt{-2 \cdot \ln\left(u_{n,0}\right)} \cdot \cos\left(2 \cdot \pi \cdot u_{n,1}\right) \\[2mm] mag_n \leftarrow VLN + VLN \cdot \sqrt{mM^2 - 1} \cdot \exp(p) \end{array} \right.$$

V, log-normal process

11.1.3.6 Interference with Specified Distribution

$distribution = 1$ distribution law specified

$$V := \left| \begin{array}{ll} V1 & \text{if } distribution = 1 \\ V2 & \text{if } distribution = 2 \\ V3 & \text{if } distribution = 3 \\ V4 & \text{if } distribution = 4 \\ V5 & \text{if } distribution = 5 \end{array} \right.$$

V, interference voltage for distribution specified

11.1.4 OUTPUT DATA

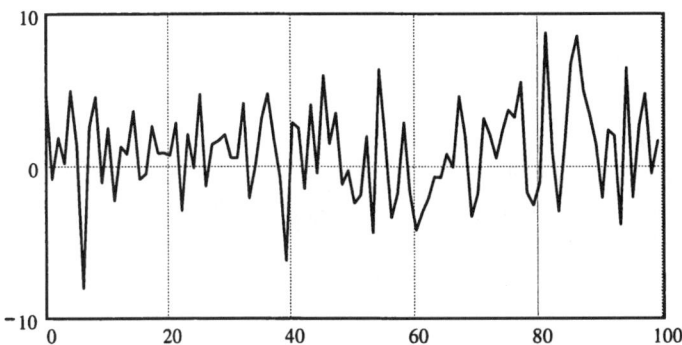

Figure 11.1 An interference magnitude (Volts or Watts depending on distribution model specified) vs. number of samples for   distribution = 1.

11.1.5 REFERENCE

[1] Pollyak, Yu. G., *Stochastic Computer Simulation,* (in Russian), Moscow: Sovetskoe Radio, 1971, Ch. 3.

11.1.6 EXAMPLES

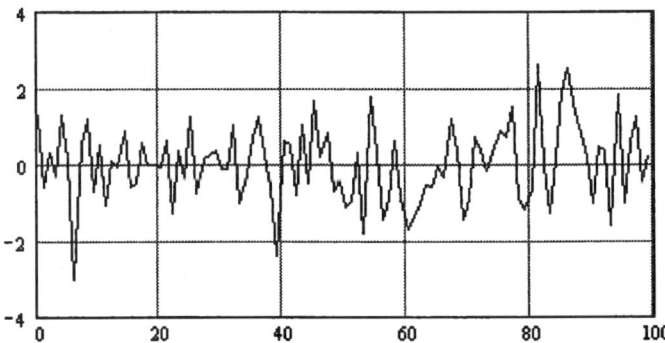

Figure 11.2 An interference magnitude (Volts) for Gaussian pdf with zero mean and rms amplitude σ = 1 V vs. number of samples (signal length to correlation interval ratio is 100).

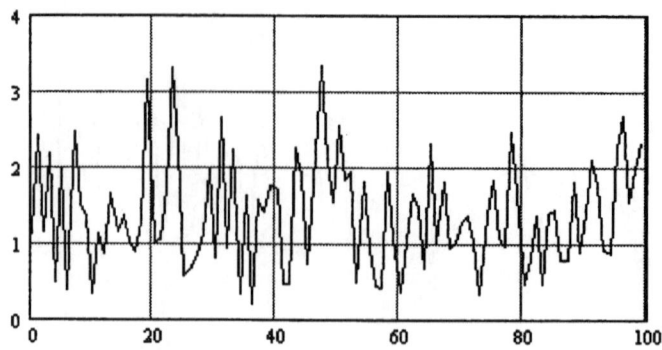

Figure 11.3 An interference magnitude (Volts) for Rayleigh pdf with rms amplitude $\sigma = 1$ V vs. number of samples (signal length to correlation interval ratio is 100).

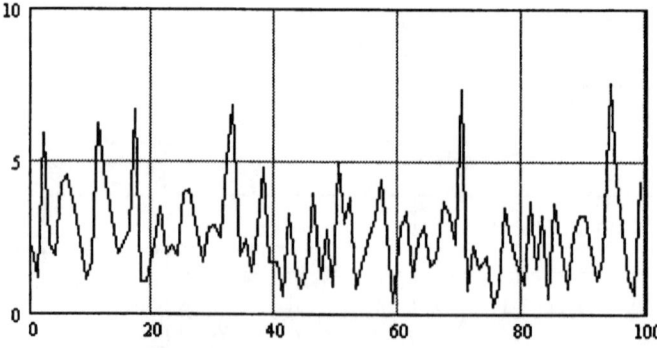

Figure 11.4 An interference magnitude (Volts) for Ricean pdf with rms amplitude $\sigma = 2$ V and $\alpha = 1$ vs. number of samples (signal length to correlation interval ratio is 100).

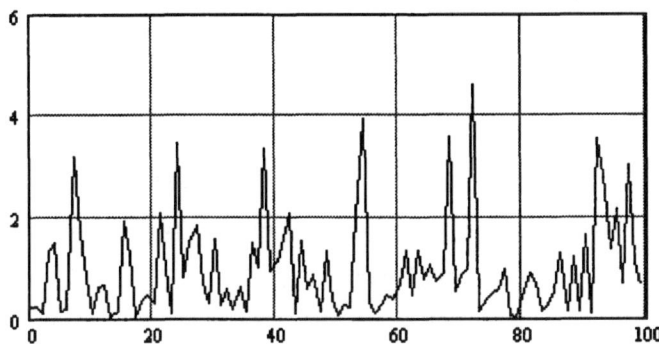

Figure 11.5 An interference magnitude (Watts) for exponential pdf with mean power
σ = 1 W vs. number of samples (signal length to correlation interval ratio
is 100).

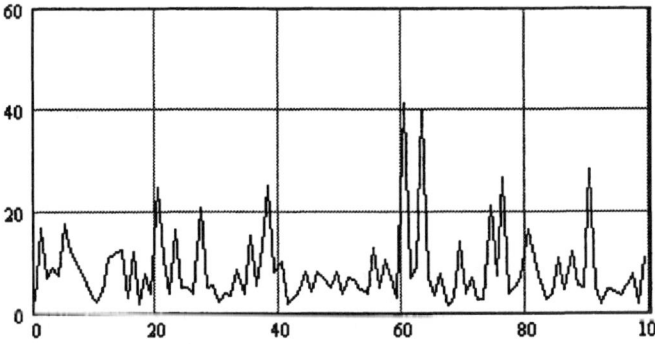

Figure 11.6 An interference magnitude (Volts) for log-normal pdf with rms amplitude
σ = 1 V, mean-to-median ratio mM = 5 vs. number of samples (signal length
to correlation interval ratio is 100).

11.2 Receiving System Noise Figure

11.2.1 DESCRIPTION

Calculates noise figure of the typical radar receiver consisting of RF input circuit, RF amplifier, mixer, IF amplifier, and analog-to-digital converter (ADC).

11.2.2 INPUT DATA

===

Parameter	Dimension and Description

===

RF Input Circuit

$LOSS_1 := 1.0$	dB, loss
$B_1 := 50$	MHz, bandwidth
$T_1 := 290$	°K, physical noise temperature of input termination used in noise figure definition

RF Amplifier

$NF_2 := 2$	dB, noise figure
$GN_2 := 30$	dB, gain
$B_2 := 50$	MHz, bandwidth

Mixer and IF Amplifier

$NF_3 := 10$	dB, noise figure
$GN_3 := 50$	dB, gain
$B_3 := 5$	MHz, bandwidth

Analog-to-Digital Converter

$N := 7$	number of bits used for signal digitization
$V := 3$	Volt, full scale of amplitudes quantized by ADC

===

11.2.3 MODEL

11.2.3.1 Constants

$$k := 1.38054 \cdot 10^{-23}$$ W*s/°K, Boltzmann's constant

$$dB(x) := 10 \cdot \log\left(\frac{x}{10^{-3}}\right)$$

conversion equation

$$ratio(x) := 10^{0.1 \cdot x}$$

conversion equation

11.2.3.2 Noise Parameters for RF Input Circuit

A. Parameters at the Input (Point 1, Figure 11.7)

$p_1 := k \cdot T_1$ $p_1 \cdot 10^{21} = 4.004$ W/Hz, noise power density

$P_1 := p_1 \cdot B_1 \cdot 10^6$ $P_1 = 2.002 \times 10^{-13}$ W, noise power

$p_{1\_dBm} := dB(p_1)$ $p_{1\_dBm} = -173.976$ dB(mW/Hz), noise power density

$P_{1\_dBm} := dB(P_1)$ $P_{1\_dBm} = -96.986$ dBm, noise power

B. Contribution of Elements Between Point 1 and Point 2 to the Equivalent Input Noise at Point 1 (Figure 11.7)

$L_1 := ratio(LOSS_1)$ $L_1 = 1.259$ input circuit loss

$p_2 := p_1 \cdot L_1$ $p_2 \cdot 10^{21} = 5.04$ W/Hz, noise power density

$P_2 := p_2 \cdot B_1 \cdot 10^6$ $P_2 = 2.52 \times 10^{-13}$ W, noise power

$p_{2\_dBm} := dB(p_2)$ $p_{2\_dBm} = -172.976$ dB(mW/Hz), noise power density

$P_{2\_dBm} := dB(P_2)$ $P_{2\_dBm} = -95.986$ dBm, noise power

C. Noise Figure Contribution from Elements Between Points 1 and Point 2, Referred to Point 1

$NF_{12} := 10 \cdot \log(L_1)$ $NF_{12} = 1$ dB, noise figure for point 1 to point 2 path

$\Delta NF_{12} := NF_{12}$ $\Delta NF_{12} = 1$ dB, noise figure contribution by front end

11.2.3.3 Noise Parameters for RF Amplifier

A. Contribution of Elements Between Point 1 and Point 3 to the Equivalent Input Noise at Point 1 (Figure 11.7)

$F_2 := \text{ratio}(NF_2)$ $F_2 = 1.585$ amplifier noise figure

$G_2 := \text{ratio}(GN_2)$ $G_2 = 1 \times 10^3$ amplifier gain

$F_{13} := L_1 \cdot F_2$ $F_{13} = 1.995$ noise figure contribution from elements
between points 1 and 3

$G_{13} := \dfrac{G_2}{L_1}$ $G_{13} = 794.328$ gain between points 1 and 3

$p_3 := p_1 \cdot F_{13} \cdot G_{13}$ $p_3 \cdot 10^{18} = 6.345$ W/Hz, noise power density

$B_{13} := \begin{vmatrix} B_1 & \text{if } B_1 \le B_2 \\ B_2 & \text{otherwise} \end{vmatrix}$ $B_{13} = 50$ MHz, the narrowest bandwidth
between points 1 and 3

$P_3 := p_3 \cdot B_{13} \cdot 10^6$ $P_3 = 3.173 \times 10^{-10}$ W, noise power

$P_{3\_dBm} := dB(p_3)$ $P_{3\_dBm} = -141.976$ dB(mW/Hz), noise power density

$P_{3\_dBm} := dB(P_3)$ $P_{3\_dBm} = -64.986$ dBm, noise power

B. Noise Figure Contribution from Elements Between Point 1 and Point 3, Referred to Point 1

$NF_{13} := 10 \cdot \log(F_{13})$ $NF_{13} = 3$ dB, noise figure for point 1 to point 3 path

$\Delta NF_{23} := NF_{13} - NF_{12}$ dB, noise figure contribution by amplifier

$\Delta NF_{23} = 2$

11.2.3.4 Noise Parameters for Mixer and IF Amplifier

A. Contribution of Elements Between Point 1 and Point 4 to the Equivalent Input Noise at Point 1 (Figure 11.7)

$F_3 := \text{ratio}(NF_3)$ $F_3 = 10$ mixer and IF amplifier noise figure

$$G_3 := \text{ratio}(GN_3) \quad G_3 = 1 \times 10^5 \qquad \text{mixer and IF amplifier gain}$$

$$F_{14} := L_1 \cdot \left(F_2 + \frac{F_3 - 1}{G_{13}} \right) \qquad \text{noise figure contribution from elements}$$
$$\text{between points 1 and 4}$$

$$F_{14} = 2.01$$

$$G_{14} := G_2 \cdot G_3 \qquad G_{14} = 1 \times 10^8 \qquad \text{gain between points 1 and 4}$$

$$p_4 := p_1 \cdot F_{14} \cdot G_{14} \qquad p_4 \cdot 10^{14} = 80.453 \qquad \text{W/Hz, noise power density}$$

$$B_{14} := B_3 \qquad B_{14} = 5 \qquad \text{MHz, bandwidth}$$

Note: bandwidth B_{14} is equal to the bandwidth of the narrowest filter (typically it is the last filter prior to analog-to-digital conversion).

$$P_4 := p_4 \cdot B_{14} \cdot 10^6 \qquad P_4 = 4.023 \times 10^{-6} \qquad \text{W, noise power}$$

$$P_{4\_dBm} := dB(p_4) \quad P_{4\_dBm} = -90.945 \qquad \text{dB(mW/Hz), noise power density}$$

$$P_{4\_dBm} := dB(P_4) \quad P_{4\_dBm} = -23.955 \qquad \text{dBm, noise power}$$

B. Noise Figure Contribution from Elements Between Point 1 and Point 4, Referred to Point 1

$$NF_{14} := 10 \cdot \log(F_{14}) \quad NF_{14} = 3.031 \quad \text{dB, noise figure for point 1 to point 4 path}$$

$$\Delta NF_{34} := NF_{14} - NF_{13} \qquad \text{dB, noise figure contribution by}$$
$$\text{mixer and IF amplifier}$$

$$\Delta NF_{34} = 0.031$$

11.2.3.5 Noise Parameters for Analog-to-Digital Converter

A. Contribution of Elements Between Point 1 and Point 5 to the Equivalent Input Noise at Point 1 (Figure 11.7)

$$\Delta V := \frac{V}{2^{N-1}} \qquad \Delta V = 0.047 \qquad \text{V, quantization level}$$

$$P_{qn} := \frac{\Delta V^2}{12} \qquad P_{qn} = 1.831 \times 10^{-4} \qquad \text{W, quantization noise power}$$

$$P_5 := P_4 + P_{qn} \qquad P_5 = 1.871 \times 10^{-4} \qquad \text{W, noise power}$$

$$p_5 := \frac{P_5}{B_{14} \cdot 10^6} \qquad p_5 = 3.743 \times 10^{-11} \qquad \text{W/Hz, noise power density}$$

$$P_{5\_dBm} := dB(p_5) \qquad P_{5\_dBm} = -74.268 \qquad \text{dB(mW/Hz), noise power density}$$

$$P_{5\_dBm} := dB(P_5) \qquad P_{5\_dBm} = -7.279 \qquad \text{dBm, noise power}$$

B. Noise Figure Contribution from Elements Between Point 1 and Point 5, Referred to Point 1

$$F_4 := \frac{P_5}{P_4} \qquad F_4 = 46.519 \qquad \text{ADC noise figure}$$

$$NF_4 := 10 \cdot \log(F_4) \qquad NF_4 = 16.676 \qquad \text{dB, ADC noise figure}$$

$$F_{15} := L_1 \cdot \left(F_2 + \frac{F_3 - 1}{G_{13}} + \frac{F_4 - 1}{G_{14}} \right) \qquad \text{noise figure contribution from elements between points 1 and 5}$$

$$F_{15} = 2.01$$

$$NF_{15} := 10 \cdot \log(F_{15}) \quad NF_{15} = 3.031 \qquad \text{dB, noise figure for point 1 to point 5 path}$$

$$\Delta NF_{45} := NF_{15} - NF_{14} \qquad \text{dB, noise figure contribution by analog-to-digital converter}$$

$$\Delta NF_{45} = 1.238 \times 10^{-6}$$

11.2.3.6 Total Receiver Noise Figure

The total receiver noise figure and contribution of various stages are given at Figure 11.7.

11.2.4 OUTPUT DATA

Receiver noise figure $NF_{15} = 3.031$ dB for input data specified

11.2.5 REFERENCE

[1] Barton, D. K., and S. A. Leonov, (eds.), *Radar Technology Encyclopedia*, Norwood, MA: Artech House, 1997, p. 287.

11.2.6 EXAMPLES

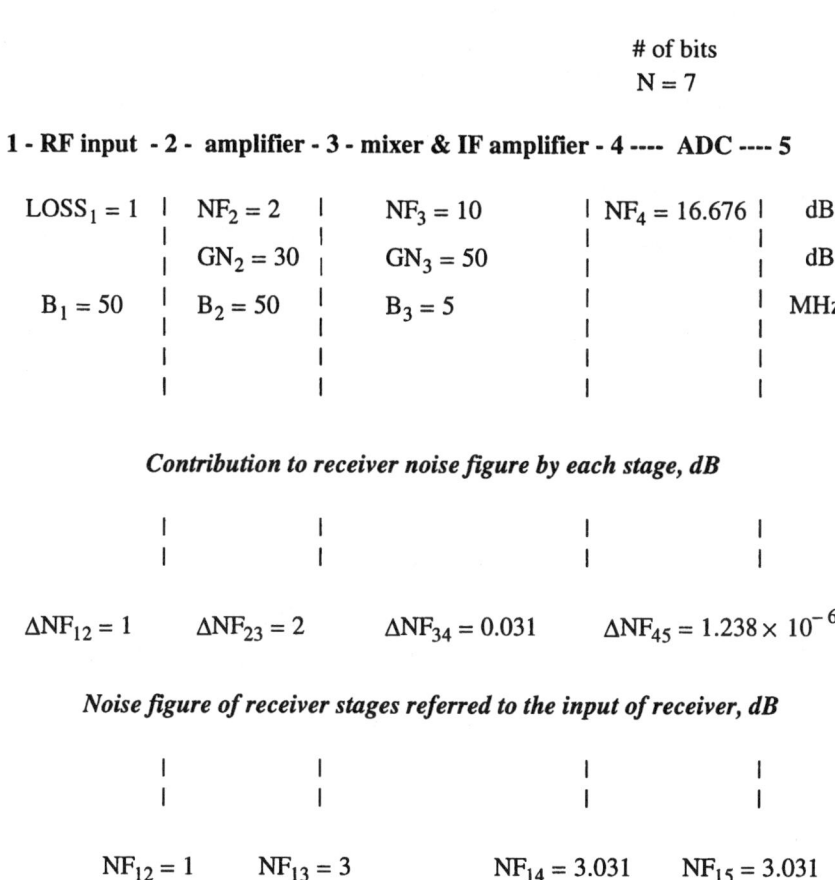

# of bits
N = 7

1 - RF input - 2 - amplifier - 3 - mixer & IF amplifier - 4 ---- ADC ---- 5

$LOSS_1 = 1$	$NF_2 = 2$	$NF_3 = 10$	$NF_4 = 16.676$	dB
	$GN_2 = 30$	$GN_3 = 50$		dB
$B_1 = 50$	$B_2 = 50$	$B_3 = 5$		MHz

Contribution to receiver noise figure by each stage, dB

$\Delta NF_{12} = 1$ $\Delta NF_{23} = 2$ $\Delta NF_{34} = 0.031$ $\Delta NF_{45} = 1.238 \times 10^{-6}$

Noise figure of receiver stages referred to the input of receiver, dB

$NF_{12} = 1$ $NF_{13} = 3$ $NF_{14} = 3.031$ $NF_{15} = 3.031$

Figure 11.7 Receiver noise figure calculation chart.

11.3 Signal-to-Interference Ratio

11.3.1 DESCRIPTION

Calculates signal-to-interference ratio for a generic pulsed radar. The following interference is covered: thermal noise, surface clutter (land and sea), volume clutter (precipitation), passive jamming (chaff), and active jamming (stand-off noise jammer). *Assumption:* unambiguous range operation when target and interference occur at the same range is considered.

11.3.2 INPUT DATA

==

Parameter	*Dimension and Description*

==

interference := 1	$1 \rightarrow$ thermal noise
	$2 \rightarrow$ land clutter
	$3 \rightarrow$ sea clutter
	$4 \rightarrow$ precipitation
	$5 \rightarrow$ noise jammer
	$6 \rightarrow$ chaff

Target Parameters

$\sigma_t := 5$	m², radar cross section of the target
$R := 20$	nm, return range for target and interference

Radar Parameters

$P_t := 10 \cdot 10^3$	W, transmitter peak power
$\tau_t := 5$	µs, transmitted pulsewidth
$\tau_r := 0.5$	µs, receive pulsewidth after processing
$\psi_3 := 1$	deg, azimuth beamwidth
$\theta_3 := 3$	deg, elevation beamwidth
$G_{t\_t} := 40$	dB, transmit antenna gain in direction to the target
$G_{r\_t} := 30$	dB, receive antenna gain in direction to the target
$f := 2000$	MHz, radar frequency
$T_s := 650$	°K, system noise temperature
$M := 1.0$	dB, receiver mismatch loss
$L_{tx} := 2.0$	dB, transmission line loss in transmit channel

$L_p := 1.3$ dB, beamshape loss

Note: typically, beamshape loss is 0 dB for tracking radar and at least 1.3 dB for search radar or tracking radar during acquisition mode.

$L_x := 8$ dB, processing loss

$G_x := 7$ dB, processing gain

$L_{oth} := 0$ dB, other losses

Note: processing gain G_x and processing loss L_x (a sum of fluctuation loss, integration loss, CFAR loss, straddling losses, etc.) in maximum detection range calculation (Blake Chart) are typically included in detectability factor D_x calculation.

$F_t := 1$ pattern-propagation factor for the radar-target path

$L_{\alpha\_t} := 0.5$ dB, two-way atmospheric attenuation loss
for the radar-target path

Note: Parameters F_t and L_{α_t} have to be calculated for range R specified above using models cited in Sections 5.3 and 8.1.

Land Clutter Parameters

$\gamma_{lc} := 0.05$ m^2/m^2, clutter mean reflectivity

$\psi_{lc} := 3$ deg, depression angle for radar-clutter cell path

$G_{t\_lc} := 10$ dB, transmit antenna gain in direction to the clutter cell

$G_{r\_lc} := 10$ dB, receive antenna gain in direction to the clutter cell

$I_{lc} := 50$ dB, improvement factor in surface clutter environment

$L_{lc} := 3$ dB, land clutter distribution loss

Note: Parameters F_{lc} and L_{α_lc} have to be calculated for range R specified above using models cited in Sections 5.3 and 8.1.

$L_{\alpha\_lc} := 0.9$ dB, two-way atmospheric attenuation loss for
the radar-clutter cell path

$F_{lc} := 1$ pattern-propagation factor for the radar-clutter cell path

Note: land clutter reflectivity parameters are typically as follows:

Terrain type	rms Roughness of the surface, σ_h	Surface reflectivity γ
Mountains	100	0.320
Urban	10	0.320
Wooded hills	10	0.100
Rolling hills	10	0.063
Farmland, desert	3	0.032
Flatland	1	0.010
Smooth surface	0.3	0.0032

Sea Clutter Parameters

$S := 4$ sea state

$\psi_{sc} := 30$ deg, depression angle for radar-clutter cell path

$G_{t\_sc} := 20$ dB, transmit antenna gain in direction to the clutter cell

$G_{r\_sc} := 20$ dB, receive antenna gain in direction to the clutter cell

$I_{sc} := 50$ dB, improvement factor in surface clutter environment

$L_{sc} := 3$ dB, sea clutter distribution loss

Note: parameters F_{sc} and L_{α_sc} have to be calculated for range R specified above using models cited in Sections 5.3 and 8.1.

$L_{\alpha\_sc} := 0.9$ dB, two-way atmospheric attenuation loss for the radar-clutter cell path

$F_{sc} := 0.75$ pattern-propagation factor for the radar-clutter cell path

Note: sea state is typically as follows:

Sea state	0	1	2	3	4	5	6	7	8	9
Beaufort wind scale number	1	2	3	4	5	6	7	8	9	10
$\sigma_h(m)$	0.003	0.027	0.090	0.210	0.420	0.720	1.14	1.70	2.43	3.33

Precipitation Clutter Parameters

$G_{t\_pc} := 40$	dB, transmit antenna gain in direction to the clutter cell
$G_{r\_pc} := 30$	dB, receive antenna gain in direction to the clutter cell
$I_{pc} := 10$	dB, improvement factor in precipitation clutter environment

Note: typically, integrated cancellation ratio is equal to 0 dB for linear polarization and 15–20 dB for circular polarization. If circular polarization is chosen, 3-dB loss should be included as L_{oth} in the Radar Parameters section of the input menu.

$ICR := 0$ dB, integrated cancellation ratio

Note: Parameters F_{pc} and L_{α_pc} have to be calculated for range R specified above using models cited in Sections 5.3 and 8.1.

$L_{\alpha\_pc} := 0.5$	dB, two-way atmospheric attenuation loss for the radar-clutter cell path
$F_{pc} := 1$	patern-propagation factor for the radar-clutter cell path

Precipitation Parameters

$r := 25$	mm/hour, precipitation rate
$a := 200$	precipitation model parameter
$b := 1.6$	precipitation model parameter
$K := 0.93$	precipitation model parameter

Note: precipitation model parameters are typically as follows:

Precipitation	a	b	K	λ (m)
Stratiform rain	200	1.60	0.930	> 0.02
	330	1.54	0.930	0.02
	570	1.00	0.930	0.0086
	280	0.80	0.930	0.06
	23	0.60	0.930	0.0032
Orographic rain	31	1.71	0.930	> 0.02
Thunderstorm	486	1.37	0.930	> 0.02
Snow (aggregate)	2000	2.00	0.197	> 0.02
Snow (dry)	500	1.6	0.197	> 0.02

Active Jamming Parameters

$P_{aj} := 10 \cdot 10^6$ W, jammer peak power

$G_{aj} := 0$ dB, jammer antenna gain in direction to the radar

$G_{r\_aj} := 0$ dB, radar receive antenna gain in direction to the jammer

$B_{aj} := 10^3$ MHz, jammer noise bandwidth

Note: parameters F_{aj} and L_{α_aj} have to be calculated for range R specified above using models cited in Sections 5.3 and 8.1.

$L_{\alpha\_aj} := 0.8$ dB, two-way atmospheric attenuation loss for the jammer-to-radar path

$F_{aj} := 1$ patern-propagation factor for the jammer-to-radar path

Passive Jamming (Chaff) Parameters

$N := 1000$ total number of dipoles

$G_{t\_pj} := 40$ dB, transmit antenna gain in direction to chaff

$G_{r\_pj} := 30$ dB, receive antenna gain in direction to chaff

Note: parameters F_{pj} and L_{α_pj} have to be calculated for range R specified above using models cited in Sections 5.3 and 8.1.

$L_{\alpha\_pj} := 0.2$ dB, two-way atmospheric attenuation loss for the chaff-to-radar path

$F_{pj} := 1$ pattern-propagation factor for the chaff-to-radar path

===

11.3.3 MODEL

11.3.3.1 Constants

$k := 1.38054 \cdot 10^{-23}$ W*s/°K, Boltzmann's constant

$c := 2.997925 \cdot 10^8$ m/s, speed of light

$z := \dfrac{\pi}{180}$ coefficient to transform degrees to radians

11.3.3.2 Total Loss

$L_{\Sigma\_t} := M + L_{tx} + L_p + L_{\alpha\_t} + L_x + L_{oth} - G_x$

$L_{\Sigma\_t} = 5.8$ dB, total loss for radar-target path

$$L_{\Sigma\_aj} := M + L_{tx} + L_p + L_{\alpha\_aj} + L_x + L_{oth} - G_x$$

$$L_{\Sigma\_aj} = 6.1 \qquad\qquad\qquad \text{dB, total loss for radar-jammer path}$$

$$L_{\Sigma\_pj} := M + L_{tx} + L_p + L_{\alpha\_pj} + L_x + L_{oth} - G_x$$

$$L_{\Sigma\_pj} = 5.5 \qquad\qquad\qquad \text{dB, total loss for radar-chaff path}$$

Note: processing gain G_x and processing loss L_x (a sum of fluctuation loss, integration loss, CFAR loss, straddling losses, etc.) in target maximum detection range calculation are typically included in detectability factor D_x value. If signal-to-interference ratio has to be compared with D_x to calculate maximum detection range in interference, make sure that corresponding losses are included only once either in D_x calculation or in signal-to-interference ratio calculation (the latter approach is chosen in this model).

11.3.3.3 Transformation of Input Data

$$\lambda := \frac{c}{f \cdot 10^6} \qquad \lambda = 0.15 \qquad\qquad \text{m, radar wavelength}$$

$$R_m := 1852 \cdot R \qquad R_m = 3.704 \times 10^4 \qquad \text{m, target and interference range}$$

$$\text{ratio}(x) := 10^{0.1 \cdot x} \qquad\qquad\qquad\qquad \text{conversion equation}$$

$$G_{t\_t\_dim} := \text{ratio}\left(G_{t\_t}\right) \qquad G_{t\_t\_dim} = 1 \times 10^4 \qquad \begin{array}{r} \text{radar transmit antenna gain} \\ \text{in direction to the target} \end{array}$$

$$G_{r\_t\_dim} := \text{ratio}\left(G_{r\_t}\right) \qquad G_{r\_t\_dim} = 1 \times 10^3 \qquad \begin{array}{r} \text{radar receive antenna gain} \\ \text{in direction to the target} \end{array}$$

$$G_{t\_lc\_dim} := \text{ratio}\left(G_{t\_lc}\right) \qquad G_{t\_lc\_dim} = 10 \qquad \begin{array}{r} \text{radar transmit antenna gain} \\ \text{in direction to land clutter cell} \end{array}$$

$$G_{r\_lc\_dim} := \text{ratio}\left(G_{r\_lc}\right) \qquad G_{r\_lc\_dim} = 10 \qquad \begin{array}{r} \text{radar receive antenna gain} \\ \text{in direction to land clutter cell} \end{array}$$

$$G_{t\_sc\_dim} := \text{ratio}\left(G_{t\_sc}\right) \qquad G_{t\_sc\_dim} = 100 \qquad \begin{array}{r} \text{radar transmit antenna gain} \\ \text{in direction to sea clutter cell} \end{array}$$

$$G_{r\_sc\_dim} := \text{ratio}\left(G_{r\_sc}\right) \qquad G_{r\_sc\_dim} = 100 \qquad \begin{array}{r} \text{radar receive antenna gain} \\ \text{in direction to sea clutter cell} \end{array}$$

$G_{t\_pc\_dim} := \text{ratio}(G_{t\_pc})$ $G_{t\_pc\_dim} = 1 \times 10^4$ radar transmit antenna gain in direction to precipitation clutter cell

$G_{r\_pc\_dim} := \text{ratio}(G_{r\_pc})$ $G_{r\_pc\_dim} = 1 \times 10^3$ radar receive antenna gain in direction to precipitation clutter cell

$G_{aj\_dim} := \text{ratio}(G_{aj})$ $G_{aj\_dim} = 1$ the jammer antenna gain in direction to the radar

$G_{r\_aj\_dim} := \text{ratio}(G_{r\_aj})$ $G_{r\_aj\_dim} = 1$ receive antenna gain in direction to the jammer

$G_{t\_pj\_dim} := \text{ratio}(G_{t\_pj})$ $G_{t\_pj\_dim} = 1 \times 10^4$ transmit antenna gain in direction to chaff

$G_{r\_pj\_dim} := \text{ratio}(G_{r\_pj})$ $G_{r\_pj\_dim} = 1 \times 10^3$ receive antenna gain in direction to chaff

$I_{lc\_dim} := \text{ratio}(I_{lc})$ $I_{lc\_dim} = 1 \times 10^5$ improvement factor for land clutter

$I_{sc\_dim} := \text{ratio}(I_{sc})$ $I_{sc\_dim} = 1 \times 10^5$ improvement factor for sea clutter

$I_{pc\_dim} := \text{ratio}(I_{pc})$ $I_{pc\_dim} = 10$ improvement factor for precipitation clutter

$ICR\_dim := \text{ratio}(ICR)$ $ICR\_dim = 1$ integrated cancellation ratio

$M\_dim := \text{ratio}(M)$ $M\_dim = 1.259$ receiver mismatch loss

$L_{tx\_dim} := \text{ratio}(L_{tx})$ $L_{tx\_dim} = 1.585$ transmission line loss in transmit channel

$L_{p\_dim} := \text{ratio}(L_p)$ $L_{p\_dim} = 1.349$ beamshape loss

$L_{oth\_dim} := \text{ratio}(L_{oth})$ $L_{oth\_dim} = 1$ other losses

$L_{lc\_dim} := \text{ratio}(L_{lc})$ $L_{lc\_dim} = 1.995$ land clutter distribution loss

$$L_{sc\_dim} := ratio(L_{sc}) \qquad L_{sc\_dim} = 1.995$$

sea clutter distribution loss

$$L_{\alpha\_lc\_dim} := ratio(L_{\alpha\_lc}) \quad L_{\alpha\_lc\_dim} = 1.23$$

atmospheric loss for radar-land clutter cell

$$L_{\alpha\_sc\_dim} := ratio(L_{\alpha\_sc}) \quad L_{\alpha\_sc\_dim} = 1.23$$

atmospheric loss for radar-sea clutter cell

$$L_{\alpha\_pc\_dim} := ratio(L_{\alpha\_pc}) \quad L_{\alpha\_pc\_dim} = 1.122$$

atmospheric loss for radar-precipitation clutter cell

$$L_{\Sigma\_t\_dim} := ratio(L_{\alpha\_pc}) \quad L_{\Sigma\_t\_dim} = 1.122$$

total loss for radar-target path

$$L_{\Sigma\_aj\_dim} := ratio(L_{\Sigma\_aj}) \quad L_{\Sigma\_aj\_dim} = 4.074$$

total loss for jammer-target path

$$L_{\Sigma\_pj\_dim} := ratio(L_{\Sigma\_pj}) \quad L_{\Sigma\_pj\_dim} = 3.548$$

total loss for chaff-target path

11.3.3.4 Signal-to-Noise Ratio

$$E_s := \frac{P_t \cdot \tau_t \cdot 10^{-6} \cdot G_{t\_t\_dim} \cdot G_{r\_t\_dim} \cdot \lambda^2 \cdot F_t^4 \cdot \sigma_t}{(4 \cdot \pi)^3 \cdot L_{\Sigma\_t\_dim} \cdot R_m^4}$$

W*s, signal energy

$$E_s \cdot 10^{20} = 1.34 \times 10^3$$

$$E_n := k \cdot T_s \qquad E_n \cdot 10^{20} = 0.897$$

W*s, thermal noise energy

$$\frac{E_s}{E_n} = 1.494 \times 10^3$$

signal-to-noise ratio

$$SIR1 := 10 \cdot \log\left(\frac{E_s}{E_n}\right) \qquad SIR1 = 31.742$$

dB, signal-to-noise ratio

11.3.3.5 Signal-to-Land Clutter Ratio

$$\sigma0 := \gamma_{lc} \cdot \sin(z \cdot \psi_{lc}) \qquad \sigma0 = 2.617 \times 10^{-3}$$

m²/m², clutter reflectivity

$$E_{lc\_in} := \frac{P_t \cdot \tau_t \cdot 10^{-6} \cdot \tau_r \cdot 10^{-6} \cdot G_{t\_lc\_dim} \cdot G_{r\_lc\_dim} \cdot \lambda^2 \cdot c \cdot z \cdot \psi_3 \cdot \sigma0 \cdot F_{lc}^4 \cdot L_{lc\_dim}}{(4 \cdot \pi)^3 \cdot 2 \cdot M\_dim \cdot L_{tx\_dim} \cdot L_{p\_dim} \cdot L_{\alpha\_lc\_dim} \cdot L_{oth\_dim} \cdot R_m^3}$$

$E_{lc\_in} \cdot 10^{20} = 0.23$

W*s, input energy of the clutter source
before doppler processing

$E_s \cdot 10^{20} = 1.34 \times 10^3$

W*s, signal energy

$\dfrac{E_s}{E_{lc\_in}} = 5.833 \times 10^3$

signal-to-clutter ratio before doppler processing

$SC := 10 \cdot \log\left(\dfrac{E_s}{E_{lc\_in}}\right)$ $SC = 37.659$

dB, signal-to-clutter ratio before
doppler processing

$E_{lc\_out} := \dfrac{E_{lc\_in}}{I_{lc\_dim}}$

W*s, output energy of the clutter source
after doppler processing

$E_{lc\_out} \cdot 10^{20} = 2.298 \times 10^{-6}$

$\dfrac{E_s}{E_{lc\_out}} = 5.833 \times 10^8$

signal-to-clutter ratio after doppler processing

$SIR2 := 10 \cdot \log\left(\dfrac{E_s}{E_{lc\_out}}\right)$ $SIR2 = 87.659$

dB, signal-to-clutter ratio after
doppler processing

11.3.3.6 Signal-to-Sea Clutter Ratio

$\gamma_{sc\_db} := 6 \cdot S - 10 \cdot \log(\lambda) - 64$

dB, scattering effectiveness of the sea

$\gamma_{sc\_db} = -31.758$

$\gamma_{sc} := 10^{\frac{\gamma_{sc\_db}}{10}}$ $\gamma_{sc} = 6.671 \times 10^{-4}$

scattering effectiveness of the sea

$\sigma 0 := \gamma_{sc} \cdot \sin(z \cdot \psi_{sc})$ $\sigma 0 = 3.336 \times 10^{-4}$

m2/m2, clutter reflectivity

$$E_{sc\_in} := \dfrac{P_t \cdot \tau_t \cdot 10^{-6} \cdot \tau_r \cdot 10^{-6} \cdot G_{t\_sc\_dim} \cdot G_{r\_sc\_dim} \cdot \lambda^2 \cdot c \cdot z \cdot \psi_3 \cdot \sigma 0 \cdot F_{sc}^4 \cdot L_{sc\_dim}}{(4 \cdot \pi)^3 \cdot 2 \cdot M\_dim \cdot L_{tx\_dim} \cdot L_{p\_dim} \cdot L_{\alpha\_sc\_dim} \cdot L_{oth\_dim} \cdot R_m^3}$$

$E_{sc\_in} \cdot 10^{20} = 0.927$

W*s, input energy of the clutter source before doppler processing

$E_s \cdot 10^{20} = 1.34 \times 10^3$

W*s, signal energy

$$\frac{E_s}{E_{sc\_in}} = 1.446 \times 10^3$$

signal-to-clutter ratio before doppler processing

$$SC := 10 \cdot \log\left(\frac{E_s}{E_{sc\_in}}\right) \qquad SC = 31.602$$

dB, signal-to-clutter ratio before doppler processing

$$E_{sc\_out} := \frac{E_{sc\_in}}{I_{sc\_dim}}$$

W*s, output energy of the clutter source after doppler processing

$E_{sc\_out} \cdot 10^{20} = 9.268 \times 10^{-6}$

$$\frac{E_s}{E_{sc\_out}} = 1.446 \times 10^8$$

signal-to-clutter ratio after doppler processing

$$SIR3 := 10 \cdot \log\left(\frac{E_s}{E_{sc\_out}}\right) \qquad SIR3 = 81.602$$

dB, signal-to-clutter ratio after doppler processing

11.3.3.7 Signal-to-Precipitation Clutter Ratio

$$\eta_v := \frac{\dfrac{\pi^5}{\lambda^4} \cdot K \cdot a \cdot 10^{-18} \cdot r^b}{ICR\_dim} \qquad \eta_v = 1.944 \times 10^{-8}$$

m²/m3, volume precipitation reflectivity

$$E_{pc\_in} := \frac{P_t \cdot \tau_t \cdot 10^{-6} \cdot G_{t\_pc\_dim} \cdot G_{r\_pc\_dim} \cdot \lambda^2 \cdot \eta_v \cdot z \cdot \theta_3 \cdot \tau_r \cdot 10^{-6} \cdot c \cdot z \cdot \psi_3 \cdot F_{pc}^4}{2 \cdot (4 \cdot \pi)^3 \cdot M\_dim \cdot L_{tx\_dim} \cdot L_{p\_dim}^2 \cdot L_{\alpha\_pc\_dim} \cdot L_{oth\_dim} \cdot R_m^2}$$

$E_{pc\_in} \cdot 10^{20} = 134.903$

W*s, input energy of the clutter source before doppler processing

$$E_s \cdot 10^{20} = 1.34 \times 10^3 \qquad\qquad \text{W*s, signal energy}$$

$$\frac{E_s}{E_{pc\_in}} = 9.935 \qquad\qquad \text{signal-to-clutter ratio before doppler processing}$$

$$SC := 10 \cdot \log\left(\frac{E_s}{E_{pc\_in}}\right) \qquad SC = 9.972 \qquad \text{dB, signal-to-clutter ratio before doppler processing}$$

$$E_{pc\_out} := \frac{E_{pc\_in}}{I_{pc\_dim}} \qquad\qquad \text{W*s, output energy of the clutter source after doppler processing}$$

$$E_{pc\_out} \cdot 10^{20} = 13.49$$

$$\frac{E_s}{E_{pc\_out}} = 99.355 \qquad\qquad \text{signal-to-clutter ratio after doppler processing}$$

$$SIR4 := 10 \cdot \log\left(\frac{E_s}{E_{pc\_out}}\right) \qquad SIR4 = 19.972 \qquad \text{dB, signal-to-clutter ratio after doppler processing}$$

11.3.3.8 Signal-to-Active Jamming Ratio

$$E_{aj} := \frac{P_{aj} \cdot G_{aj\_dim} \cdot G_{r\_aj\_dim} \cdot \lambda^2 \cdot F_{aj}^2}{(4 \cdot \pi)^2 \cdot B_{aj} \cdot 10^6 \cdot L_{\Sigma\_aj\_dim} \cdot R_m^2}$$

$$E_{aj} \cdot 10^{20} = 2.546 \times 10^4 \qquad\qquad \text{W*s, jammer power spectral density}$$

$$E_s \cdot 10^{20} = 1.34 \times 10^3 \qquad\qquad \text{W*s, signal energy}$$

$$\frac{E_s}{E_{aj}} = 0.053 \qquad\qquad \text{signal-to-jamming ratio}$$

$$SIR5 := 10 \cdot \log\left(\frac{E_s}{E_{aj}}\right) \qquad SIR5 = -12.786 \qquad \text{dB, signal-to-jamming ratio}$$

11.3.3.9 Signal-to-Passive Jamming Ratio

$$\sigma_d := 0.18 \cdot \lambda^2 \cdot N \qquad \sigma_d = 4.044 \qquad \text{m2, RCS of chaff dipoles}$$

$$E_{pj} := \frac{P_t \cdot \tau_t \cdot 10^{-6} \cdot G_{t\_pj\_dim} \cdot G_{r\_pj\_dim} \cdot \lambda^2 \cdot \sigma_d \cdot F_{pj}^{4}}{(4 \cdot \pi)^3 \cdot L_{\Sigma\_pj\_dim} \cdot R_m^{4}}$$

$$E_{pj} \cdot 10^{20} = 342.84 \qquad \text{W*s, chaff return power spectral density}$$

$$E_s \cdot 10^{20} = 1.34 \times 10^3 \qquad \text{W*s, signal energy}$$

$$\frac{E_s}{E_{pj}} = 3.909 \qquad \text{signal-to-jamming ratio}$$

$$SIR6 := 10 \cdot \log\left(\frac{E_s}{E_{pj}}\right) \qquad SIR6 = 5.921 \qquad \text{dB, signal-to-jamming ratio}$$

11.3.3.10 Signal-to-Interference Ratio

$$SIR := \begin{vmatrix} SIR1 & \text{if interference} = 1 \\ SIR2 & \text{if interference} = 2 \\ SIR3 & \text{if interference} = 3 \\ SIR4 & \text{if interference} = 4 \\ SIR5 & \text{if interference} = 5 \\ SIR6 & \text{if interference} = 6 \end{vmatrix}$$

dB, signal-to-interference ratio

11.3.4 OUTPUT DATA

Signal-to-interference ratio $SIR = 31.742$ for   interference = 1
and input data specified.

11.3.5 REFERENCE

[1] Barton, D. K., and S. A. Leonov, (eds.), *Radar Technology Encyclopedia*, Norwood, MA: Artech House, 1997, p. 423.

11.4 System Noise Temperature

11.4.1 DESCRIPTION

Calculates system noise temperature as the sum of antenna noise temperature, the receiving-transmission line noise temperature and receiver noise temperature. *Assumption*: Barton's model for atmospheric attenuation loss is used (see Section 8.1).

11.4.2 INPUT DATA

===

Parameter	Dimension and Description

===

$f := 400$	MHz, radar frequency
$k_\alpha := 0.01$	parameter of atmospheric attenuation loss

Note: parameter k_α is frequency-dependent. See Section 8.1 for the table of the parameters k_α for various frequency bands in clear and precipitation.

$\theta := 1$	deg, elevation angle to the point of radiation
space_noise := 1	1 → normal solar and galactic noise
	2 → minimum solar and galactic noise
	3 → maximum solar and galactic noise
$R0_{km} := 500$	km, sky noise path through atmosphere
$L_a := 0.75$	dB, dissipative loss within antenna
$T_{tr} := 290$	°K, the receiving-transmission-line thermal temperature
$L_r := 3$	dB, the receiving-transmission-line loss
$F_n := 5$	dB, receiver noise figure

===

11.4.3 MODEL

11.4.3.1 Constants

$T_0 := 290$	°K, conventional reference temperature
$z := \dfrac{\pi}{180}$	coefficient to transform degrees to radians

11.4.3.2 Galactic and Solar Noise Temperature

$$k_g := \begin{vmatrix} 3 \cdot 10^8 & \text{if space\_noise} = 1 \\ 5 \cdot 10^7 & \text{if space\_noise} = 2 \\ 1.8 \cdot 10^9 & \text{if space\_noise} = 3 \end{vmatrix}$$

parameter depending on space radiation intensity

$$T_g := \frac{k_g}{f^{2.5}} + 5 \qquad T_g = 98.75$$

°K, solar and galactic noise temperature

11.4.3.3 Noise Temperature Due to Atmospheric Attenuation

$$\theta_{eff} := z \cdot \theta + \frac{2.5 \cdot 10^{-4}}{z \cdot \theta + 0.028}$$

rad, effective elevation angle

$$R_{eff} := \frac{3.0}{\sin(\theta_{eff})}$$

km, effective range

$$L1_\alpha := k_\alpha \cdot R_{eff} \cdot \left(1 - \exp\left(-\frac{R0_{km}}{R_{eff}}\right)\right)$$

dB, one-way atmospheric attenuation loss

$$L1_\alpha = 1.279$$

$$L_\alpha := 10^{\frac{L1_\alpha}{10}} \qquad L_\alpha = 1.342$$

two-way atmospheric attenuation loss

$$T_{pa} := T_0 \cdot \left(1 - \frac{1}{\sqrt{L_\alpha}}\right) \qquad T_{pa} = 39.696$$

°K, noise temperature due to atmospheric attenuation

11.4.3.4 Sky Temperature

$$T_{sky} := T_g + T_{pa} \qquad T_{sky} = 138.446$$

°K, sky temperature

11.4.3.5 Antenna Noise Temperature

$$L_{a\_dim} := 10^{\frac{L_a}{10}} \qquad L_{a\_dim} = 1.189$$

antenna loss

$$T_a := \frac{(0.88 \cdot T_{sky} - 254)}{L_{a\_dim}} + T_0 \qquad T_a = 178.795 \qquad °\text{K, antenna noise temperature}$$

11.4.3.6 Transmission Line Noise Temperature

$$L_{r\_dim} := 10^{\frac{L_r}{10}} \qquad L_{r\_dim} = 1.995 \qquad \text{receiving-transmission-line loss}$$

$$T_r := T_{tr} \cdot (L_{r\_dim} - 1) \qquad T_r = 288.626 \qquad °\text{K , transmission line noise temperature}$$

11.4.3.7 Receiver Noise Temperature

$$F_{n\_dim} := 10^{\frac{F_n}{10}} \qquad F_{n\_dim} = 3.162 \qquad \text{receiver noise figure}$$

$$T_e := T_0 \cdot (F_{n\_dim} - 1) \qquad T_e = 627.061 \qquad °\text{K, receiver noise temperature}$$

11.4.3.8 System Noise Temperature

$$T_s := T_a + T_r + L_{r\_dim} \cdot T_e \qquad T_s = 1.719 \times 10^3 \qquad °\text{K, system noise temperature}$$

11.4.4 OUTPUT DATA

System noise temperature is $\quad T_s = 1.719 \times 10^3 \quad$ °K for input data specified.

11.4.5 REFERENCES

[1] Barton, D. K., and S. A. Leonov, (eds.), *Radar Technology Encyclopedia*, Norwood, MA: Artech House, 1997, pp. 437, 438.
[2] Skolnik M., (ed.), *Radar Handbook*, New York: McGraw-Hill, 1990, pp. 2.26– 2.31.

11.5 MTI, Limitations to Performance

11.5.1 DESCRIPTION

Calculates the maximum attainable improvement factor in the presence of equipment instabilities and clutter motion for MTI radar. *Assumptions:* ground-based surveillance radar, pulse compression, quadrature channels, Gaussian clutter spectrum centered around zero.

11.5.2 INPUT DATA

==

Parameter	Dimension and Description

==

Radar Parameters

$f := 1000$	MHz, radar frequency
$R_{ins} := 50$	nm, instrumented range
$PRF := 1000$	Hz, pulse repetition frequency
$\tau := 50$	μs, transmit pulsewidth
$K := 300$	pulse compression ratio
$n := 50$	number of pulses integrated
$v := 1.1$	PRF stagger ratio (maximum-to-minimum PRF)

Clutter Parameters

$\sigma_{vc} := 1$	m/s, rms clutter spectrum velocity spread

Canceler Parameters

$N := 4$	number of pulses employed by the canceler

Equipment Instabilities

$\Delta A := 0.1$	%, relative pulse amplitude variation
$\Delta f\_p := 5$	Hz, interpulse frequency variation (power-oscillator transmitter, e.g., magnetron-based)
$\Delta \psi := 0.05$	deg, interpulse phase variation
$\Delta f\_s := 0.05$	Hz, STALO frequency drift
$\Delta f\_c := 0.01$	Hz, COHO frequency drift
$\Delta t\_t := 10^{-9}$	s, pulse timing jitter

$\Delta t\_w := 10^{-9}$ s, pulsewidth jitter

$\Delta t\_adc := 0.1 \cdot 10^{-9}$ s, analog-to-digital converter (ADC) sampling time jitter

$M := 10$ # of bits in ADC

===

11.5.3 MODEL

11.5.3.1 Constants

$z := \dfrac{\pi}{180}$ coefficient to transform degrees to radians

$c := 2.997925 \cdot 10^{8}$ m/s, speed of light

11.5.3.2 Clutter Motion and Scan Modulation Effects

$\lambda := \dfrac{c}{f \cdot 10^{6}}$ $\lambda = 0.3$ m, wavelength

$\sigma_{cm} := \dfrac{2 \cdot \sigma_{vc}}{\lambda}$ $\sigma_{cm} = 6.671$ Hz, spectral spread of the clutter due to its motion

$\sigma_{sm} := \dfrac{\sqrt{\ln(2)} \cdot PRF}{\pi \cdot n}$ $\sigma_{sm} = 5.3$ Hz, spectral spread of the clutter due to scan modulation

$\sigma_{c} := \sqrt{\sigma_{cm}^{2} + \sigma_{sm}^{2}}$ $\sigma_{c} = 8.52$ Hz, spectral spread of the clutter due to motion and scan modulation

$I_0 := \dfrac{2^{N-1}}{(N-1)!} \cdot \left(\dfrac{PRF}{2 \cdot \pi \cdot \sigma_c} \right)^{2 \cdot (N-1)}$ improvement factor limit due to clutter motion and scan modulation

$I\_db_0 := 20 \cdot \log(I_0)$ $I\_db_0 = 155.062$ dB, improvement factor

11.5.3.3 Pulse-to-Pulse PRF Staggering

$I_1 := \dfrac{2.5 \cdot n}{v - 1}$ improvement factor limit due to PRF staggering

$I\_db_1 := 20 \cdot \log(I_1)$ $I\_db_1 = 61.938$ dB, improvement factor

11.5.3.4 Equipment Instabilities

$$I_2 := \frac{1}{\Delta A \cdot 0.01}$$

improvement factor limit due to pulse amplitude variation

$$I\_db_2 := 20 \cdot \log(I_2) \qquad I\_db_2 = 60 \qquad \text{dB, improvement factor}$$

$$I_3 := \frac{1}{\pi \cdot \Delta f\_p \cdot \tau \cdot 10^{-6}}$$

improvement factor limit due to interpulse frequency variation (power-oscillator transmitter only)

$$I\_db_3 := 20 \cdot \log(I_3) \qquad I\_db_3 = 62.098 \qquad \text{dB, improvement factor}$$

$$I_4 := \frac{1}{\Delta \psi \cdot z}$$

improvement factor limit due to interpulse phase variation

$$I\_db_4 := 20 \cdot \log(I_4) \qquad I\_db_4 = 61.183 \qquad \text{dB, improvement factor}$$

$$R_m := R_{ins} \cdot 1852 \qquad R_m = 9.26 \times 10^4 \qquad \text{m, instrumented range}$$

$$I_5 := \frac{1}{2 \cdot \pi \cdot \Delta f\_s \cdot 2 \cdot \dfrac{R_m}{c}}$$

improvement factor limit due to STALO frequency drift

$$I\_db_5 := 20 \cdot \log(I_5) \qquad I\_db_5 = 74.241 \qquad \text{dB, improvement factor}$$

$$I_6 := \frac{1}{2 \cdot \pi \cdot \Delta f\_c \cdot 2 \cdot \dfrac{R_m}{c}}$$

improvement factor limit due to COHO frequency drift

$$I\_db_6 := 20 \cdot \log(I_6) \qquad I\_db_6 = 88.22 \qquad \text{dB, improvement factor}$$

$$I_7 := \frac{\tau \cdot 10^{-6}}{\sqrt{2} \cdot \Delta t\_t \cdot \sqrt{K}}$$

improvement factor limit due to pulse timing jitter

$$I\_db_7 := 20 \cdot \log(I_7) \qquad I\_db_7 = 66.198 \qquad \text{dB, improvement factor}$$

$$I_8 := \frac{\tau \cdot 10^{-6}}{\Delta t\_w \cdot \sqrt{K}}$$ improvement factor limit due to pulsewidth jitter

$I\_db_8 := 20 \cdot \log(I_8)$ $I\_db_8 = 69.208$ dB, improvement factor

$$I_9 := \frac{\tau \cdot 10^{-6}}{\Delta t\_adc \cdot K}$$ improvement factor limit due to ADC jitter

$I\_db_9 := 20 \cdot \log(I_9)$ $I\_db_9 = 64.437$ dB, improvement factor

$$I_{10} := (2^M - 1) \cdot \sqrt{1.5}$$ improvement factor limit due to ADC
                                              quantization noise

$I\_db_{10} := 20 \cdot \log(I_{10})$ $I\_db_{10} = 61.958$ dB, improvement factor

$i := 0 .. 10$ cycle

$$I\_p_i := 10^{\frac{I\_db_i}{10}}$$ improvement factor conversion

$$I := \frac{1}{\sum_i \frac{1}{I\_p_i}}$$ total sum of improvement factors

$I\_db_\Sigma := 10 \cdot \log(I)$ dB, total maximum attainable improvement factor
                             due to clutter motion, antenna rotation and equipment
                                                                   limitations

11.5.4 OUTPUT DATA

Improvement factor (dB): $I\_db_\Sigma = 53.551$ for input data specified.

11.5.5 REFERENCES

[1] Barton, D. K., and S. A. Leonov, (eds.), *Radar Technology Encyclopedia*, Norwood, MA: Artech House, 1997, p. 281.
[2] Skolnik, M., *Introduction to Radar Systems*, New York: McGraw-Hill, 1980, p. 129.
[3] Skolnik, M., (ed.), *Radar Handbook*, New York: McGraw-Hill, 1990, pp. 15.45.
[4] Schleher, D. C., *MTI and Pulsed Doppler Radar*, Norwood, MA: Artech House, 1991, Ch. 3.

11.5.6 EXAMPLES

MTI Limitations Budget

Radar Parameters	
Frequency (MHz)	$f = 1 \times 10^3$
Instrumented Range (nm):	$R_{ins} = 50$
Pulse Repetition Frequency (Hz):	$PRF = 1 \times 10^3$
Transmit Pulsewidth (μs):	$\tau = 50$
Pulse Compression Ratio:	$K = 300$
Number of Pulses Integrated:	$n = 50$
Number of Pulses Employed by the Canceler:	$N = 4$

Contributing Factor	Value	Improvement Factor (dB)
Clutter Motion and Scan Modulation: Spectral Spread (Hz):	$\sigma_c = 8.52$	$I\_db_0 = 155.062$
PRF Staggering	$v = 1.1$	$I\_db_1 = 61.938$
Pulse Amplitude Variation (%):	$\Delta A = 0.1$	$I\_db_2 = 60$
Interpulse Frequency Variation (Hz):	$\Delta f\_p = 5$	$I\_db_3 = 62.098$
Interpulse Phase Variation (deg):	$\Delta \psi = 0.05$	$I\_db_4 = 61.183$
STALO Frequency Drift (Hz):	$\Delta f\_s = 0.05$	$I\_db_5 = 74.241$
COHO Frequency Drift (Hz):	$\Delta f\_c = 0.01$	$I\_db_6 = 88.22$
Pulse Timing Jitter (s):	$\Delta t\_t = 1 \times 10^{-9}$	$I\_db_7 - 66.198$
Pulsewidth Jitter (s):	$\Delta t\_w = 1 \times 10^{-9}$	$I\_db_8 = 69.208$
ADC Jitter (s):	$\Delta t\_adc = 1 \times 10^{-10}$	$I\_db_9 = 64.437$
ADC Quantization Noise: Number of Bits:	$M = 10$	$I\_db_{10} = 61.958$
TOTAL:		$I\_db_\Sigma = 53.551$

11.6 Subclutter Visibility

11.6.1 DESCRIPTION

Models subclutter visibility and improvement factor as a function of a target velocity. *Assumption*: Gaussian clutter spectrum, coherent integration of pulses, constant PRF (no staggering), unlimited dynamic range.

11.6.2 INPUT DATA

==

Parameter	Dimension and Description

==

frequency := 1000	MHz, radar frequency
f_s := 5	MHz, digital sampling rate
rpm := 20	rpm, antenna rotation rate
Θ := 1	deg, beamwidth
f_p := 1000	Hz, pulse repetition frequency
N := 3	number of pulses for coherent integration
K := 3	number of filters in a doppler bank

$$c := \begin{pmatrix} 0.85 & 0.85 + 0.85j & 0.85 \\ -0.85 \cdot j & -3 - 3 \cdot j & 0.85 \cdot j \\ -0.85 & 3 + 3 \cdot j & -0.85 \\ 0.85 \cdot j & -0.85 - 0.85 \cdot j & -0.85 \cdot j \end{pmatrix}$$

doppler filter coefficients

I_lim := 50	dB, equipment limitations for improvement factor (see Section 11.5)
V_{xc} := 10	dB, clutter visibility factor
v_{rc} := 0	m/s, clutter radial velocity
σ_v := 5	m/s, rms velocity spread due to clutter motion
Vmax := 1000	knots, maximum velocity to be plotted
$M := 2^{10}$	number of discrete velocity samples in response
$M = 1.024 \times 10^3$	

==

11.6.3 MODEL

11.6.3.1 Input Data Transformation

$$T_p := \frac{1}{f_p} \qquad\qquad T_p = 1 \times 10^{-3} \qquad\qquad \text{s, pulse repetition interval}$$

$$\lambda := \frac{2.997925 \cdot 10^8}{\text{frequency} \cdot 10^6} \qquad\qquad \lambda = 0.3 \qquad\qquad \text{m, radar wavelength}$$

$$m := 0 \, .. \, M - 1 \qquad\qquad\qquad\qquad\qquad\qquad\qquad \text{cycle}$$

$$v_m := V\text{max} \cdot \frac{m}{M} \qquad\qquad\qquad\qquad \text{knots, current velocity sample}$$

$$f_m := 2 \cdot \frac{v_m}{\lambda} \cdot \frac{1852}{3600} \qquad\qquad \text{Hz, current doppler frequency sample}$$

$$f\_max := 2 \cdot \frac{V\text{max}}{\lambda} \cdot \frac{1852}{3600} \qquad\qquad \text{Hz, maximum doppler frequency}$$

$$f\_max = 3.432 \times 10^3$$

$$p := \text{floor}\left(\frac{f\_max}{f_p}\right)$$

number of filter response

$$\text{filter\_peaks} := p + 1 \qquad \text{filter\_peaks} = 4 \qquad \begin{array}{l}\text{maxima occurring at } 0 - V_{max}\end{array}$$

11.6.3.2 Clutter Spectrum and Power

$$f0 := \frac{2 \cdot v_{rc}}{\lambda} \qquad\qquad f0 = 0 \qquad\qquad \text{Hz, clutter central frequency}$$

$$t_0 := \frac{\Theta}{\text{rpm} \cdot \dfrac{360}{60}} \qquad t_0 = 8.333 \times 10^{-3} \qquad\qquad \text{s, time-on-target}$$

$$\sigma_{c\_sm} := \frac{1.178}{\sqrt{2} \cdot \pi \cdot t_0} \qquad\qquad \text{Hz, spectrum spread by scanning modulation}$$

$$\sigma_{c\_sm} = 31.817$$

$$\sigma_{v\_sm} := \frac{\sigma_{c\_sm} \cdot \lambda}{2}$$

m/s, rms velocity spread by scanning modulation

$$\sigma_{v\_sm} = 4.769$$

$$\sigma_v = 5$$

m/s, rms velocity spread by clutter motion

$$\sigma_{v\Xi} := \sqrt{\sigma_v{}^2 + \left(\sigma_{v\_sm}\right)^2}$$

m/s, resultant rms velocity spread

$$\sigma_{v\Xi} = 6.91$$

$$\sigma_c := \frac{2 \cdot \sigma_{v\Xi}}{\lambda} \qquad \sigma_c = 46.098$$

Hz, resultant clutter spectrum spread

$$u := 0 .. p$$

cycle

$$S\_clutter_m := \sum_u \exp\left[-\frac{\left[f_m - \left(f0 + u \cdot f_p\right)\right]^2}{2 \cdot \left(\sigma_c\right)^2}\right]$$

discrete clutter spectrum

$$S\_clutter\_db_m := 10 \cdot \log\left(S\_clutter_m + 10^{-10}\right)$$

dB, discrete spectrum

$$z := 0 .. 1$$

cycle

$$S\_clutter\_d_m := \sum_z \exp\left[-\frac{\left[f_m - \left(f0 + z \cdot f_p\right)\right]^2}{2 \cdot \left(\sigma_c\right)^2}\right]$$

discrete clutter spectrum function

$$S\_clutter(f) := \sum_z \exp\left[-\frac{\left[f - \left(f0 + z \cdot f_p\right)\right]^2}{2 \cdot \left(\sigma_c\right)^2}\right]$$

continuous clutter spectrum function

$$P\_c := \int_0^{f_p} \left(S\_clutter(f) + 10^{-10}\right) df$$

clutter power

$$P\_c = 115.549$$

$$P\_c\_db := 10 \cdot \log(P\_c) \qquad P\_c\_db = 20.628 \qquad \text{dB, clutter power}$$

11.6.3.3 Filter Bank Frequency Response

$$n := 0..N-1 \qquad k := 0..K-1 \qquad \text{cycles}$$

$$c\_rms_{n,k} := \frac{c_{n,k}}{\sqrt{\displaystyle\sum_n \left(\left|c_{n,k} + 10^{-10}\right|\right)^2}} \qquad \begin{array}{l}\text{filter coefficients normalized}\\ \text{to noise bandwidth}\end{array}$$

$$c\_max_{n,k} := \frac{c_{n,k}}{\displaystyle\sum_n \left(\left|c_{n,k} + 10^{-10}\right|\right)} \qquad \begin{array}{l}\text{filter coefficients normalized}\\ \text{to maximum}\end{array}$$

Filter # 1 responses (discrete and continuous)

$$R1\_rms_m := \sum_n \left(c\_rms_{n,0} \cdot \exp\left(j \cdot 2 \cdot \pi \cdot n \cdot T_p \cdot f_m\right) + 10^{-10}\right)$$

$$R1\_max_m := \sum_n \left(c\_max_{n,0} \cdot \exp\left(j \cdot 2 \cdot \pi \cdot n \cdot T_p \cdot f_m\right) + 10^{-10}\right)$$

$$R1(f) := \sum_n \left(c\_max_{n,0} \cdot \exp\left(j \cdot 2 \cdot \pi \cdot n \cdot T_p \cdot f\right)\right)$$

Filter # 2 responses (discrete and continuous)

$$R2\_rms_m := \sum_n \left(c\_rms_{n,1} \cdot \exp\left(j \cdot 2 \cdot \pi \cdot n \cdot T_p \cdot f_m\right) + 10^{-10}\right)$$

$$R2\_max_m := \sum_n \left(c\_max_{n,1} \cdot \exp\left(j \cdot 2 \cdot \pi \cdot n \cdot T_p \cdot f_m\right) + 10^{-10}\right)$$

$$R2(f) := \sum_n \left(c\_max_{n,1} \cdot \exp\left(j \cdot 2 \cdot \pi \cdot n \cdot T_p \cdot f\right)\right)$$

Filter # 3 responses (discrete and continuous)

$$R3\_rms_m := \sum_n \left(c\_rms_{n,\,2} \cdot \exp\left(j \cdot 2 \cdot \pi \cdot n \cdot T_p \cdot f_m\right) + 10^{-10} \right)$$

$$R3\_max_m := \sum_n \left(c\_max_{n,\,2} \cdot \exp\left(j \cdot 2 \cdot \pi \cdot n \cdot T_p \cdot f_m\right) + 10^{-10} \right)$$

$$R3(f) := \sum_n \left(c\_max_{n,\,2} \cdot \exp\left(j \cdot 2 \cdot \pi \cdot n \cdot T_p \cdot f\right) \right)$$

Filter gains normalized to noise bandwidth

$G1\_rms_m := 20 \cdot \log\left(\left| R1\_rms_m \right| \right)$ dB, filter # 1 gain

$G2\_rms_m := 20 \cdot \log\left(\left| R2\_rms_m \right| \right)$ dB, filter # 2 gain

$G3\_rms_m := 20 \cdot \log\left(\left| R3\_rms_m \right| \right)$ dB, filter # 3 gain

Filter gains normalized to maximum

$G1\_max_m := 20 \cdot \log\left(\left| R1\_max_m \right| \right)$ dB, filter # 1 gain

$G2\_max_m := 20 \cdot \log\left(\left| R2\_max_m \right| \right)$ dB, filter # 2 gain

$G3\_max_m := 20 \cdot \log\left(\left| R3\_max_m \right| \right)$ dB, filter # 3 gain

$G\_rms_m :=$
$\left|\begin{array}{l} par\_in \leftarrow -1000 \\ par_0 \leftarrow G1\_rms_m \\ par_1 \leftarrow G2\_rms_m \\ par_2 \leftarrow G3\_rms_m \\ \text{for } ind \in 0.. K - 1 \\ \quad par\_in \leftarrow par_{ind} \text{ if } par_{ind} \geq par\_in \\ MAX \leftarrow par\_in \\ MAX \end{array}\right.$

$$G\_av := \frac{1}{M-1} \cdot \sum_m G\_rms_m \qquad G\_av = 2.903 \qquad \text{dB, average gain}$$

$$Gav_m := G\_av$$

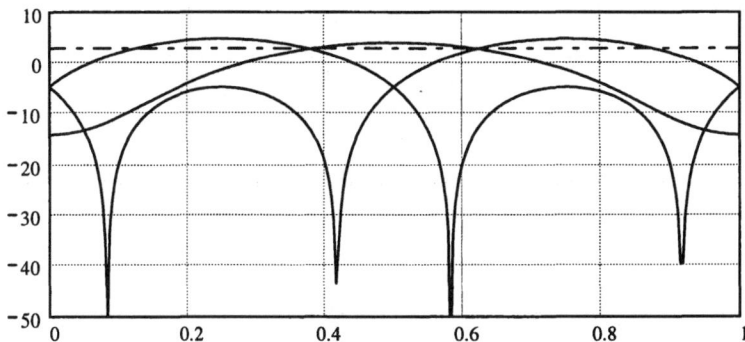

Figure 11.8 Filter bank responses normalized to noise bandwidth (dB) vs. doppler frequency relative to PRF (dashed line is an average gain).

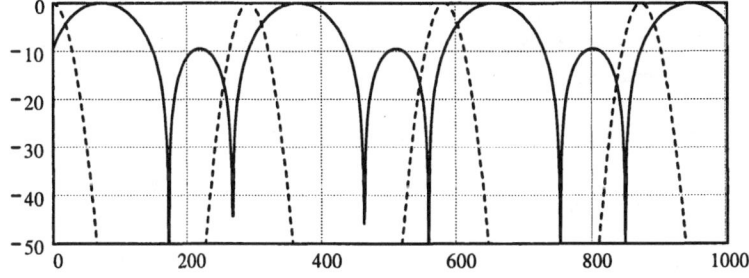

Figure 11.9 Filter # 1 response (solid) overlaid on the clutter spectrum (dashed) over velocity in knots.

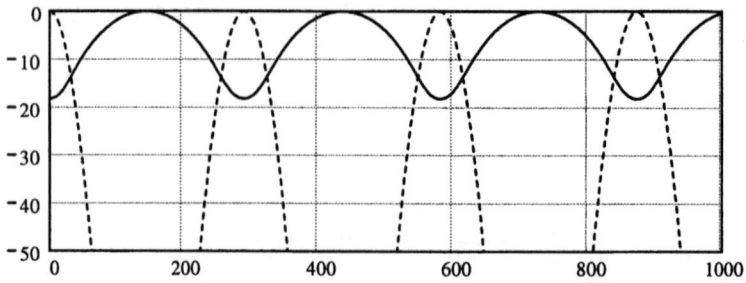

Figure 11.10 Filter # 2 response (solid) overlaid on the clutter spectrum (dashed) over velocity in knots.

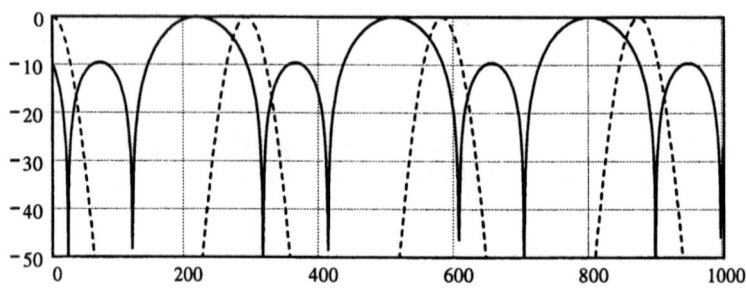

Figure 11.11 Filter # 3 response (solid) overlaid on the clutter spectrum (dashed) over velocity in knots.

11.6.3.4 Clutter Power in Filter # 2 (Notch Filter)

$$F2\_res\_d_m := \left(\left|R2\_max_m\right|\right)^2 \cdot S\_clutter\_d_m + 10^{-10}$$

$$res\_F2_m := 10 \cdot \log\left(F2\_res\_d_m\right) \qquad \text{dB, discrete} \atop \text{clutter residue}$$

$$F2\_res(f) := \left(\left|R2(f)\right|\right)^2 \cdot S\_clutter(f) \qquad \text{clutter residue} \atop \text{(continuous function)}$$

$$P\_c\_F2 := \int_0^{f_p} F2\_res(f)\, df \qquad \text{clutter residue power}$$

P_c_F2 = 2.176

$$\text{clutter\_res\_F2\_db} := 10 \cdot \log\left(\frac{\text{P\_c\_F2}}{\text{P\_c}}\right)$$ dB, clutter residue power

clutter_res := −clutter_res_F2_db   a sign change

clutter_res = 17.251   dB, clutter residue power

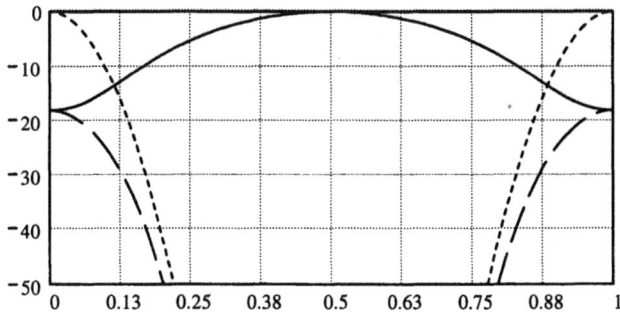

Figure 11.12 Clutter residue (dashed) in filter # 2 (solid line is filter # 2 frequency response, dotted line is the clutter spectrum) vs. doppler frequency relative to pulse repetition frequency.

11.6.3.5 Clutter Attenuation

$$CA := \frac{1}{\dfrac{1}{10^{\frac{I\_lim}{10}}} + \dfrac{1}{10^{\frac{\text{clutter\_res}}{10}}}}$$ clutter attenuation

$$CA\_db := 10 \cdot \log(CA)$$ dB, clutter attenuation

CA_db = 17.249

11.6.3.6 Improvement Factor and Subclutter Visibility

$$IMP_m := \left(G2\_rms_m + CA\_db\right)$$ dB, improvement factor for filter # 2

$$SCV_m := IMP_m - V_{xc}$$ dB, subclutter visibility for filter # 2

11.6.4 OUTPUT DATA

For input data specified:

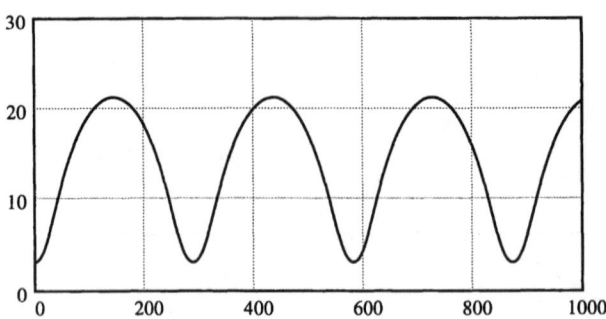

Figure 11.13 Improvement factor (dB) vs. target velocity in knots.

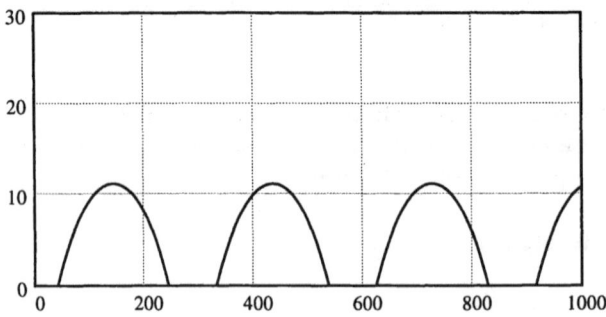

Figure 11.14 Subclutter visibility (dB) vs. target velocity in knots.

11.6.5 REFERENCES

[1] Barton, D. K., and S. A. Leonov, (eds.), *Radar Technology Encyclopedia*, Norwood, MA: Artech House, 1997, pp. 79–87.

[2] Schleher, D. C., *MTI and Pulsed Doppler Radar*, Norwood, MA: Artech House, 1991, p. 108.

[3] Skolnik, M., *Introduction to Radar Systems*, New York: McGraw-Hill, 1980, pp. 119–126.

APPENDIXES

For the list of references cited in Appendixes, see page 147.

Appendix 1 **Notations and Special Functions Used in the Handbook**

1 *For an arbitrary function* $A(t)$:

$\overline{A}(t)$ is a complex representation of a function $A(t)$ (with the exception of the random events where $\overline{A}$ is a complement of an event A); $\text{Re}\{\overline{A}(t)\}$ is the real part of the function $\overline{A}(t)$, $\text{Im}\{\overline{A}(t)\}$ is the imaginary part of the function $\overline{A}(t)$;

$\overline{A}^{*}(t)$ is a complex conjugate; $\dot{A}(t)$ is a derivative of the first order, $\ddot{A}(t)$ is a derivative of the second order, and so forth; $\hat{K}$ is an estimate of an arbitrary parameter K; *conv* or $\otimes$ are convolution signs; $C_n^k = n!/[k!(n-k)!]$.

2 *Gamma function* $\Gamma(z)$:

$\Gamma(z)$ is defined as the solution of the equation:

$$\Gamma \cdot z + 1 = z \cdot \Gamma(z), \text{ where } \Gamma(1) = 1$$

If Re $z > 0$:

$$\Gamma(z) = \int_0^1 \left(\ln\frac{1}{u} \right)^{z-1} du$$

For integer $n = 1, 2, 3,...$

$$\Gamma(1) = 1, \quad \Gamma(2) = 1, \quad \Gamma(n) = (n-1)!, \quad n = 3, 4,...$$

$$\Gamma\left(n + \frac{1}{2} \right) = \sqrt{\pi} \frac{1 \cdot 3 \cdot 5 \cdot ...(2n-1)}{2^n}$$

$$\Gamma\left(-n + \frac{1}{2} \right) = \sqrt{\pi} \frac{(-2)^n}{1 \cdot 3 \cdot 5 \cdot ...(2n-1)}, \quad n = 1, 2,...$$

$$\Gamma\left(\frac{1}{2} \right) = \sqrt{\pi}, \quad \Gamma\left(-\frac{1}{2} \right) = -2\sqrt{\pi}$$

3 *Incomplete beta function* $B_x(a,b)$:

$$B_x(a,b) = \int_{-\infty}^{\infty} x^{a-1}(1-x)^{b-1}\,dx$$

4 *Function of a parabolic cylinder* $D_p(z)$:

The function $D_p(z)$ is the solution of the differential equation:

$$\frac{d^2\omega}{dz^2} + \left(p + \frac{1}{2} - \frac{1}{4}z^2\right)\omega = 0$$

where p is a parameter. For integer $p = n = 0, 1, 2,\ldots$

$$D_n(z) = 2^{-\frac{n}{2}} e^{-z^2/4} H_n\left(z/\sqrt{2}\right)$$

where $H_n(u) = (-1)^n e^{z^2} \dfrac{d^n}{dz^n}\left(e^{-z^2}\right)$

5 *Error function* $\Phi(x)$:

$$\Phi(x) = \frac{1}{\sqrt{2\pi}} \int_{-\infty}^{x} e^{-t^2/2}\,dt$$

$$\Phi\left(\frac{x-m}{\sigma}\right) = \frac{1}{\sqrt{2\pi}\sigma} \int_{-\infty}^{x} \exp\left[-\frac{(t-m)^2}{2\sigma^2}\right]dt$$

$$erf(x) = \frac{2}{\sqrt{\pi}} \int_{0}^{x} \exp(-t^2)\,dt$$

6 *Error function of a complex argument* $W(z)$:

$$W(z) = e^{-z^2}\left(1 + \frac{2j}{\pi}\int_{0}^{z} e^{t^2}\,dt\right)$$

7 _Bessel functions_ $J_0(z)$ and $I_0(z)$:

$$J_0(a\vartheta) = \frac{1}{\pi} \int_{-a}^{a} \frac{e^{j\vartheta t}}{\sqrt{a^2 - t^2}} dt$$

$$I_0(a\vartheta) = J_0(ja\vartheta) = \frac{1}{2\pi} \int_0^{2\pi} e^{-ja\vartheta \cos t} dt$$

8 _The confluent hypergeometric function_ $_1F_1(\alpha, \gamma, x)$:

$$_1F_1(\alpha, \gamma, x) = 1 + \frac{\alpha}{\gamma} + \frac{\alpha(+1)}{\gamma(\gamma+1)} \cdot \frac{x^2}{2!} + \frac{\alpha(\alpha+1)\cdot(\alpha+2)}{\gamma(\gamma+1)\cdot(\gamma+2)} \cdot \frac{x^3}{3!} + \dots$$

9 _The function_ cnorm (x) returns the integral from $-\infty$ to x of the standard normal distribution (x must be real):

$$cnorm(\sqrt{2}u) = erf(u)$$

10 _The function_ qnorm (x, m, σ) returns the inverse cumulative normal distribution with mean m and standard deviation σ .

The function $qnorm[(1-P),0,1] = -\sqrt{2}u$, where u is the solution of the equation $erf(u)$ for $u = 2P-1$.

11 _The function_ dbinom (I, M, p) returns the probability density for the negative binomial distribution with size M and probability p:

$$dbinom(i, M, p) = \frac{M!}{i!M-i!} p^i (1-p)^{M-i}$$

12 _The functions_ js(n,x) and ys(n,x) return the value of spherical Bessel functions of the first and second kinds, correspondingly, of order n that is the solution of equation:

$$x^2 \frac{d^2}{dx} y + 2x \frac{d^2}{dx} y + \left[x^2 - n(n+1) \right] y = 0$$

Appendix 2 **The Major Features of Delta Function**

1 Delta function is an *even function* of its argument:

$$\delta(t - t_0) = \delta(t_0 - t)$$

2 *Selectivity* feature: delta function selects the value of the signal $s(t)$ only in the moment $t = t_0$:

$$s(t) \cdot \delta(t - t_0) = s(t_0) \cdot \delta(t - t_0)$$

3 The variation of timescale:

$$\delta(at) = \frac{1}{|a|} \cdot \delta(t) \quad (\text{e.g., } \delta(\omega) = \frac{1}{2\pi} \delta(f))$$

4 Power and energy of delta function is finite at any infinite interval T:

$$P = \frac{1}{T} \int_{-\frac{T}{2}}^{T/2} \delta^2(t) dt = \lim_{\Delta t \to 0} \frac{1}{T} \int_{-T/2}^{T/2} \left(\frac{1}{\Delta t}\right)^2 dt = \lim_{\Delta t \to 0} \frac{1}{T \cdot \Delta t}$$

5 The second power of delta function is equal to delta function:

$$\delta^2(t) = \delta(t)$$

6 Delta function is always *orthogonal* at noncoincident moments of time:

$$\lim_{T \to \infty} \int_{-T/2}^{T/2} \delta(t - t_1) \cdot \delta(t - t_2) dt = \begin{cases} 0, & t_1 \neq t_2 \\ 1, & t_1 = t_2 \end{cases}$$

7 *Filtration* feature: the following equation is valid for any signal that is continuous and limited in point $t = t_0$

$$\int_{t_a}^{t_b} s(t) \cdot \delta(t - t_0) dt = \begin{cases} s(t_0), & t_a < t_0 < t_b \\ \frac{1}{2} s(t_0), & t_0 = t_a, & t_0 = t_b \\ 0, & t_0 < t_a, & t_0 > t_b \end{cases}$$

Appendix 3 **Signal Models Based on Elementary Time-Domain Functions**

1 *Truncated Signal* $s_1(t) = s(t) \cdot \sigma(t - t_0)$

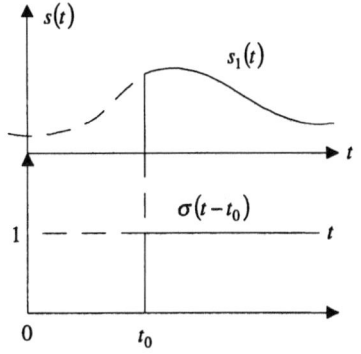

Figure A3.1 The truncated signal model.

2 *Finite Signal* $s_1 = s(t)[\sigma(t - t_a) - \sigma(t - t_b)]$

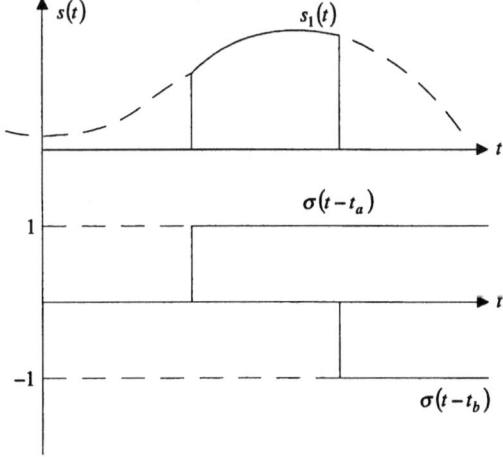

Figure A3.2 The finite signal model.

3 *Rectangular Pulse* $\quad s_r(t) = \sigma\left(t + \dfrac{\tau}{2}\right) - \sigma\left(t - \dfrac{\tau}{2}\right)$

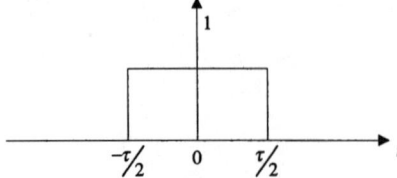

Figure A3.3 The rectangular pulse model.

Appendix 4 The Decomposition Models of the Signals (after [2, pp. 518–525])

1 *Trigonometric* system of functions (interval $T = \dfrac{2\pi}{\omega_0}$):

$$s(t)= \frac{a_0}{2}+\sum_{n=1}^{\infty} A_n \cdot \cos(n\omega_0 t + \varphi_n) \qquad a_0 = \frac{2}{T}\int_{-T/2}^{T/2} s(t)dt$$

2 *Legendre polynomials* (interval $-1 < t < 1$):

$$s(t)= C_0 \cdot P_0(t)+ C_1 \cdot P_1(t)+ ...C_n \cdot P_n(t)+ ...$$

$$C_n = \frac{2n+1}{2}\int_{-1}^{1} s(t)P_n(dt) \qquad P_n(t)= \frac{1}{2^n n!}\frac{d^n}{dt^n}(t^2-1)^n$$

3 *Chebyshev polynomials* (interval $-1 < t < 1$):

$$s(t)= C_0 +\sum_{n=1}^{\infty} C_n \cdot T_n(t)$$

$$C_0 = \frac{1}{\pi}\int_{-1}^{1}\frac{s(t)}{\sqrt{1-t^2}}dt \qquad C_n = \frac{2}{\pi}\int_{-1}^{1}\frac{s(t)T_n(t)}{\sqrt{1-t^2}}dt$$

$$T_n(t)= \frac{(-2)^n \cdot n!}{(2n)!}\sqrt{1-t^2}\frac{d^n}{dt^n}\left(\sqrt{1-t^2}\right)^{2n-1}$$

4 *Laquerre polynomials* (interval $0 < t < \infty$):

$$s(t)= \sum_{n=0}^{\infty} C_n \cdot l_n(t)$$

$$C_n = \int_{0}^{\infty} s(t)l_n(t)dt \qquad l_n(t)= \sqrt{\rho(t)}\cdot L_n(t); \qquad \rho(t)= e^{-t}$$

$$L_n(t) = \frac{e^t}{n!} \cdot \frac{d^n}{dt^n}\left(t^n e^{-t}\right) \quad t \geq 0$$

5 *Hermitian polynomials*:

$$s(t) = \sum_{n=0}^{\infty} C_n \cdot \varphi_n(t)$$

$$C_n = \int_{-\infty}^{\infty} s(t)\varphi_n(t)dt$$

$$\varphi_n(t) = \frac{H_n(t) \cdot e^{-t^2/2}}{\sqrt{2^n \sqrt{\pi} n!}}$$

$$H_n(t) = (-1)^n \cdot e^{t^2} \frac{d^n}{dt^n}\left(e^{-t^2}\right)$$

Appendix 5 **Basic Features of the Discrete Fourier Transform** (after [1, pp. 12,13])

1 *Linearity*:   If $X(k) = DFT[x(n\Delta t)]$ and $Y(k) = DFT[y(n\Delta t)]$,   then $DFT[ax(n\Delta t) + by(n\Delta t)] = aX(n) + bY(k)$, where a, b are arbitrary constants.

2 *Shift*: If $X(k) = DFT[x(n\Delta t)]$, and $y(n\Delta t)$ is obtained from $x(n\Delta t)$ by shifting it by n_0 counts, then $Y(k) = DFT[y(n\Delta t)] = X(k) \cdot W_N^{n_0 k}$

3 *Symmetry*: $\mathrm{Re}[X(k)] = \mathrm{Re}[X(N-k)]$, $\mathrm{Im}[X(k)] = -\mathrm{Im}[X(N-k)]$, $|X(k)| = |X(N-k)|$, $\arg X(k) = -\arg X(N-k)$

4 *Convolution*:   If $X(k) = DFT[x(n\Delta t)]$ and $Y(k) = DFT[\dot{y}(n\Delta t)]$,   if

$$u(n\Delta t) = \sum_{l=0}^{N-1} x(l\Delta t) \cdot y[(n-l)\Delta t], \text{ then } DFT[u(n\Delta t)] = X(k) \cdot Y(k) = U(k)$$

If $u(n\Delta t) = x(n\Delta t) \cdot y(n\Delta t)$, then:

$$DFT[u(n\Delta t)] = U(k) = \frac{1}{N} \sum_{l=0}^{N-1} X(l)Y(k-l), \quad k = 0,...N-1$$

5 *Reciprocity*: The inverse DFT can be calculated via the direct DFT as

$$x(n\Delta t) = P^*(n\Delta t), \text{ where } P(n\Delta t) = \frac{1}{N}\left(\sum_{k=0}^{N-1} X^*(k)W_N^{nk}\right), \quad n = 0,..., N-1$$

Appendix 6 **The Fast Fourier Transform Models** (after [1, pp. 14–21])

1 *Time-decimation FFT (base 2)*. The sequence $x(n\Delta t)$ with length $N = 2^\gamma$ is divided into two $N/2$-points sequences, and then correspondingly into $N/4$-points ones, $N/8$-points ones, etc. The steps of transformation $\gamma = \log_2 N$.

$$X(k) = \sum_{n=0}^{\frac{N}{2}-1} x(2n\Delta t)W_{N/2}^{nk} + W_N^k \sum_{n=0}^{\frac{N}{2}-1} x[(2n+1)\Delta t]W_{N/2}^{nk}$$

2 *Frequency-decimation FFT (base 2)*. The sequence $x(n\Delta t)$ is divided into $x_1(n\Delta t) = x(n\Delta t)$, and $x_2(n\Delta t) = x\left[\left(n + \dfrac{N}{2}\right)\Delta t\right]$, then even and odd DFT are performed separately:

$$X(2k) = \sum_{n=0}^{\frac{N}{2}-1} [x_1(n\Delta t) + x_2(n\Delta t)]W_{N/2}^{nk}$$

$$X(2k+1) = \sum_{n=0}^{\frac{N}{2}-1} [(x_1(n\Delta t) - x_2(n\Delta t)]W_N^n W_{N/2}^{nk}$$

3 *Algorithm with rotation*. If $= N = N_1 \cdot N_2$ where N_1, N_2 are any positive integers, the N-point DFT is reduced to $N_1\,N_2$ − point and $N_2\,N_1$ − point DFT and N multiplications by rotation factors W_N^l :

$$k = k_1 + k_2 N_2 \quad n = n_1 + n_2 N_1 \quad n_1, k_2 = 0,...,N_1 - 1 \quad n_2, k_1 = 0,...,N_2 - 1$$

$$X(k_1 + k_2 N_2) = \sum_{n_1=0}^{N_1-1}\left[\left\{\sum_{n_2=0}^{N_2-1} x[(n_1 + n_2 N_1)\Delta t]W_{N_2}^{k_1 n_2}\right\}W_N^{k_1 n_1}\right]W_{N_1}^{k_2 n_1}$$

4 *Algorithm with arbitrary base*. FFT model with any base different from 2 can be obtained by using the algorithm with rotation The most common bases are 4, 8, 16. The formulas to compare the required number of operations for FFT with bases 2, 4, 8, and 16 are given in Table A6.1.

Table A6.1

The comparison of number of operations required for FFT algorithms with different bases (assumed that a basic operation N-point DFT is performed by the algorithm with a minimum number of operations)

Algorithm	Brief description	Number of real multiplications	Number of real summations
Base 2.	1. Computation of $\left(\dfrac{N}{2}\right)\gamma$	0	$2N\gamma$
$N = 2^{\gamma}$,	2-point DFT		
$\gamma \geq 1$ is an	2. Rotation	$(2\gamma - 4)N + 4$	$(\gamma - 2)N + 2$
integer	3. Complete transform $(2\gamma - 4)N + 4$		$(3\gamma - 2)N + 2$
Base 4.	1. Computation of $\left(\dfrac{N}{4}\right)\cdot\dfrac{\gamma}{2}$	0	$2N\gamma$
$N = \left(2^{2}\right)^{\gamma/2}$	4-point DFT		
$\gamma/2 \geq 1$ is	2. Rotation	$(3\gamma/2 - 4)N + 4$	$(3\gamma/4 - 2)N + 2$
an integer	3. Complete transform $(3\gamma/2 - 4)N + 4$		$(2.75\gamma/4 - 2)N + 2$
Base 8.	1. Computation of $\left(\dfrac{N}{8}\right)\cdot\dfrac{\gamma}{3}$	$N\gamma/6$	$13N\gamma/6$
$N = \left(2^{3}\right)^{\gamma/3}$	8-point DFT		
$\gamma/3 \geq 1$ is	2. Rotation	$(7\gamma/6 - 4)N + 4$	$(7\gamma/12 - 2)N + 2$
an integer	3. Complete transform $(4\gamma/3 - 4)N + 4$		$(2.75\gamma - 4)N + 4$
Base 16.	1. Computation of $\left(\dfrac{N}{16}\right)\cdot\dfrac{\gamma}{4}$	$5N\gamma/8$	$37N\gamma/16$
$N = \left(2^{4}\right)^{\gamma/4}$	16-point DFT		
$\gamma/4 \geq 1$ is	2. Rotation	$(15\gamma/16 - 4)N + 4$	$(15\gamma/32 - 2)N + 2$
an integer	3. Complete transform $(25\gamma/16 - 4)N + 4$		$(89\gamma/32 - 2)N + 2$

Appendix 7 **Z-Transform of Some Common Functions** (after [1, p. 8], [3, p. 290])

$x(n\Delta t)$	$Z\{x(n\Delta t)\}$
1	$z/(z-1)$
$(-1)^n$	$1/(1+z^{-1})$
n	$z^{-1}(1-z^{-1})^2$
n^2	$(z^{-1}+z^{-2})(1-z^{-1})^3$
a^n	$1/(1-az^{-1})$
na^{n-1}	$z^{-1}/(1-az^{-1})^2$
$e^{s\tau}$	$z/(z-e^\tau)$
$a^n \sin(n\tau)$	$(az^{-1}\sin\tau)/(1-2az^{-1}\cos\tau+a^2z^{-2})$
$a^n \cos(n\tau)$	$(1-az^{-1}\cos\tau)/(1-2az^{-1}\cos\tau+a^2z^{-2})$
$ch(s\tau)$	$[z(z-ch\tau)]/(z^2-2zch\tau+1)$
$sh(s\tau)$	$(zsh\tau)/(z^2-2zch\tau+1)$

Appendix 8 **Active Device Noise**

1 _Noise factor F_ is the ratio of noise power output of actual device to power output of source noise (thermal noise power of the source resistance). When expressed in decibels, noise factor is called a *noise figure (NF)*.
Basic models are:

$$F = \frac{P_{no}}{P_{ni}} = \frac{P_{si}/P_{ni}}{P_{so}/P_{no}} \quad \text{if } P_{so} = P_{si}$$

$$F = \frac{\left(V_{si}/V_{ni}\right)^2}{\left(V_{so}/V_{no}\right)^2} \quad \text{if } R_i = R_o$$

$$F = \frac{\left(V_{no}\right)^2}{4kTBR_s A}$$

$$NF = 10\log F$$

$$F = F_1 + \frac{F_2 - 1}{G_1} + ... \frac{F_m - 1}{G_1 G_2 ... G_m}$$

$$G_m = A_m^2 R_{im}/R_{om}$$

where P_{no} is output noise power, P_{ni} is input noise power, P_{so}, P_{si} are output and input signal powers, $V_{so}, V_{si}, V_{no}, V_{ni}$ are output and input signal and noise voltages, R_0, R_i are input and output (load) impedances, k is Boltzmann's constant, T is absolute temperature, B is noise bandwidth, R_s is source resistance, A is voltage gain, F_m, G_m are noise factor and available power gain of each cascade in a cascaded network.

2 _Equivalent input noise voltage_ is:

$$V_{ne} = \left[4kTBR_s + V_n^2 + \left(I_n R_s\right)^2\right]^{1/2}$$

where V_{ne} is equivalent input noise voltage of a device, V_n is voltage of device noise when $R_s = 0$, I_n is current of device noise when $R_s \neq 0$.

3 *The equivalent noise temperature* T_e is the increase in source resistance temperature necessary to produce the observed noise power at the output of the device. Basic models are:

$$T_e = 290(F - 1)$$

$$T_e = 290\left(10^{NF/10} - 1\right)$$

$$T_e = \frac{V_n^2 + (I_n R_s)^2}{4kBR_s}$$

$$T_{e_\Sigma} = T_{e_1} + \frac{T_{e_2}}{G_1} + \ldots \frac{T_{e_m}}{G_1 G_2 \ldots G_m}$$

Appendix 9 **Recursive and Nonrecursive Filter Frequency Responses** (after [1, p. 55])

1 *Recursive filter:*

Complex response: $H\left(e^{j\omega\Delta t}\right)=\dfrac{\displaystyle\sum_{l=0}^{N-1}b_l e^{-jl\omega\Delta t}}{1+\displaystyle\sum_{i=1}^{M-1}a_i e^{-ji\omega\Delta t}}$

Amplitude response: $A(\omega)=\left|H\left(e^{j\omega\Delta t}\right)\right|=\sqrt{\dfrac{\displaystyle\sum_{m=0}^{N-1}\sum_{k=0}^{N-1}b_m b_k \cos(m-k)\omega\Delta t}{\displaystyle\sum_{p=o}^{M-1}\sum_{s=0}^{M-1}a_p a_s \cos(p-s)\omega\Delta t}}$ $a_0=1$

Phase response: $\Phi(\omega)=\arg\left[H\left(e^{j\omega\Delta t}\right)\right]=-\arctan A+\arctan B$

$A=\dfrac{\displaystyle\sum_{l=0}^{N-1}b_l \sin(l\omega\Delta t)}{\displaystyle\sum_{l=0}^{N-1}b_l \cos(l\omega\Delta t)}$ $B=\dfrac{\displaystyle\sum_{i=0}^{M-1}a_i \sin(i\omega\Delta t)}{\displaystyle\sum_{l=0}^{M-1}a_i \cos(i\omega\Delta t)}$

2 *Nonrecursive filter*

Complex response: $H\left(e^{j\omega\Delta t}\right)=\displaystyle\sum_{l=0}^{N-1}b_l e^{-jl\omega\Delta t}$

Amplitude response: $A(\omega)=\left|H\left(e^{j\omega\Delta t}\right)\right|=\sqrt{\displaystyle\sum_{m=0}^{N-1}\sum_{n=0}^{N-1}b_m b_k \cos(m-k)\omega\Delta t}$

Phase response: $\Phi(\omega)=\arg\left[H\left(e^{j\omega\Delta t}\right)\right]=-\arctan(A/B)$

$A=\displaystyle\sum_{l=0}^{N-1}b_l \sin(l\omega\Delta t)$ $B=\displaystyle\sum_{l=0}^{N-1}b_l \cos(l\omega\Delta t)$

Appendix 10 **The Transfer Functions of a Pair of Cascaded Filters** (after [1, p. 51])

1 *Sequential connection* $H_\Sigma(z) = H_1(z) \times H_2(z)$

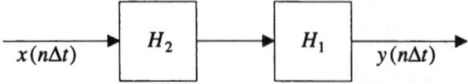

2 *Parallel connection* $H_\Sigma(z) = H_1(z) + H_2(z)$

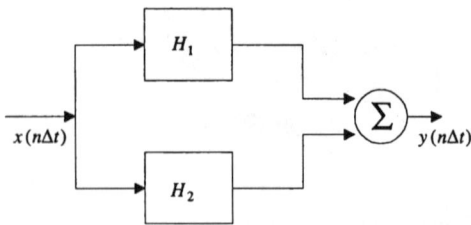

3 *A feedback loop* $H_\Sigma(z) = \dfrac{H_1(z)}{1 - H_1(z)H_2(z)}$

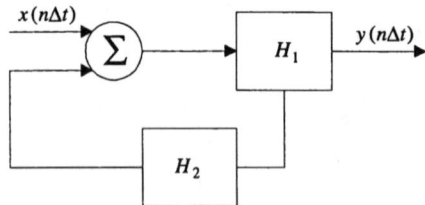

Appendix 11 **The Distribution Laws for the Discrete Random Variables** (after [4, pp. 30–33])

1 *Binomial (Bernoulli) Distribution*

Variable variation range: $k = 0,1,...n$

Distribution function: $P_n(k) = C_n^k p^k (1-p)^{n-k}$

Parameters: n, p

Characteristic function: $\left[1 + p\left(e^{j\vartheta} - 1\right)\right]^n$

2 *Binomial (Negative) Distribution*

Variable variation range: $k = n, n+1,...$ or $k = 0,1,2...$

Distribution function: $P(k) = C_{k-1}^{n-1} \cdot p^n \div (1-p)^{k-n}$ or $P(k) = C_{k+n}^{n-1} \cdot p^n \cdot (1-p)^k$

Parameters: n, p

Characteristic function: $p^n \cdot \left[1 - (1-p)e^{j\vartheta}\right]^{-n}$

3 *Geometric (Farry) Distribution* (a partial case of Pascal distribution function when $n = 1$)

Variable variation range: $k = 0,1,2...$

Distribution function: $P_n(k) = p(1-p)^k$

Parameters: p

Characteristic function: $p\left[1 - (1-p)e^{j\vartheta}\right]^{-1}$

4 *Hypergeometric Distribution*

Variable variation range: $k = 0,1,2...$ $\min(M,n)$

Distribution function: $P_n(k) = \dfrac{C_M^k\, C_{N-M}^{n-k}}{C_N^n}$

Parameters: N, M, n

Characteristic function: $\displaystyle\sum_{k=0}^{\min(M,n)} \dfrac{C_M^k\, C_{N-M}^{n-k}}{C_N^n} \cdot e^{jk\vartheta}$

5 Pascal Distribution

Variable variation range: $k = 0, 1, 2, \ldots$

Distribution function: $P_n(k) = C_{k+n-1}^{n-1} \cdot p^n \cdot (1-p)^k$

Parameters: p, n, k

6 Poisson Distribution

Variable variation range: $: k = 0, 1, 2, \ldots$

Distribution function: $P_n(k) = \dfrac{\lambda^k}{k!} \cdot e^{-\lambda}$

Parameter: λ

Characteristic function: $e^{\lambda\left(e^{j\vartheta}-1\right)}$

7 Polya Distribution

Variable variation range: $: k = 0, 1, 2, \ldots$ if $\alpha = 0$; $k = 1, 2, \ldots$ if $\alpha > 0$

Distribution function: $P(k) = P_0 \cdot \left(\dfrac{\lambda}{1+\alpha\cdot\lambda}\right)^k \times \dfrac{1\cdot(1+\alpha)..[1+(k-1)\cdot\alpha]}{k!}$

$\alpha \geq 0,\ \lambda > 0$

$$P_0 = P(0) = (1+\alpha\cdot\lambda)^{-1/\alpha}$$

Parameters: α, λ

Characteristic function: $\left(1+\alpha\lambda\left(1-e^{j\vartheta}\right)\right)^{-1/\alpha}$

8 *Polynomial Distribution*

Variable variation range: $k_1 = 0,1,...n \quad k_2 = 0,1,...n \ ... \ k_m = 0,1,...n \ ; \ \sum_{i=1}^{m} k_i = n$

Distribution function: $P_n\left(k_1,k_2,...k_m\right) = \dfrac{n!}{k_1!k_2!...k_m!} \, p_1^{k_1}...p_m^{k_m}$

$$p_1 + p_2 + ... + p_m = 1$$

Parameters: n and any $n-1$ values from $p_1, p_2,...p_m$

9 *Uniform Distribution*

Variable variation range: $k = 1, 2, ... n$

Distribution function: $P(k) = \dfrac{1}{n}$

Parameter: n

Characteristic function: $\dfrac{e^{j\vartheta}\left(1-e^{j\vartheta n}\right)}{n\left(1-e^{j\vartheta}\right)}$

Appendix 12 **The Distribution Laws for the Continuous Random Variables** (after [4, pp. 33–47])

1 *Arcsin Distribution*

Variable variation range: $-a < x < a$

Distribution function: $F_1(x) = \begin{cases} 0, & x \le -a \\ \dfrac{1}{2} + \dfrac{1}{\pi} \cdot \arcsin \dfrac{x}{a} & -a < x < a \\ 1 & x \ge a \end{cases}$

Density function: $f_1(x) = \dfrac{1}{\pi\sqrt{a^2 - x^2}}$

Characteristic function: $\dfrac{1}{\pi} \displaystyle\int_{-a}^{a} \dfrac{e^{j\vartheta x}}{\sqrt{a^2 - x^2}} dx$

Parameters: a

2 *Beta Distribution*

Variable variation range: $0 < x < 1$

Distribution function: $F_1(x) = \begin{cases} 0, & x \le 0 \\ \dfrac{Bx(a,b)}{B(a,b)} & 0 < x < 1 \\ 1 & x \ge 1 \end{cases}$

Density function: $f_1(x) = \dfrac{1}{B(a,b)} \cdot x^{a-1} \cdot (1-x)^{b-1}$

$B(a,b) = \Gamma(a) \cdot \Gamma(b)/\Gamma(a+b)$; $B_x(a,b) =$ incomplete beta function

Characteristic function: $\dfrac{\Gamma(a+b)}{\Gamma(a)} \times \displaystyle\sum_{m=0}^{\infty} \dfrac{(j\vartheta)^m}{m!} \dfrac{\Gamma(a+m)}{\Gamma(a+b+m)}$

Parameters: $a > 0, b > 0$

3 *Chi Distribution*

Variable variation range: $0 < x < \infty$

Distribution function: $F_1(x) = \begin{cases} 0 & x \le 0 \\[2ex] \dfrac{\Gamma\left(\dfrac{n}{2}, \dfrac{x^2}{2}\right)}{\Gamma\left(\dfrac{n}{2}\right)} & x > 0 \end{cases}$

Density function: $f_1(x) = \dfrac{1}{2^{\frac{n}{2}-1}\Gamma\left(\dfrac{n}{2}\right)} \cdot x^{n-1} \cdot e^{-x^2/2}$

Characteristic function: $\dfrac{\Gamma(n)}{2^{\frac{n}{2}-1} \cdot \Gamma\left(\dfrac{n}{2}\right)} \times e^{-\vartheta^2/4} \cdot D_{-n}\left(-j\vartheta\right)$

$D_p(z)$—function of parabolic cylinder (see Appendix 1)

Parameters: n

4 *Chi-squared Distribution*

Variable variation range: $0 < x < \infty$

Distribution function: $F_1(x) = \begin{cases} 0 & x \le 0 \\[2ex] \dfrac{\Gamma\left(\dfrac{n}{2}, \dfrac{x}{2}\right)}{\Gamma\left(\dfrac{n}{2}\right)} & x > 0 \end{cases}$

Density function: $f_1(x) = \dfrac{1}{2^{\frac{n}{2}}\Gamma\left(\dfrac{n}{2}\right)} \cdot x^{\frac{n}{2}-1} \cdot e^{-x/2}$

Characteristic function: $\left(1 - 2j\vartheta\right)^{-n/2}$

Parameters: n

5 Erlang Distribution

Variable variation range: $0 < x < \infty$

Distribution function: $F_1(x) = \begin{cases} 0 & x \leq 0 \\ \dfrac{\Gamma[(k+1), \lambda x]}{\Gamma(k+1)} & x > 0 \end{cases}$

Density function: $f_1(x) = \dfrac{\lambda^{k+1}}{\Gamma(k+1)} \cdot x^k \cdot e^{-\lambda x}$

Characteristic function: $\left(1 - \dfrac{j\vartheta}{\lambda}\right)^{-(k+1)}$

Parameters: λ, k (k is integer)

6 Exponential Distribution

Variable variation range: $0 < x < \infty$

Distribution function: $F_1(x) = \begin{cases} 0 & x \leq 0 \\ 1 - e^{-\lambda x} & x > 0 \end{cases}$

Density function: $f_1(x) = \lambda \cdot e^{-\lambda x}$

Characteristic function: $\dfrac{\lambda}{\lambda - j\vartheta}$

Parameters: λ

7 Exponential Distribution, Double

Variable variation range: $-\infty < x < \infty$

Distribution function: $F_1(x) = \exp\left(-c \cdot e^{-\alpha x}\right)$

Density function: $f_1(x) = c \cdot \alpha \cdot \exp\left(-\alpha x - c \cdot e^{-\alpha x}\right)$

Parameters: $c > 0, \alpha > 0$

8 *Exponential Distribution, Power*

Variable variation range: $0 < x < \infty$

Distribution function: $F_1(x) = \begin{cases} 0 & x \le 0 \\ \dfrac{\Gamma[(m+1), x]}{\Gamma(m+1)} & x > 0 \end{cases}$

Density function: $f_1(x) = \dfrac{x^m}{m!} e^{-x}$

Characteristic function: $(1 - j\vartheta)^{-(m+1)}$

Parameters: m

9 *Fisher Distribution (F-Distribution)*

Variable variation range: $0 < x < \infty$

Density function:

$$f_1(x) = \frac{\Gamma\left(\dfrac{n_1 + n_2}{2}\right)}{\Gamma\left(\dfrac{n_1}{2}\right) \cdot \Gamma\left(\dfrac{n_2}{2}\right)} \times \left(\frac{n_1}{n_2}\right)^{n_1/2} \times x^{(n_1/2)-1} \times \left(1 + \frac{n_1}{n_2} \cdot x\right)^{-\frac{n_1 + n_2}{2}}$$

Parameters: n_1, n_2

10 *Gamma Distribution*

Variable variation range: $0 < x < \infty$

Distribution function: $F_1(x) = \begin{cases} 0 & x \le 0 \\ \dfrac{\Gamma\left[(\alpha+1), \dfrac{x}{\beta}\right]}{\Gamma(\alpha+1)} & x > 0 \end{cases}$

Density function: $f_1(x) = \dfrac{1}{\beta^{\alpha+1} \cdot \Gamma(\alpha+1)} \cdot x^{\alpha} \cdot e^{-x/\beta}$

Characteristic function: $(1 - j\vartheta\beta)^{-(\alpha+1)}$

Parameters: $\alpha > -1, \beta > 0$

11 *Gaussian (Normal) Distribution*

Variable variation range: $-\infty < x < \infty$

Distribution function: $F_1(x) = \dfrac{1}{\sigma\sqrt{2\pi}} \displaystyle\int_{-\infty}^{x} e^{-\frac{(t-m)^2}{2\sigma^2}} \, dt = \Phi\!\left(\dfrac{x-m}{\sigma}\right)$

Density function: $f_1(x) = \dfrac{1}{\sigma\sqrt{2\pi}} \exp\!\left[-\dfrac{(x-m)^2}{2\sigma^2}\right]$

Characteristic function: $e^{jm\vartheta - \frac{\sigma^2\vartheta^2}{2}}$

Parameters: m, σ

12 *Gaussian (Standard) Distribution*

Variable variation range: $-\infty < x < \infty$

Distribution function: $F_1(x) = \dfrac{1}{\sqrt{2\pi}} \displaystyle\int_{-\infty}^{x} e^{-t^2/2} \, dt = \Phi(x)$

Density function: $f_1(x) = \dfrac{1}{\sqrt{2\pi}} \cdot e^{-x^2/2}$

Characteristic function: $e^{-\vartheta^2/2}$

Parameters: $m = 0, \sigma = 1$

13 *Gaussian (Log-Normal) Distribution*

Variable variation range: $0 < x < \infty$

Distribution function: $F_1(x) = \begin{cases} 0 & x \leq 0 \\ \Phi\left(\dfrac{\log x - m}{\sigma}\right) & x > 0 \end{cases}$

Density function: $f_1(x) = \dfrac{\log e}{x\sigma\sqrt{2\pi}} \cdot \exp\left(-\dfrac{(\log x - m)^2}{2\sigma^2}\right)$

Parameters: m, σ

14 *Gaussian (Truncated) Distribution*

Variable variation range: $a < x < b$

Distribution function: $F_1(x) = \begin{cases} 0 & x \leq a \\ c\left[\Phi\left(\dfrac{x-m}{\sigma}\right) - \Phi\left[\dfrac{a-m}{\sigma}\right]\right], & a < x < b \\ 1 & x > b \end{cases}$

Density function: $f_1(x) = \dfrac{c}{\sigma\sqrt{2}} \exp\left[-\dfrac{(x-m)^2}{2\sigma^2}\right]$

$$c = \left[\Phi\left(\dfrac{b-m}{\sigma}\right) - \Phi\left(\dfrac{a-m}{\sigma}\right)\right]^{-1}$$

Characteristic function: $c \times \exp(z) \times \left[\Phi\left[\left(\dfrac{b-m-u}{\sigma}\right) - \Phi\left(\dfrac{a-m-u}{\sigma}\right)\right]\right]$

$$z = jm\vartheta - \dfrac{\vartheta^2\sigma^2}{2} \quad u = j\vartheta\sigma^2$$

Parameters: m, σ, a, b

15 *Cauchy Distribution*

Variable variation range: $-\infty < x < \infty$

Distribution function: $F_1(x) = \dfrac{1}{2} + \dfrac{1}{\pi} a \tan\left(\dfrac{x - x_0}{h}\right)$

Density function: $f_1(x) = \dfrac{1}{\pi} \cdot \dfrac{h}{h^2 + (x - x_0)^2}$

Characteristic function: $e^{jx_0\vartheta - h|\vartheta|}$

Parameters: h, x_0

16 *Laplace Distribution*

Variable variation range: $-\infty < x < \infty$

Distribution function: $F_1(x) = \begin{cases} \dfrac{1}{2} e^{-\lambda(x-\mu)}, & -\infty < x < \mu \\ 1 - \dfrac{1}{2} e^{-\lambda(x-\mu)}, & \mu < x < \infty \end{cases}$

Density function: $f_1(x) = \dfrac{\lambda}{2} e^{-\lambda|x-\mu|}$

Characteristic function: $\dfrac{\lambda^2 \cdot e^{j\vartheta\mu}}{\lambda^2 + \vartheta^2}$

Parameters: λ, μ

17 *M-Distribution (Nakagami)*

Variable variation range: $0 < x < \infty$

Distribution function: $F_1(x) = \begin{cases} 0 & x \le 0 \\ \dfrac{\Gamma\left(m, \dfrac{mx^2}{\sigma^2}\right)}{\Gamma(m)} & x > 0 \end{cases}$

Density function: $f_1(x) = \Gamma\left(\dfrac{2}{m}\right) \cdot \left(\dfrac{m}{\sigma^2}\right)^m \cdot x^{2m-1} \times \exp\left(-\dfrac{m}{\sigma^2} \cdot x^2\right)$ $m \geq 1/2$

Characteristic function: $\dfrac{\Gamma(2m)}{2^{m-1}\Gamma(m)} \times e^{\frac{-\vartheta^2 \sigma^2}{8m}} \times D_{-2m}\left(-\dfrac{j\vartheta\sigma}{\sqrt{2m}}\right)$

$D_p(z)$—function of parabolic cylinder (see Appendix 1)

Parameters: m, σ

18 Maxwell Distribution

Variable variation range: $0 < x < \infty$

Distribution function: $F_1(x) = \begin{cases} 0 & x \leq 0 \\ \dfrac{2}{\sqrt{\pi}}\Gamma\left(\dfrac{3}{2}, \dfrac{x^2}{2\sigma^2}\right) & x > 0 \end{cases}$

Density function: $f_1(x) = \sqrt{\dfrac{2}{\pi}} \cdot \dfrac{1}{\sigma^3} x^2 e^{-x^2/2\sigma^2}$

Characteristic function: $(1 - \sigma^2\vartheta^2) \cdot W\left(\dfrac{\sigma\vartheta}{\sqrt{2}}\right) + j\sigma\vartheta\sqrt{\dfrac{2}{\pi}}$

$W(z)$ is a probability integral (see Appendix 1)

Parameters: σ

19 The Modulus of a Gaussian Variable

Variable variation range: $0 < x < \infty$

Distribution function: $F_1(x) = \begin{cases} 0 & x \leq 0 \\ \Phi\left(\dfrac{x-m}{\sigma}\right) + \Phi\left(\dfrac{x+m}{\sigma}\right) - 1 & x > 0 \end{cases}$

Density function: $f_1(x) = \dfrac{1}{\sigma\sqrt{2\pi}} \times \{\exp(A) + \exp(B)\}$

$$A = -\frac{(x-m)^2}{2\sigma^2} \quad B = -\frac{(x+m)^2}{2\sigma^2}$$

Characteristic function: $e^{-\frac{\sigma^2\vartheta^2}{2}} \times \left[e^{j\vartheta m}\Phi\left(\frac{m}{\sigma}+j\vartheta\sigma\right) + e^{-j\vartheta m}\Phi\left(-\frac{m}{\sigma}+j\vartheta\sigma\right) \right]$

Parameters: m, σ

20 *The Modulus of a Multidimensional Vector*

Variable variation range: $0 < x < \infty$

Distribution function: $F_1(x) = \begin{cases} 0 & x \le 0 \\ \dfrac{\Gamma\left(\dfrac{n}{2},\dfrac{x^2}{2\sigma^2}\right)}{\Gamma\left(\dfrac{n}{2}\right)} & x > 0 \end{cases}$

Density function: $f_1(x) = \dfrac{2x^{n-1}\cdot e^{-x^2/2\sigma^2}}{\left(2\sigma^2\right)^{n/2}\cdot\Gamma\left(\dfrac{n}{2}\right)}$

Characteristic function: $\dfrac{\Gamma(n)}{2^{\frac{n}{2}-1}\Gamma\left(\dfrac{n}{2}\right)} \exp\left(-\dfrac{\vartheta^2\sigma^2}{4}\right) D_{-n}(j\vartheta\sigma)$

Parameters: n, σ

21 *Rayleigh Distribution*

Variable variation range: $0 < x < \infty$

Distribution function: $F_1(x) = \begin{cases} 0 & x \le 0 \\ 1 - e^{-x^2/2\sigma^2} & x > 0 \end{cases}$

Density function: $f_1(x) = \dfrac{x}{\sigma^2} e^{-\frac{x^2}{2\sigma^2}}$

Characteristic function: $1 + ja\vartheta \sqrt{\dfrac{\pi}{2}} W\left(\dfrac{a\vartheta}{\sqrt{2}}\right)$

Parameters: σ

22 Ricean (Generalized Rayleigh) Distribution

Variable variation range: $0 < x < \infty$

Distribution function: $F_1(x) = \begin{cases} 0 & x \leq 0 \\ e^{-\frac{a^2}{2\sigma^2}} \displaystyle\sum_{k=0}^{\infty} \dfrac{x}{(k!)^2} \left(\dfrac{a^2}{2\sigma^2}\right)^k \Gamma\left(k+1, \dfrac{x^2}{2\sigma^2}\right) & x > 0 \end{cases}$

Density function: $f_1(x) = \dfrac{x}{2\sigma^2} \exp\left(-\dfrac{x^2+a^2}{2\sigma^2}\right) I_0\left(\dfrac{ax}{\sigma^2}\right)$

Parameters: a, σ

23 Ch-squared Distribution

Variable variation range: $-\infty < x < \infty$

Distribution function: $F_1(x) = \dfrac{1}{2} + \dfrac{1}{2} th(ax)$

Density function: $f_1(x) = \dfrac{a}{2 ch^2(ax)}$

Characteristic function: $\dfrac{\vartheta\pi}{2a \, sh\left(\dfrac{\vartheta\pi}{2a}\right)}$

Parameters: a

24 *Simpson (Triangular) Distribution*

Variable variation range: $a < x < b$

Distribution function:
$$F_1(x) = \begin{cases} 0 & -\infty < x < a \\ \dfrac{2(x-a)^2}{(b-a)^2} & a < x < \dfrac{a+b}{2} \\ 1 - \dfrac{2(b-x)^2}{(b-a)^2} & \dfrac{a+b}{2} < x < b \\ 1 & b < x < \infty \end{cases}$$

Density function:
$$f_1(x) = \begin{cases} 0 & -\infty < x < a \\ \dfrac{4(x-a)}{(b-a)^2} & a < x < \dfrac{b+a}{2} \\ \dfrac{4(b-x)}{(b-a)^2} & \dfrac{a+b}{2} < x < b \\ 0 & b < x < \infty \end{cases}$$

Characteristic function: $-\dfrac{4}{\vartheta^2(b-a)^2} \times \left(e^{j\vartheta\frac{b}{2}} - e^{j\vartheta\frac{a}{2}} \right)^2$

Parameters: a, b

25 *Student Distribution (t-distribution)*

Variable variation range: $-\infty < x < \infty$

Distribution function:
$$F_1(x) = \frac{\Gamma\left(\dfrac{k+1}{2}\right)}{\sqrt{k\pi}\,\Gamma\left(\dfrac{k}{2}\right)} \times \int_{\infty}^{x} \left(1 + \frac{t^2}{k}\right)^{-\frac{k+1}{2}} dt$$

Density function:
$$f_1(x) = \frac{\Gamma\left(\dfrac{k+1}{2}\right)}{\sqrt{k\pi}\,\Gamma\left(\dfrac{k}{2}\right)} \cdot \left(1 + \frac{x^2}{k}\right)^{-\frac{k+1}{2}}$$

Characteristic function: $\Gamma\left(\dfrac{k+1}{2}\right)\sqrt{k\pi}\Gamma\left(\dfrac{k}{2}\right)\displaystyle\int_{-\infty}^{\infty}\left(1+\dfrac{t^2}{k}\right)^{-\frac{k+1}{2}}\cdot e^{j\vartheta t}\,dt$

Parameters: k

26 Uniform Distribution

Variable variation range: $a < x < b$

Distribution function: $F_1(x)=\begin{cases} 0 & x < a \\ \dfrac{x-a}{b-a} & a \le x \le b \\ 1 & x > b \end{cases}$

Density function: $f_1(x)=\dfrac{1}{b-a}$

Characteristic function: $\dfrac{e^{jb\vartheta}-e^{-ja\vartheta}}{j\vartheta(b-a)}$

Parameters: a, b

27 Weibull Distribution

Variable variation range: $0 < x < \infty$

Distribution function: $F_1(x)=\begin{cases} 0 & x \le 0 \\ 1-\exp\!\left(-c\cdot x^\alpha\right) & x > 0 \end{cases}$

Density function: $f_1(x)=c\cdot\alpha\cdot x^{\alpha-1}\cdot\exp\!\left(-c\cdot x^a\right)$

Characteristic function: $1+j\vartheta\displaystyle\int_0^{\infty} e^{j\vartheta x-cx^\alpha}\,dx$

Parameters: $c > 0$, $\alpha > 0$

Appendix 13 **Integral Forms for Gaussian Distribution** (after [5, p. 83])

Typically, the following integral forms of the Gaussian distributions are used:

$$1 \ \ \Phi(x) = \frac{1}{\sqrt{2\pi}} \int_{-\infty}^{x} e^{-t^2/2} dt \qquad 2 \ \ \hat{\Phi}(x) = \frac{\rho}{\sqrt{\pi}} \int_{-\infty}^{x} e^{-\rho^2 t^2} dt$$

$$3 \ \ \Phi_0(x) = \frac{1}{\sqrt{2\pi}} \int_{0}^{x} e^{-t^2/2} dt \qquad 4 \ \ \hat{\Phi}_0(x) = \frac{\rho}{\sqrt{\pi}} \int_{0}^{x} e^{-\rho^2 t^2} dt$$

$$5 \ \ F(x) = \frac{2}{\sqrt{2\pi}} \int_{0}^{x} e^{-t^2/2} dt \qquad 6 \ \ \hat{F}(x) = \frac{2\rho}{\sqrt{\pi}} \int_{0}^{x} e^{-\rho^2 t^2} dt$$

$$7 \ \ erf(x) = \frac{2}{\sqrt{\pi}} \int_{0}^{x} e^{-t^2} dt$$

The correspondence between different forms is as follows:

$$1 \ \ \Phi(x) = \hat{\Phi}\left(\frac{x}{\rho\sqrt{2}}\right) = \Phi_0(x) + 0.5 = \hat{\Phi}_0\left(\frac{x}{\rho\sqrt{2}}\right) + 0.5 = \frac{1}{2}[1 + F(x)]$$

$$= \frac{1}{2}\left[1 + \hat{F}\left(\frac{x}{\rho\sqrt{2}}\right)\right] = \frac{1}{2}\left[1 + erf\left(\frac{x}{\sqrt{2}}\right)\right]$$

$$2 \ \ \hat{\Phi}(x) = \Phi(\rho\sqrt{2}x) = \Phi_0(\rho\sqrt{2}x) + 0.5 = \hat{\Phi}_0(x) + 0.5 = \frac{1}{2}\left[1 + F(\rho\sqrt{2}x)\right]$$

$$= \frac{1}{2}[1 + \hat{F}(x)] = \frac{1}{2}[1 + erf(\rho x)]$$

$$3 \ \ \Phi_0(x) = \Phi(x) - 0.5 = \hat{\Phi}\left(\frac{x}{\rho\sqrt{2}}\right) - 0.5 = \hat{\Phi}_0\left(\frac{x}{\rho\sqrt{2}}\right) = \frac{1}{2}F(x)$$

$$= \frac{1}{2}\hat{F}\left(\frac{x}{\rho\sqrt{2}}\right) = \frac{1}{2}erf\left(\frac{x}{\sqrt{2}}\right)$$

$$4 \ \ \hat{\Phi}_0(x) = \Phi(\rho\sqrt{2}x) - 0.5 = \hat{\Phi}(x) - 0.5 = \Phi_0(\rho\sqrt{2}x) = \frac{1}{2}F(\rho\sqrt{2}x) = \frac{1}{2}\hat{F}(x)$$

$$= \frac{1}{2}erf(\rho x)$$

5 $F(x) = 2\Phi(x) - 1 = 2\hat{\Phi}\left(\dfrac{x}{\rho\sqrt{2}}\right) - 1 = 2\Phi_0(x) = 2\hat{\Phi}_0\left(\dfrac{x}{\rho\sqrt{2}}\right)$

$= \hat{F}\left(\dfrac{x}{\rho\sqrt{2}}\right) = erf\left(\dfrac{x}{\sqrt{2}}\right)$

6 $\hat{F}(x) = 2\Phi\left(\rho\sqrt{2}x\right) - 1 = 2\hat{\Phi}(x) - 1 = 2\Phi_0\left(\rho\sqrt{2}x\right) = 2\hat{\Phi}_0(x) = F\left(\rho\sqrt{2}x\right)$

$= erf\left(\rho x\right)$

7 $erf(x) = 2\Phi\left(\sqrt{2}x\right) - 1 = 2\hat{\Phi}\left(\dfrac{x}{\rho}\right) - 1 = 2\Phi_0\left(\sqrt{2}x\right) = 2\hat{\Phi}_0\left(\dfrac{x}{\rho}\right) = F\left(\sqrt{2}x\right)$

$= \hat{F}\left(\dfrac{x}{\rho}\right)$

Appendix 14 **Moments of Random Variables**

1 *Initial Moment of Order k*:

Discrete variable: $m_k = M\left(X^k\right) = \sum_{n=1}^{N} x_n^k p_n$

Continuous variable: $m_k = \int_{-\infty}^{\infty} x^k f_1(x) dx$

2 *Absolute Initial Moment of Order k*:

Discrete variable: $\beta_k = M\left(|X|^k\right) = \sum_{n=1}^{N} |x_n|^k p_n$

Continuos variable: $\beta_k = \int_{-\infty}^{\infty} |x|^k f_1(x) dx$

3 *Factorial Initial Moment of Order k* *****:

$m_{[k]} = M\left(X^{[k]}\right) = \sum_{n=1}^{N} x_n^{[k]} \cdot p_n$

$m_{[k]} = \int_{-\infty}^{\infty} x^{[k]} f_1(x) dx$

4 *Central Moment of Order k:*

Discrete variable: $M_k = M\left(X_0^k\right) = \sum_{n=1}^{N} (x_n - m_x)^k \cdot p_n$

Continuos variable: $M_k = \int_{-\infty}^{\infty} (x - m_x)^k f_1(x) dx$

5 *Absolute Central Moment of Order k:*

Discrete variable: $B_k = M\left(|X_0|^k\right) = \sum_{n=1}^{N} |x_n - m_x|^k \cdot p_n$

Continuous variable: $B_k = \int\limits_{-\infty}^{\infty} |x - m_x|^k f_1(x) dx$

6 Factorial Central Moment of Order k *:

Discrete variable: $M_{[k]} = M\left(X_0^{[k]}\right) = \sum\limits_{n=1}^{N} (x_n - m_x)^{[k]} \cdot p_n$

Continuous variable: $M_{[k]} = \int\limits_{-\infty}^{\infty} (x - m_x)^{[k]} f_1(x) dx$

---

* $z^{[k]} = z(z-1)(z-2)...(z-k+1)$

Appendix 15 **Moments of Discrete Random Variables** (after [4, p. 52])

1 *Binomial Distribution*

$$m = np; \quad \sigma^2 = npq; \quad \gamma_1 = \frac{q-p}{\sqrt{npq}}; \quad \gamma_2 = \frac{1-6pq}{npq}$$

$$M_{k+1} = pq \cdot \left(nkM_{k-1} + \frac{dM_k}{dp} \right); \quad q = 1-p$$

2 *Hypergeometric Distribution*

$$m = n\frac{M}{N}; \quad \sigma^2 = \frac{M(N-M)n(N-n)}{N^2(N-1)}; \quad \gamma_1 = \frac{(N-2M)(N-2n)\sqrt{N-1}}{(N-2)\sqrt{M(N-M)n(N-n)}}$$

$$M_4 = \frac{M(N-M)n(N-n)}{N^4(N-1)\cdot(N-2)\cdot(N-3)}$$
$$\times \left\{ N^3(N+1) - 6N^2n(N-n) + 3M(N-M)\left[N^2(n-2) - Nn^2 + 6n(N-n) \right] \right\}$$

3 *Poisson Distribution*

$$m = \lambda; \quad \sigma^2 = \lambda; \quad \gamma_1 = \frac{1}{\sqrt{\lambda}}; \quad \gamma_2 = \frac{1}{\lambda}; \quad M_{k+1} = \lambda_k M_{k-1} + \lambda \cdot \frac{dM_k}{d\lambda}$$

$$m_{k+1} = m_k + \lambda \cdot \sum_{n=0}^{k-1} C_n^{k-1} \cdot m_{k-n}$$

4 *Polya Distribution*

$$m = \lambda; \quad \sigma^2 = \lambda(1+\alpha\lambda); \quad \gamma_1 = \frac{1+2\alpha\lambda}{\sqrt{\lambda(1+\alpha\lambda)}}; \quad \gamma_2 = 6\alpha + \frac{1}{\lambda(1+\alpha\lambda)}$$

5 *Uniform Distribution*

$$m = \frac{n+1}{2}; \quad \sigma^2 = \frac{n^2-1}{12}; \quad \gamma_1 = 0; \quad \gamma_2 = -1.2 + \frac{4}{n^2-1}$$

Appendix 16 **Moments of Continuous Random Variables** (after [4, pp. 54, 55])

1 *Arcsin Distribution*

$$m = 0; \quad \sigma^2 = \frac{1}{2}a^2$$

2 *Beta Distribution*

$$m = \frac{a}{a+b}; \quad \sigma^2 = \frac{ab}{(a+b)^2(a+b+1)}; \quad m_k = \frac{\Gamma(a+b)\Gamma(a+k)}{\Gamma(a)\Gamma(a+b+k)}$$

3 *Chi-Squared Distribution*

$$m = n; \quad \sigma^2 = 2n$$

4 *Exponential Distribution*

$$m = \lambda; \quad \sigma^2 = \frac{1}{\lambda^2}; \quad \gamma_1 = 2; \quad \gamma_2 = 6; \quad m_k = k!\lambda^{-k}; \quad m_{k+1} = \frac{k+1}{\lambda}m_k$$

5 *Gamma Distribution*

$$m = (\alpha+1)\beta; \quad \sigma^2 = (\alpha+1)\beta^2; \quad \gamma_1 = \frac{2}{\sqrt{\alpha+1}}; \quad \gamma_2 = \frac{6}{\alpha+1}$$

$$m_{k+1} = (\alpha+k+1)\cdot\beta\cdot m_k$$

6 *Gaussian Distribution*

$$m = m; \quad \sigma^2 = \sigma^2; \quad \gamma_1 = 0; \quad \gamma_2 = 0; \quad m_k = k!\sum_{i=0}^{[k/2]}A_i; \quad M_{2k} = \frac{(2k)!}{k!}\left(\frac{\sigma^2}{2}\right)^k$$

$$M_{2k+1} = 0; \quad A_i = \frac{m^{k-2i}}{(k-2i)!}\frac{\left(\dfrac{\sigma^2}{2}\right)^i}{i!}; \quad [k/2] = \text{floor}\left(\frac{k}{2}\right)$$

7 *Gaussian Distribution, Standard*

$$m = 0; \; \sigma^2 = 1; \; \gamma_1 = 0; \; \gamma_2 = 0; \; m_{2k} = M_{2k} = \frac{(2k)}{k!} \left(\frac{1}{2}\right)^k; \; m_{2k+1} = M_{2k+1} = 0$$

8 *Laplace Distribution*

$$m = \mu; \; \sigma^2 = \frac{2}{\lambda^2}; \; \gamma_1 = \frac{3\sqrt{2}}{2}; \; \gamma_2 = 3; \; m_k = k! \sum_{i=0}^{[k/2]} A_i; \; M_{2k} = (2k)! \lambda^{-2k}$$

$$M_{2k+1} = 0; \; A_i = \frac{M^{k-2i}}{(k-2i)!} \lambda^{-2i}$$

9 *Maxwell Distribution*

$$m = 2\sqrt{\frac{2}{\pi}}\sigma; \; \sigma^2 = \left(3 - \frac{8}{\pi}\right) \cdot \sigma^2$$

10 *Modulus of a Gaussian Variable*

$$m = 2\left[m\Phi\left(\frac{m}{\sigma}\right) + \sigma\Phi\left[\frac{m}{\sigma}\right]\right] = m_1; \; \sigma^2 = \sigma^2 + m^2 - m_1^2$$

11 *Rayleigh Distribution*

$$m = \sigma\sqrt{\frac{\pi}{2}}; \; \sigma^2 = \sigma^2\left(2 - \frac{\pi}{2}\right); \; \gamma_1 = (\pi - 3)\sqrt{\frac{\pi}{2}}; \; \gamma_2 = 5 - \frac{3}{4}\pi^2$$

$$m_k = \left(\sigma\sqrt{2}\right)^k \Gamma\left(\frac{k+2}{2}\right); \; M_k = \sum_{i=0}^{k}(-1)^i C_k^i \cdot m^i m_{k-i}$$

12 *Simpson Distribution*

$$m = \frac{a+b}{2}; \; \sigma^2 = \frac{(b-a)^2}{6}$$

13 *Student Distribution*

$$m = 0 \; (k > 1); \; \sigma^2 = \frac{k}{k-2} \; (k > 2)$$

14 *Uniform Distribution*

$$m = (a+b)/2; \; \sigma^2 = \frac{(b-a)^2}{12}; \; \gamma_1 = 0; \; \gamma_2 = -1.2; \; m_k = \frac{b^{k+1} - a^{k+1}}{(b-a)(k+1)}$$

$$m_{k+1} = m_k \cdot \frac{\left(b^{k+2} - a^{k+2}\right)(k+1)}{\left(b^{k-1} - a^{k-1}\right)(k+2)}$$

15 *Weibull Distribution*

$$m = c^{-1/\alpha} \Gamma\left(1 + \frac{1}{\alpha}\right); \; \sigma^2 = c^{-2/\alpha} \times \left[\Gamma\left(1 + \frac{2}{\alpha}\right) - \Gamma^2\left(1 + \frac{1}{\alpha}\right)\right]; \; \gamma_1 = \frac{M_3}{\sigma^3}$$

$$\gamma_2 = \frac{M_4}{\sigma^4} - 3; \; m_k = c^{-k/\alpha} \Gamma\left(1 + \frac{k}{\alpha}\right); \; M_k = \sum_{i=0}^{k} (-1)^i \times C_k^i m^i m_{k-i}$$

Appendix 17 **The Functional Transform of Random Variables**

1 *Linear Transform* $(Y = aX + b)$

Initial function: $f_1(x)$

Transformed function: $f_1(y) = \dfrac{1}{|a|} f_1\left(\dfrac{y - b}{a} \right)$

2 *Square Transform* $(Y = X^2)$

Initial function: $f_1(x)$

Transformed function: $f_1(y) = \begin{cases} \dfrac{1}{2\sqrt{y}} \left[f_1\left(\sqrt{y}\right) + f_1\left(-\sqrt{y}\right) \right] & y > 0 \\ 0 & , \quad y < 0 \end{cases}$

3 *Square Transform of Gaussian Variable*

Initial function: $f_1(x) = \dfrac{1}{\sigma\sqrt{2\pi}} \cdot e^{-\frac{(x-a)^2}{2\sigma^2}}$

Transformed function: $f_1(y) = \dfrac{1}{\sigma\sqrt{2\pi y}} e^{-\frac{y^2 + a^2}{2\sigma^2}} ch\left(\dfrac{a\sqrt{y}}{\sigma^2} \right), \quad y > 0$

4 *Square Transform of Gaussian Variable with Zero Mean*

Initial function: $f_1(x) = \dfrac{1}{\sigma\sqrt{2\pi}} \cdot e^{-\frac{x^2}{2\sigma^2}}$

Transformed function: $f_1(y) = \dfrac{1}{\sigma\sqrt{2\pi y}} \cdot e^{-\frac{y}{2\sigma^2}}, \quad y > 0$

5 *Square Transform of Rayleigh Variable*

Initial variable: $f_1(x) = \dfrac{x}{\sigma^2} \cdot e^{-\frac{x^2}{2\sigma^2}}$ (Rayleigh pdf)

Transformed variable: $f_1(y) = \dfrac{1}{2\sigma^2} \cdot \exp\left(-\dfrac{y}{2\sigma^2}\right)$ $y > 0$ (exponential pdf)

6 *Exponential Transform* ($Y = e^X$)

Initial function: $f_1(x) = \dfrac{1}{\sigma\sqrt{2\pi}} \cdot e^{-\frac{(x-a)^2}{2\sigma^2}}$

Transformed function: $f_1(y) = \dfrac{1}{y\sqrt{2\pi\sigma^2}} \exp\left[-\dfrac{(\ln y - a)^2}{2\sigma^2}\right]$, $y \geq 0$

7 *Detector of Power* ($a > 0$) $Y = \begin{cases} (X - x_0)^a, & X \geq x_0 \\ 0 & , & X < x_0 \end{cases}$

Initial function: $f_1(x)$

Transformed function:

$$f_1(y) = \delta(y) \int_{-\infty}^{x_0} f_1(x)dx + \left[a \cdot y^{\frac{a-1}{a}}\right]^{-1} \cdot f_1\left(x_0 + y^{1/a}\right) \quad y \geq 0$$

8 *Linear Detector* ($a = 0$) $Y = \begin{cases} X - x_0, & X \geq x_0 \\ 0 & , & X < x_0 \end{cases}$

Initial function: $f_1(x)$

Transformed function: $f_1(y) = \delta(y) \cdot \Phi\left(\dfrac{x_0}{\delta}\right) + \dfrac{1}{\sigma\sqrt{2\pi}} e^{-\frac{(y+x_0)^2}{2\sigma^2}}$ $y \geq 0$

9 *Harmonic Oscillation with Random Phase* ($Y = a \cdot \sin X$)

Initial function: $f_1(x)$

Transformed function:

$$f_1(y) = \frac{1}{a\sqrt{1-\left(\dfrac{y}{a}\right)^2}} \sum_{k=-\infty}^{\infty} f_1\left[\pi k + (-1)^k \arcsin\frac{y}{a}\right] \quad |y| < a$$

10 *Harmonic Oscillation with Random Phase Distributed Uniformly at* ($0, 2\pi$)

Initial function: $f_1(x) = \begin{cases} \dfrac{1}{2\pi}, & |x| \le \pi \\ 0, & |x| > \pi \end{cases}$

Transformed function: $f_1(y) = \dfrac{1}{\pi a\sqrt{1-\left(\dfrac{y}{a}\right)^2}}, \quad |y| < a$

Appendix 18 **Probability Density Function of Sum, Difference, Product, and Quotient of Two Random Variables**

1 _Sum_ $(Y = X_1 + X_2)$

Dependent variables: $f_1(y) = \int\limits_{-\aleph}^{\infty} f_2(y - x_2, x_2)dx_2 = \int\limits_{-\infty}^{\infty} f_2(y - x_1, x_1)dx_1$

Independent variables:

$f_1(y) = \int\limits_{-\infty}^{\infty} f_1(y - x_2) \cdot f_1(x_2)dx_2 = \int\limits_{-\infty}^{\infty} f_1(y - x_1) \cdot f(x_1)dx_1$

2 _Difference_ $(Y = X_1 - X_2)$

Dependent variables: $f_1(y) = \int\limits_{-\infty}^{\infty} f_2(y + x_2, x_2)dx_2 = \int\limits_{-\infty}^{\infty} f_2(x_1 - y, x_1)dx_1$

Independent variables:

$f_1(y) = \int\limits_{-\infty}^{\infty} f_1(y + x_2) \cdot f(x_2)dx_2 = \int\limits_{-\infty}^{\infty} f_1(x_1 - y) \cdot f_1(x_1)dx_1$

3 _Product_ $(Y = X_1 \cdot X_2)$

Dependent variables: $f_1(y) = \int\limits_{-\infty}^{\infty} f_2\left(\dfrac{y}{x_2}, x_2\right)\dfrac{1}{|x_2|}dx_2 = \int\limits_{-\infty}^{\infty} f_2\left(\dfrac{y}{x_1}, x_1\right) \cdot \dfrac{1}{|x_1|}dx_1$

Independent variables:

$f_1(y) = \int\limits_{-\infty}^{\infty} f_1\left(\dfrac{y}{x_2}\right) \cdot f_1(x_2)\dfrac{1}{|x_2|}dx_2 = \int\limits_{-\infty}^{\infty} f_1\left(\dfrac{y}{x_1}\right) \cdot f_1(x_1) \cdot \dfrac{1}{|x_1|} \cdot dx_1$

4 _Quotient_ $(Y = \dfrac{X_1}{X_2})$

Dependent variables: $f_1(y) = \int\limits_{-\infty}^{\infty} f_2(y \cdot x_2, x_2) \cdot |x_2|dx_2 = \int\limits_{-\infty}^{\infty} f_2\left(\dfrac{x_1}{y}, x_1\right)\left|\dfrac{x_1}{y^2}\right|dx_1$

Independent variables:

$$f_1(y) = \int_{-\infty}^{\infty} f_1(y \cdot x_2) f_1(x_2) |x_2| dx = \int_{-\infty}^{\infty} f_1\left(\frac{x_1}{y}\right) \cdot f_1(x_1) \left|\frac{x_1}{y^2}\right| dx_1$$

Appendix 19 **Simulation Algorithms for Common Distributions**

1 *Gaussian Distribution*

Density function: $f(z) = \dfrac{1}{\sqrt{2\pi}\,\sigma_z} \cdot \exp\left[-\dfrac{(z - m_z)^2}{2\sigma_z^2} \right]$

Simulation algorithm: $z_0 = \sqrt{-2 \cdot \ln(z1_u)} \cdot \cos(2\pi\, z2_u)$

or

$z_0 = \sqrt{-2 \cdot \ln(z1_u)} \cdot \sin(2\pi\, z2_u)$

$z_N = m_z + \sigma_z \cdot z_0$

2 *Rayleigh Distribution*

Density function: $f(z) = \dfrac{z}{\sigma_z} \cdot \exp\left(-\dfrac{z^2}{2\sigma_z^2} \right) \quad z \geq 0$

Simulation algorithm: $z_R = \sigma_z \cdot \sqrt{-2 \cdot \ln(z_u)}$

3 *Ricean Distribution*

Density function:

$$f(z) = \dfrac{z}{\sigma_z} \cdot \exp\left(-\dfrac{z^2 + a^2}{2\sigma_z^2} \right) \cdot I_0\left(\dfrac{az}{\sigma_z^2} \right) \quad z \geq 0$$

Simulation algorithm:

$$z_{RN} = \sqrt{a^2 - 2\sigma_z^2 \cdot \ln(z1_u) - 2a\sigma_z \sqrt{-2 \cdot \ln(z1_u)} \cdot \cos(2\pi\, z2_u)}$$

4 *Exponential Distribution*

Density function: $f(z) = \lambda \cdot \exp(-\lambda z)$

Simulation algorithm: $z_E = \dfrac{-\ln(z_u)}{\lambda}$

5 *Log-Normal Distribution*

Density function: $f(z) = \dfrac{1}{z\sqrt{2\pi}\,\sigma_z} \cdot \exp\left[-\dfrac{(\ln(z) - m_z)^2}{2\sigma_z^2}\right]$

Simulation algorithm: $z_L = \exp(z_N)$

6 *Chi-Squared Distribution* (with $2k$ degrees of freedom)

Density function: $f(z) = \dfrac{1}{2^k \cdot \Gamma(k)} \cdot z^{k-1}\, e^{-\frac{z}{2}}$

Simulation algorithm: $z_C = -2\ln\left(\displaystyle\prod_{i=1}^{k} zi_u\right)$

7 *Student Distribution* (with n degrees of freedom)

Density function: $f(z) = \dfrac{\Gamma\left(\dfrac{n+1}{2}\right)}{\sqrt{n\pi}\cdot\Gamma\left(\dfrac{n}{2}\right)\cdot\left(1+\dfrac{z^2}{n}\right)^{\frac{n+1}{2}}}$

Simulation algorithm: $z_S = \dfrac{z_0}{\sqrt{z_C}}\cdot\sqrt{n}$ for z_C
with $2k = n$

Appendix 20 Basic Simulation Algorithms to Generate Random Variables with Gaussian Distribution (after [11, p. 60])

Algorithm # 1

$$\varepsilon_1 = \sqrt{-2\ln\gamma_1} \cdot \cos(2\pi\gamma_2); \quad \varepsilon_2 = \sqrt{-2\ln\gamma_1} \cdot \sin(2\pi\gamma_2)$$

$\gamma_1 \in UN(0,1)$ and $\gamma_2 \in UN(0,1)$ are independent variables with uniform distribution at the interval $(0,1)$. *Note:* the algorithm is sensitive to correlation between γ_1 and γ_2.

Algorithm # 2

$$\varepsilon = \begin{cases} \Phi^{-1}(\gamma), & 0.5 < \gamma < 1 \\ -\Phi^{-1}(1-\gamma), & 0 < \gamma \le 0.5 \end{cases}$$

$\gamma \in UN(0,1)$, $\Phi^{-1}(\gamma)$ is an approximation of the function reciprocal to:

$$\Phi(y) = \int_{-\infty}^{y} \frac{1}{\sqrt{2\pi}} \exp\left(-t^2/2\right) dt$$

Typical approximation is:

$$\Phi^{-1}(\gamma) = \frac{C_1 + C_2 \cdot \vartheta}{1 + C_3 \cdot \vartheta + C_4 \cdot \vartheta^2}$$

$C_1 = 2.30753$; $C_2 = 0.27061$; $C_3 = 0.99229$; $C_4 = 0.04481$; $\vartheta = \sqrt{-2\ln\gamma}$

Appendix 21 **Simulation Algorithms Based on a Rigorous Solution for the Reciprocal Functions** (after [12, p. 15])

1 *Rayleigh Distribution*

Density function: $f_1(y) = \dfrac{y}{\sigma^2} e^{-y^2/2\sigma^2}$ $y \geq 0$

Simulation algorithm: $y = \sigma\sqrt{-2 \cdot \ln(1 - x_u)} = \sigma\sqrt{-2 \cdot \ln x_u}$

(variables $1 - x_u$ and x_u have the same pdf)

2 *Exponential Distribution*

Density function: $f_1(y) = \lambda \cdot e^{-\lambda y}$ $y \geq 0$

Simulation algorithm: $y = -\dfrac{1}{\lambda} \cdot \ln x_u$

3 *Cauchy Distribution*

Density function: $f_1(y) = \dfrac{a}{\pi\left[a^2 + (y - b)^2\right]}$

Simulation algorithm: $y = a \cdot \tan\left[\pi\left(x_u - \dfrac{1}{2}\right)\right] + b$

4 *Distribution with Probability Density Function*

$f_1(y) = r^2 \cdot y^2 \left(1 + r^2 \cdot y^2\right)^{-3/2}$ $y \geq 0$

Simulation algorithm: $y = \dfrac{1}{r}\sqrt{\dfrac{1}{x_u^2} - 1}$

5 *Distribution with Probability Density Function*

$f_1(y) = \alpha^{-1} \cdot y \cdot \left(\alpha^2 - y^2\right)^{-1/2}$ $0 < y \leq \alpha$

Simulation algorithm: $y = \alpha \cdot \sqrt{1 - x_u^2}$

6 Combination of Two Cauchy Distributions

Density function: $f_1(y) = \dfrac{\alpha}{2\pi}\left[\dfrac{1}{\alpha^2 + (y + \beta)^2} + \dfrac{1}{\alpha^2 + (y - \beta)^2}\right]$

Simulation algorithm: $y = \alpha \cdot \tan\left[\pi\left(x_u - \dfrac{1}{2}\right)\right] + \gamma_\beta$

γ_β is a variable that takes two values only: $+\beta$ or $-\beta$ with equal probability.

7 Combination of Two Gaussian Distributions

Density function: $f_1(y) = \dfrac{1}{2\alpha\sqrt{2\pi}} \cdot \left\{ e^{-\frac{(y+\beta)^2}{2\alpha^2}} + e^{-\frac{(y-\beta)^2}{2\alpha^2}} \right\}$

Simulation algorithm: $y = \alpha \cdot x_N + \gamma_\beta$

$x_N \in N(0,1)$ is a normal variable, γ_β is defined above.

Appendix 22 **Simulation Algorithms Derived by Means of Gamma Distribution** (after [12, p. 17])

1 *Gamma Distribution*

Density function: $f_1(x) = \dfrac{\beta^a}{\Gamma(\alpha)} x^{\alpha-1} \cdot e^{-\beta x}$, $\quad x > 0$

Simulation algorithm: $x = -\ln\left(\displaystyle\prod_{k=1}^{n} \dfrac{x k_u}{\beta}\right)$; $\alpha = n = $ integer, $x \in \Gamma(\alpha, \beta)$

2 μ *-th Degree of a Variable with Gamma Distribution*

Density function: $f_1(x) = \dfrac{\beta^a}{|\mu|\Gamma(a)} \cdot x^{\frac{\alpha}{\mu}-1} \cdot \exp\left(-\beta \cdot x^{1/\mu}\right)$

Simulation algorithm: $x = z^\mu$, $\quad z \in \Gamma(\alpha, \beta)$

For $\mu = \dfrac{1}{2}$ density function is $f_1(x) = \dfrac{2 \cdot \beta^a}{\Gamma(a)} \cdot x^{2\alpha-1} \cdot e^{-\beta x^2}$ and
simulation algorithm is $x = \sqrt{z}$, $\quad z \in \Gamma(\alpha, \beta)$

3 *F-Distribution*

Density function: $f_1(x) = \dfrac{\Gamma(\alpha_1 + \alpha_2) \cdot \beta^{\alpha_1}}{\Gamma(a_1)\Gamma(\alpha_2)} \cdot \dfrac{x^{\alpha_1-1}}{(1+\beta x)^{\alpha_1+\alpha_2}}$

Simulation algorithm: $x = x_1/x_2$,

where $x_1 \in \Gamma(\alpha_1, \beta_1)$ $\quad x_2 \in \Gamma(\alpha_2, \beta_2)$

4 μ *-th Degree of a Variable with F-Distribution*

Density function: $f_1(x) = \dfrac{\Gamma(\alpha_1 + \alpha_2)\beta^{\alpha_1}}{|\mu|\Gamma(a_1) \cdot \Gamma(\alpha_2)} \cdot \dfrac{x^{\alpha_1/\mu-1}}{\left(1+\beta x^{1/\mu}\right)^{\alpha_1+\alpha_2}}$

Simulation algorithm: $x = (x_1/x_2)^\mu$

For $\mu = \dfrac{1}{2}$ density function is $f_1(x) = \dfrac{2\Gamma(\alpha_1 + \alpha_2) \cdot \beta^{\alpha_1}}{\Gamma(a_1) \cdot \Gamma(\alpha_2)} \cdot \dfrac{x^{2\alpha_1 - 1}}{\left(1 + \beta x^2\right)^{\alpha_1 + \alpha_2}}$ and

simulation algorithm is $x = \sqrt{x_1/x_2}$

5 Beta Distribution

Density function: $f_1(x) = \dfrac{\Gamma(p+m)}{\Gamma(p)\Gamma(m)} x^{p-1} \cdot (1-x)^{m-1}, \quad 0 \le x \le 1$

Simulation algorithm: $x = \dfrac{x_1}{\left(x_1 + x_N\right)}$,

where x_1 is defined above and $x_N \in N(0,1)$ is a normal variable.

Appendix 23 Common Correlation Functions and Spectra of the Stationary Random Processes

---

Correlation Function	Spectral Density

$$K(\tau)=\sigma^2 \cdot e^{-\alpha|\tau|}$$

$$S(\omega)=\frac{\sigma^2}{\pi} \cdot \frac{a}{\omega^2+\alpha^2}$$

$$K(\tau)=\sigma^2 \cdot e^{-\alpha|\tau|} \cdot \cos \beta\tau$$

$$S(\omega)=\frac{\sigma^2\alpha}{\pi} \cdot \frac{\omega^2+\alpha^2+\beta^2}{\left(\omega^2-\beta^2-\alpha^2\right)^2+4\alpha^2\omega^2}$$

$$K(\tau)=\sigma^2 \cdot e^{-\alpha|\tau|} \cdot \cos \beta\tau$$
$$\times\left(\cos \beta\tau+\frac{a}{\beta}\sin \beta|\tau|\right)$$

$$S(\omega)=\frac{2\sigma^2\alpha}{\pi} \cdot \frac{\alpha^2+\beta^2}{\left(\omega^2-\beta^2-\alpha^2\right)^2+4\alpha^2\omega^2}$$

$$K(\tau)=\sigma^2 \cdot e^{-\alpha^2\tau^2} \cdot \cos \beta\tau$$

$$S(\omega)=\frac{\sigma^2}{4\alpha\sqrt{\pi}} \cdot \left[e^{-\frac{(\omega+\beta)^2}{4\alpha^2}}+e^{-\frac{(\omega-\beta)^2}{4\alpha^2}}\right]$$

$$K(\tau)=a_0 \cdot \delta(\tau)+a_1 \cdot \delta^{(2)}(\tau)$$
$$+...a_n \cdot \delta^{(2n)}(\tau)$$

$$S(\omega)=\frac{1}{2\pi}\left[\begin{array}{l} a_0-\omega^2 a_1+\omega^4\alpha_2+... \\ +(-1)^n \omega^{2n} a_n \end{array}\right]$$

Appendix 24 **The Derivatives of the Random Process**

1 *A Generic Random Process*

Covariation : $B_{\xi^{(n)}}(t_1, t_2) = \dfrac{\partial^{2n} B_\xi(t_1, t_2)}{\partial t_1^n \partial t_2^n}$

Mutual covariation : $B_{\xi^{(k)}\xi^{(l)}}(t_1, t_2) = \dfrac{\partial^{k+l} K_\xi(t_1, t_2)}{\partial t_1^k \partial t_2^l}$

Correlation: $K_{\xi^{(n)}}(t_1, t_2) = \dfrac{\partial^{2n} K_\xi(t_1, t_2)}{\partial t_1^n \partial t_2^n}$

Mutual correlation: $K_{\xi^{(k)}\xi^{(l)}}(t_1, t_2) = \dfrac{\partial^{k+l} K_\xi(t_1, t_2)}{\partial t_1^k \partial t_2^l}$

2 *A Stationary Random Process*

Covariation: $B_{\xi^{(n)}}(\tau) = (-1)^n \dfrac{d^{2n} B_\xi(\tau)}{d\tau^{2n}}$

Mutual covariation: $B_{\xi^{(k)}\xi^{(l)}}(\tau) = (-1)^k \dfrac{d^{k+l} B_\xi(\tau)}{d\tau^{k+l}}$

Correlation: $K_{\xi^{(n)}}(\tau) = (-1)^n \dfrac{d^{2n} K_\xi(\tau)}{d\tau^{2n}}$

Mutual correlation: $K_{\xi^{(k)}\xi^{(l)}}(\tau) = (-1)^k \dfrac{d^{k+l} B_\xi(\tau)}{d\tau^{k+l}}$

Spectral density: $S_{\xi^{(n)}}(\omega) = \omega^{2n} \cdot S_{\xi^0}(\omega)$

Appendix 25 A Gaussian Random Process

1 A n-*Dimensional Generic Process*

Density function:

$$f_n\left(x_1, x_2, \ldots x_n; t_1, \ldots t_n\right) = \frac{1}{\sqrt{(2\pi)^n \Delta}}$$

$$\times \exp\left\{-\frac{1}{2\Delta} \cdot \sum_{\mu=1}^{n} \sum_{\gamma=1}^{n} \Delta_{\mu\gamma} \left[x_\mu - m_\xi\left(t_\mu\right)\right] \cdot \left[x_\gamma - m_\xi\left(t_\gamma\right)\right]\right\}$$

Characteristic function:

$$\theta_n\left(j\vartheta_1, j\vartheta_2, \ldots j\vartheta_n; t_1, \ldots t_n\right)$$

$$= \exp\left[j\sum_{\mu=1}^{n} m_\xi\left(t_\mu\right) \cdot \vartheta_\mu + \frac{1}{2} j^2 \sum_{\mu=1}^{n} \sum_{\gamma=1}^{n} K_\xi\left(t_\mu, t_\gamma\right)\vartheta_\mu \vartheta_\gamma\right]$$

2 A n-*Dimensional Stationary Process*

Density function:

$$f_n\left(x_1, \ldots x_n\right) = \frac{1}{\sigma^n \sqrt{(2\pi)^n D}}$$

$$\times \exp\left\{-\frac{1}{2\sigma^2 D} \cdot \sum_{\mu=1}^{n} \sum_{\gamma=1}^{n} D_{\mu\gamma}\left(x_\mu - m\right) \cdot \left(x_\gamma - m\right)\right\}$$

Characteristic function:

$$\theta_n\left(j\vartheta_1, \ldots j\vartheta_n\right)$$

$$= \exp\left[j \cdot m \cdot \sum_{\mu=1}^{n} \vartheta_\mu + \frac{1}{2} j^2 \sigma^2 \sum_{\mu=1}^{n} \sum_{\gamma=1}^{n} R\left(\tau_{\mu\gamma}\right)\vartheta_\mu \vartheta_\gamma\right]$$

3 A Two-Dimensional Stationary Process

Density function:

$$f_2(x_1, x_2) = \frac{1}{2\pi\sigma^2 \sqrt{1 - R^2(\tau)}}$$

$$\times \exp\left\{-\frac{1}{2\sigma^2[1 - R^2(\tau)]} \times \left[(x_1 - m)^2 - 2R(\tau)(x_1 - m)(x_2 - m) + (x_2 - m)^2\right]\right\}$$

Characteristic function:

$$\theta_2(j\vartheta_1, \ldots j\vartheta_2)$$

$$= \exp\left\{jm(\vartheta_1 + \vartheta_2) - \frac{1}{2}\sigma^2\left[\vartheta_1^2 + 2R(\tau)\vartheta_1\vartheta_2 + \vartheta_2^2\right]\right\}$$

4 A One-Dimensional Stationary Process

Density function: $f_1(x) = \dfrac{1}{\sigma\sqrt{2\pi}} \exp\left[-\dfrac{(x - m)^2}{2\sigma^2}\right]$

Characteristic function: $e^{jm\vartheta - \frac{\sigma^2\vartheta^2}{2}}$

Appendix 26 **The Models of Common Pulses in Time and Frequency Domains** (after [4, pp. 212, 213])

1 *Rectangular Pulse*

Pulse: $s_0(t) = \begin{cases} A_m, & |t| \leq \tau/2 \\ 0 & |t| > \tau/2 \end{cases}$

Spectrum: $F_1(\omega, \tau) = A_m \tau \dfrac{\sin\left(\dfrac{\omega\tau}{2}\right)}{\dfrac{\omega\tau}{2}}$

2 *Triangular Pulse*

Pulse: $s_0(t) = \begin{cases} \dfrac{A_m}{\tau}(t+\tau), & -\tau \leq t \leq 0 \\ -\dfrac{A_m}{\tau}(t-\tau), & 0 \leq t \leq \tau \\ 0 & |t| > \tau \end{cases}$

Spectrum: $F_1(\omega, \tau) = A_m \tau \left(\dfrac{\sin\dfrac{\omega\tau}{2}}{\dfrac{\omega\tau}{2}} \right)$

3 *Trapezoid Pulse*

Pulse: $s_0(t) = \begin{cases} \dfrac{A_m}{\alpha\tau}\left[t + (1+\alpha)\dfrac{\tau}{2}\right], & -(1+\alpha)\dfrac{\tau}{2} \leq t \leq -(1-\alpha)\dfrac{\tau}{2} \\ A_m, & |t| \leq (1-\alpha)\dfrac{\tau}{2} \\ -\dfrac{A_m}{\alpha\tau}\left[t - (1+\alpha)\dfrac{\tau}{2}\right], & (1-\alpha)\dfrac{\tau}{2} \leq t \leq (1+\alpha)\dfrac{\tau}{2} \\ 0 & |t| > (1+\alpha)\dfrac{\tau}{2} \end{cases}$

Spectrum: $F_1(\omega,\tau) = A_m \dfrac{\sin\left(\dfrac{\omega\tau}{2}\right)\sin\left(\dfrac{\alpha\omega\tau}{2}\right)}{\left(\dfrac{\omega\tau}{2}\right)\left(\dfrac{\alpha\omega\tau}{2}\right)}$

4 *Gaussian Pulse*

Pulse: $s_0(t) = A_m \exp\left(-4\ln 2\dfrac{t^2}{\tau^2}\right), \quad -\infty < t < \infty$

Spectrum: $F_1(\omega,\tau) = A_m\tau\sqrt{\dfrac{\pi}{4\ln 2}}\exp\left(-\dfrac{\omega^2\tau^2}{16\ln 2}\right)$

5 *Error Function Pulse*

Pulse: $s_0(t) = \dfrac{A_m}{2}\left[erf\,\dfrac{\sqrt{\pi}}{\alpha}\left(\dfrac{t}{\tau}+\dfrac{1}{2}\right) - erf\,\dfrac{\sqrt{\pi}}{\alpha}\left(\dfrac{t}{\tau}-\dfrac{1}{2}\right)\right], \quad -\infty < t < \infty$

$erfx = \dfrac{2}{\sqrt{\pi}}\displaystyle\int_0^x e^{-u^2}\,du$

Spectrum: $F_1(\omega,\tau) = A_m\tau\dfrac{\sin\left(\dfrac{\omega\tau}{2}\right)}{\left(\dfrac{\omega\tau}{2}\right)}\exp\left(-\dfrac{\alpha^2\omega^2\tau^2}{4\pi}\right)$

Appendix 27 The Spectral Densities for Common Pulsed Random Processes (after [4, pp. 214, 215])

1 *The process* $\xi_1(t)$ *with the following properties:*

A_γ and τ_γ do not depend on ϑ_γ; $f_2(A_\gamma, \tau_\gamma) = f_2(A, \tau)$; $f_1(\vartheta_\gamma) = f_1(\vartheta)$;

$A_\gamma, \tau_\gamma, \vartheta_\gamma$ are mutually independent

$$\text{Spectral density: } S(\omega) = \frac{1}{m_\vartheta} \left[\begin{array}{l} M\left\{A^2 |F_1(\omega,\tau)|^2\right\} + 2M\left\{AF_1(\omega,\tau)\right\} \times \\ M\left\{AF_1^*(\omega,\tau)\right\} \text{Re}\dfrac{\theta_\vartheta(\omega)}{1-\theta_\vartheta(\omega)} \end{array} \right] + 2\pi m_\xi^2 \delta(\omega)$$

2 *The process* $\xi_2(t)$ *with the following properties:*

in addition to previous assumptions for process $\xi_1(t)$: $f_2(A,\tau) = f_1(A) \cdot f_1(\tau)$

Spectral density:

$$S(\omega) = \frac{1}{m_\vartheta} \left[M\left\{A^2\right\} + 2M^2\{A\} \cdot \text{Re}\frac{\theta_\vartheta(\omega)}{1-\theta_\vartheta(\omega)} \right] \times M\left\{F_1(\omega,\tau)^2\right\}$$

$$+ \frac{2\pi}{m_\vartheta^2} [M\{A\} \cdot M\{F_1(0,\tau)\}]^2 \cdot \delta(\omega)$$

3 *The process* $\xi_3(t)$ *with the following properties:*

in addition to previous assumptions : $A_\gamma = A_0$ = const (m_τ, m_Δ, m_A are mathematical expectations of pulse duration, time between pulses and amplitude)

Spectral density:

$$S(\omega) = \frac{2A_0^2}{\omega^2(m_\tau + m_\Delta)} \text{Re}\frac{[1-\theta_\tau(\omega)][1-\theta_\Delta(\omega)]}{1-\theta_\tau(\omega)\theta_\Delta(\omega)} + A_0^2 \left(\frac{m_\tau}{m_\tau + m_\Delta}\right)^2 \cdot 2\pi\delta(\omega)$$

4 *The process* $\xi_4(t)$ *with the following properties:*

$\vartheta = \vartheta_0$ = const; $\tau = \tau_0$ = const; A_γ are random and correlated variables;

$\xi_4(t)$ is a stationary process: m_ξ = const, $K_\xi(\tau) = D_\xi \cdot R_\xi(\tau)$ (amplitude-pulse modulation case)

$$S(\omega)= 2\pi m_\xi^2 \cdot \vartheta_0^{-2}\left|F_1(\omega,\tau_0)\right|^2$$

Spectral density:
$$\times \sum_{k=-\infty}^{\infty} \delta\left(\omega-\frac{2\pi k}{\vartheta_0}\right)+\frac{D_\xi}{\vartheta_0}\left|F_1(\omega,\tau_0)\right|^2 \sum_{k=-\infty}^{\infty} R_\xi(k\vartheta_0)\cdot e^{j\omega k \vartheta_0}$$

5 _The process_ $\xi_5(t)$ _with the following properties:_
in addition to previous assumptions for process $\xi_4(t):\vartheta_0 \gg \tau_c$ (τ_c is correlation interval for $\xi_5(t)$)
Spectral density:

$$S(\omega)= 2\pi m_\xi^2 \vartheta_0^{-2}\left|F_1(\omega,\tau)\right|^2$$
$$\times \sum_{k=-\infty}^{\infty} \delta\left(\omega-\frac{2\pi k}{\vartheta_0}\right)+\frac{D_\xi}{\vartheta_0}\cdot\left|F_1(\omega,\tau)\right|^2 =S_d(\omega)\cdot \sum_{k=-\infty}^{\infty}\delta\left(\omega-\frac{2\pi k}{\vartheta_0}\right)+S_c(\omega)$$

$S_d(\omega)$ is the discrete spectral density portion, $S_c(\omega)$ is the continuous spectral density portion.

Appendix 28 Probability Density Functions and Spectral Densities for Common Pulse Amplitude Distributions (after [4, p. 217])

1 *Sinusoidal Distribution*

Density function: $f(x) = \begin{cases} \left[\pi \sqrt{x_0^2 - (x - m_x)^2} \right]^{-1}, & |x - m_x| < x_0 \\ 0 & |x - m_x| > x_0 \end{cases}$

Mean: m_x

Variance: $\dfrac{1}{2} x_0^2$

Spectral density:

$$\frac{2\pi}{\vartheta_0^2} \cdot m_x^2 \left| F_1(\omega, \tau_0) \right|^2 \cdot \delta(\omega - n\omega_0) + \frac{x_0^2}{2\vartheta_0} \left| F_1(\omega, \tau_0) \right|^2$$

2 *Gaussian Distribution*

Density function: $f_x = \dfrac{1}{\sigma\sqrt{2\pi}} \exp\left[-\dfrac{(x - m_x)^2}{2\sigma^2} \right] \quad -\infty < x < \infty$

Mean: m_x

Variance: σ^2

Spectral density:

$$\frac{2\pi}{\vartheta_0^2} \cdot m_x^2 \left| F_1(\omega, \tau_0) \right|^2 \cdot \delta(\omega - n\omega_0) + \frac{\sigma^2}{\vartheta_0} \cdot \left| F_1(\omega, \tau_0) \right|^2$$

3 *Uniform Distribution*

Density function: $f(x) = \begin{cases} \dfrac{1}{2x_0}, & |x - m_x| \leq x_0 \\ 0 & |x - m_x| > x_0 \end{cases}$

Mean: m_x

Variance: $\dfrac{1}{3}x_0^2$

Spectral density:

$$\frac{2\pi}{\vartheta_0^2} \cdot m_x^2 \left| F_1(\omega,\tau_0) \right|^2 \cdot \delta(\omega - n\omega_0) + \frac{x_0^2}{2\vartheta_0} \left| F_1(\omega,\tau_0) \right|^2$$

4 *The Sum of Two Delta Functions*

Density function: $f_1 \delta(x - x_1) + (1 - f_1) \delta(x - x_2)$

Mean: $f_1(x_1 - x_2) + x_2$

Variance: $f_1(1 - f_1) \times (x_1 - x_2)^2$

Spectral density:

$$\frac{2\pi}{\vartheta_0^2} [f_1(x_1 - x_2) + x_2]^2 \left| F_1(\omega,\tau_0) \right|^2 \cdot \delta(\omega - n\omega_0)$$

$$+ \frac{f_1(1 - f_1)(x_1 - x_2)^2}{\vartheta_0} \left| F_1(\omega,\tau_0) \right|^2$$

Appendix 29 Spectral Densities for the Stationary Pulse Trains (after [4, p. 223, 223])

1 Pulse Position Modulation (general case)

Assumptions: pulse duration $\tau_i = \tau_0 = \text{const}$; amplitude A_γ and edge shift ε_k are random, stationary and independent variables; pulses do not overlap:

$$\tau_0 + \varepsilon_{max} < \frac{\vartheta_0}{2}$$

Spectral density:
$$S(\omega) = \frac{1}{\vartheta_0} \left| F_1(\omega, \tau_0) \right|^2 \cdot M\{A^2\} - m_A^2 \left| \theta_\varepsilon(\omega) \right|^2$$

$$+ 2\pi \cdot \frac{m_A^2}{\vartheta} \left| \theta_\varepsilon(\omega) \right|^2 \cdot \sum_{k=-\infty}^{\infty} \delta\left(\omega - \frac{2\pi k}{\omega_0} \right)$$

$$\theta_\varepsilon(\omega) = M\{\exp(j\omega\varepsilon)\}$$

2 Pulse-Position Modulation (constant amplitudes $A_k = A_0$)

Spectral density: $S(\omega) = S_d(\omega) \cdot \sum_{k=-\infty}^{\infty} \delta\left(\omega - \frac{2\pi k}{\vartheta_0} \right) + S_c(\omega)$

$$S_d(\omega) = 2\pi A_0^2 \vartheta_0^{-2} \left| F_1(\omega, \tau_0) \right|^2 \cdot \left| \theta_\varepsilon(\omega) \right|^2 ; \quad S_c(\omega) = A_0^2 \vartheta_0^{-1} \left\{ \left| F_1(\omega, \tau_0) \right|^2 - \left| \theta_\varepsilon(\omega) \right|^2 \right\}$$

3 Single-Sided Pulsewidth Modulation

Assumptions: amplitudes $A_0 = \text{const}$; pulsewidths are random and independent variables; pulses do not overlap

Spectral density: $S(\omega) = S_d(\omega) \cdot \sum_{k=-\infty}^{\infty} \delta\left(\omega - \frac{2\pi k}{\vartheta_0} \right) + S_c(\omega)$

$$S_d(\omega) = 2\pi A_0^2 \vartheta_0^{-2} \left| M\{F_1(\omega, \tau)\}^2 \right|$$

$$S_c(\omega) = A_0^2 \vartheta_0^{-1} \left\{ M\left\{ \left| F_1(\omega, \tau) \right|^2 \right\} - \left| M\{F_1(\omega, \tau)\} \right|^2 \right\}$$

4 *Double-Sided Pulsewidth Modulation*

Assumptions: amplitudes A_0 = const; the interval between middle points of two pulses ϑ_0 = const; pulsewidth are random and independent variables; pulses do not overlap.

Spectral density: $S(\omega) = S_d(\omega) \cdot \sum\limits_{k=-\infty}^{\infty} \delta\left(\omega - \frac{2\pi k}{\vartheta_0}\right) + S_c(\omega)$

$$S_d(\omega) = 2\pi A_0^2 \vartheta_0^{-2} \cdot \left| M\left\{F_1(\omega,\tau) \cdot e^{j\omega\tau/2}\right\}\right|^2$$

$$S_c(\omega) = A_0^2 \vartheta_0^{-1} \left\{ M\left\{|F_1(\omega,\tau)|^2\right\} - \left| M\left\{F_1(\omega,\tau) \cdot ^{j\omega\tau/2}\right\}\right|^2 \right\}$$

Appendix 30 Spectral Densities of the Stationary Pulse Trains for Various Types of Modulation and Distribution of Pulse Parameters
(after [4, pp. 219-223])

Spectral density: $S(\omega) = S_d(\omega) \cdot \sum_{k=-\infty}^{\infty} \delta\left(\omega - \frac{2\pi k}{\vartheta_0}\right) + S_c(\omega)$

PULSE-POSITION MODULATION
(random parameter is time shift of the pulse edge ε)

1 *Sinusoidal Distribution*

$$S_d(\omega) = \frac{2\pi}{\vartheta_0^2} A_0^2 \cdot J_0^2(\omega x_0) |F_1(\omega, \tau_0)|^2$$

$$S_c(\omega) = \frac{A_0^2 |F_1(\omega, \tau_0)|^2}{\vartheta_0} \left[1 - J_0^2(\omega x_0)\right]$$

2 *Gaussian Distribution*

$$S_d(\omega) = \frac{2\pi}{\vartheta_0^2} A_0^2 e^{-\sigma^2 \omega_0^2} |F_1(\omega, \tau_0)|^2$$

$$\frac{A_0^2 |F_1(\omega, \tau_0)|^2}{\vartheta_0} \left[1 - e^{-\sigma^2 \omega^2}\right]$$

3 *Uniform Distribution*

$$S_d(\omega) = \frac{2\pi}{\vartheta_0^2} A_0^2 \left[\frac{\sin(\omega x_0)}{\omega x_0}\right]^2 \cdot |F_1(\omega, \tau_0)|^2$$

$$+ \frac{A_0^2 |F_1(\omega, \tau_0)|^2}{\vartheta_0} \left[1 - \left(\frac{\sin(\omega x_0)}{\omega x_0}\right)^2\right]$$

4 *Sum of Two Delta Functions*

$$S_d(\omega) = \frac{2\pi}{\vartheta_0^2}\left[1 - 4f_1(1 - f_1)\sin^2\left(\frac{\omega x_1 - \omega x_2}{2}\right)\right] \times A_0^2\left|F_1(\omega,\tau_0)\right|^2$$

$$+ \frac{A_0^2\left|F_1(\omega,\tau_0)\right|^2}{\vartheta_0} \cdot 4f_1(1 - f_1)\cdot\sin^2\left(\frac{\omega x_1 - \omega x_2}{2}\right)$$

SINGLE-SIDED PULSEWIDTH MODULATION
(random parameter is pulsewidth τ)

1 *Sinusoidal Distribution*

$$S_d(\omega) = \frac{2\pi A_0^2}{\omega^2\vartheta_0^2}\left[1 + J_0^2(\omega x_0) - 2J_0(\omega x_0)\cos\omega\tau\right]$$

$$S_c(\omega) = \frac{A_0^2}{\omega^2\vartheta_0}\left[1 - J_0^2(\omega x_0)\right]$$

2 *Gaussian Distribution*

$$S_d(\omega) = \frac{2\pi A_0^2}{\omega^2\vartheta_0^2}\left[1 + e^{-\sigma^2\omega^2} - 2e^{-\frac{1}{2}\sigma^2\omega^2}\cos\omega\tau\right]$$

$$S_c(\omega) = \frac{A_0^2}{\omega^2\vartheta_0}\left[1 - e^{-\sigma^2\omega^2}\right]$$

3 *Uniform Distribution*

$$S_d(\omega) = \frac{2\pi A_0^2}{\omega^2\vartheta_0^2}\left[1 + \frac{\sin^2(\omega x_0)}{(\omega x_0)^2} - 2\frac{\sin(\omega x_0)}{\omega x_0}\cos\omega\tau\right]$$

$$S_c(\omega) = \frac{A_0^2}{\omega^2\vartheta_0}\left[1 - \frac{\sin^2(\omega x_0)}{(\omega x_0)^2}\right]$$

4 *Sum of Two Delta Functions*

$$S_d(\omega) = \frac{4\pi A_0^2}{\omega^2 \vartheta_0^2} \left\{ 1 - f_1 + f_1^2 - f_1 \cos(\omega x_1) - (1 - f_1) \cdot \cos(\omega x_2) \right]$$

$$S_c(\omega) = \frac{2A_0^2}{\omega^2 \vartheta_0} f_1(1 - f_1)\{1 - \cos[\omega(x_1 - x_2)]\}$$

DOUBLE-SIDED PULSEWIDTH MODULATION
(random parameter is pulsewidth τ)
RECTANGULAR PULSE

1 *Sinusoidal Distribution*

$$S_d(\omega) = 2\pi A_0^2 \left(\frac{\tau}{\vartheta_0}\right)^2 \cdot \frac{\sin^2\left(\frac{\omega\tau}{2}\right)}{\left(\frac{\omega\tau}{2}\right)^2} \cdot J_0^2\left(\frac{\omega x_0}{2}\right)$$

$$S_c(\omega) = \frac{2A_0^2}{\omega^2 \vartheta_0}\left[1 - J_0(\omega x_0) - 2J_0^2\left(\frac{\omega x_0}{2}\right)\sin^2\left(\frac{\omega\tau}{2}\right)\right]$$

2 *Gaussian Distribution*

$$S_d(\omega) = 2\pi A_0^2 \left(\frac{\tau}{\vartheta_0}\right)^2 \cdot \frac{\sin^2\left(\frac{\omega\tau}{2}\right)}{\left(\frac{\omega\tau}{2}\right)^2} e^{-\frac{\sigma^2\omega^2}{4}}$$

$$S_c(\omega) = \frac{2A_0^2}{\omega^2 \vartheta_0^2}\left[1 - e^{-\frac{\sigma^2\omega^2}{2}} \cdot \cos(\omega\tau) - 2e^{-\frac{\sigma^2\omega^2}{4}} \cdot \sin^2\left(\frac{\omega\tau}{2}\right)\right]$$

3 *Uniform Distribution*

$$S_d(\omega) = \pi A_0^2 \left(\frac{\tau}{\vartheta_0}\right)^2 \cdot \frac{\sin^2\left(\frac{\omega\tau}{2}\right)\sin^2\left(\frac{\omega x_0}{2}\right)}{\left(\frac{\omega\tau}{2}\right)^2\left(\frac{\omega x_0}{2}\right)^2}$$

$$S_c(\omega) = \frac{2A_0^2}{\omega^2 \vartheta_0} \left[1 - \frac{\sin(\omega\, x_0)}{\omega\, x_0} \cos(\omega\tau) - 2 \frac{\sin^2\left(\frac{\omega\, x_0}{2}\right)}{\left(\frac{\omega\, x_0}{2}\right)^2} \cdot \sin^2\left(\frac{\omega\tau}{2}\right) \right]$$

4 *Sum of Two Delta Functions*

$$S_d(\omega) = \frac{8\pi A_0^2}{\omega^2 \vartheta_0^2} \left[f_1 \sin\left(\frac{\omega\, x_1}{2}\right) + (1 - f_1)\sin\left(\frac{\omega\, x_2}{2}\right) \right]^2$$

$$S_c(\omega) = \frac{4A_0^2}{\omega^2 \vartheta_0} f_1(1 - f_1)\left[\sin\left(\frac{\omega\, x_1}{2}\right) - \sin\left(\frac{\omega\, x_2}{2}\right) \right]^2$$

DOUBLE-SIDED PULSEWIDTH MODULATION
(random parameter is pulsewidth τ)
GAUSSIAN PULSE

1 *Gaussian Distribution*

$$S_d(\omega) = \frac{\pi A_0^2 \tau^2}{2\vartheta_0^2 \ln 2} \left(\frac{8 \ln 2}{\sigma^2 \omega^2 + 8 \ln 2} \right)^3 \exp\left(-\frac{\omega^2 \tau^2}{\sigma^2 \omega^2 + 8 \ln 2} \right)$$

$$S_c(\omega) = \frac{\pi A_0^2 \tau^2}{4\vartheta_0 \ln 2} \left[u - \frac{1}{\left(1 + \dfrac{\sigma^2 \omega^2}{8 \ln 2}\right)^3} \exp\left(-\frac{\omega^2 \tau^2}{8 \ln 2 + 2\sigma^2 \omega^2} \right) \right]$$

$$u = \frac{1 + \dfrac{\sigma^2}{\tau^2}\left(1 + \dfrac{\sigma^2 \omega^2}{4 \ln 2}\right)}{\left(1 + \dfrac{\sigma^2 \omega^2}{4 \ln 2}\right)^{5/2}} \exp\left(-\frac{\omega^2 \tau^2}{8 \ln 2 + 2\sigma^2 \omega^2} \right)$$

2 *Uniform Distribution*

$$S_d(\omega) = \frac{2\pi}{\vartheta_0}\left[\frac{A_0^2\pi 16\ln 2}{\omega^4 x_0^2}\exp\left(-\frac{\omega^2\tau^2 + \omega^2 x_0^2}{8\ln 2}\right)sh^2\left(\frac{\omega^2\,\tau\,x_0}{8\ln 2}\right)\right]$$

$$S_c(\omega) = \frac{\pi A_0^2}{2\omega^2\vartheta_0 x_0}\left\{\begin{array}{l}2\exp\left[-b^2\left(\tau^2 + x_0^2\right)\right]\cdot\left[\tau\cdot sh\left(2b^2\tau\,x_0\right) - x_0 ch\left(2b^2\tau\,x_0\right)\right]+ \\[2mm] +\dfrac{\sqrt{\pi}}{2b}\left[erf\left(\dfrac{\tau + x_0}{b}\right) - erf\left(\dfrac{\tau - x_0}{b}\right)\right]- \\[2mm] -\dfrac{4}{b^2 x_0}\exp\left[-b^2\left(\tau^2 + x_0^2\right)\right]\times sh^2\left(b^2\tau\,x_0\right)\end{array}\right\},$$

$$b^2 = \frac{\omega^2}{8\ln 2}$$

3 *Sum of Two Delta Functions*

$$S_d(\omega) = \frac{2\pi}{\vartheta_0^2}\cdot\frac{\pi A_0^2}{4\ln 2}\left[f_1 x_1 e^{-\frac{\omega^2 x_1^2}{16\ln 2}} + (1 - f_1)x_2\cdot e^{-\frac{\omega^2 x_2^2}{16\ln 2}}\right]$$

$$S_c(\omega) = \frac{A_0^2}{\vartheta_0^2}\cdot\frac{\pi}{4\ln 2}f_1(1 - f_1)\left[x_1\cdot e^{\frac{-\omega^2 x_1}{16\ln 2}} - x_2\cdot e^{\frac{-\omega^2 x_2^2}{16\ln 2}}\right]^2$$

Appendix 31 Simulation Algorithms Based on Recurrent Methods for the Stationary Processes with Common Correlation Functions
(after [12, p. 51])

1 *The process* $\xi_1(t)$ *with correlation function* $K_1(\tau) = \sigma_\xi^2 \cdot e^{-\alpha|\tau|}$

Simulation algorithm: $\xi_k = a_1 \cdot \xi_{k-1} + b_1 \cdot \varepsilon_k$

Parameters: $a_1 = e^{-\alpha \Delta t}$; $b_1 = \sigma_\xi \cdot \sqrt{1 - e^{-2\alpha \Delta t}}$

2 *The process* $\xi_2(t)$ *with correlation function* $K_2(\tau) = \sigma_\xi^2 \cdot e^{-\alpha|\tau|} \cdot \cos \beta\tau$

Simulation algorithm: $\xi_k = a_1\xi_{k-1} + a_2\xi_{k-2} + b_1 \cdot \varepsilon_k + b_2 \cdot \varepsilon_{k-1}$

$a_1 = 2 \cdot e^{-\alpha \Delta t} \cdot \cos(\beta \Delta t)$; $a_2 = -e^{-2\alpha \Delta t}$; $b_1 = \sqrt{-c_0/\vartheta_1}$; $b_2 = -\vartheta_1 \cdot b_1$

$\vartheta_1 = -\chi + sign(\chi) \cdot \sqrt{\chi^2 - 1}$; $\chi = c_1/(2c_0)$; $c_0 = a_{12}R_{12} - R_{11} \cdot a_{22}$;

$c_1 = R_{11}(a_{22})^2 - 2R_{12}a_{12}a_{22} + R_{22}(a_{12})^2 + R_{11}$; $a_{11} = \rho s_1$; $a_{12} = (\rho/\beta) \cdot \sin \beta_\Delta$;

$a_{21} = -\rho\alpha\overline{\alpha}(1 + \overline{\beta}^2)\sin \beta_\Delta$; $a_{22} = \rho s_2$;

$R_{11} = 1 - \rho^2 + 2\overline{\alpha}^2 \rho^2 \sin^2 \beta_\Delta \left(\sqrt{1 + \overline{\beta}^2} - 1 \right)$;

$R_{12} = R_{21} = \alpha\rho^2 \left[\cos \beta_\Delta + \sin \beta_\Delta \left(\sqrt{\alpha^2 + 1} - \overline{\alpha} \right) \right]^2 - \alpha$;

$R_{22} = \alpha^2 \left\{ \rho^2 \left[2s_1^2 \left(\sqrt{1 + \overline{\beta}^2} - 1 \right) - 1 - \overline{\beta}^2 \right] + 1 + \left(1 - \sqrt{1 + \overline{\beta}^2} \right)^2 \right\}$

3 *The process* $\xi_3(t)$ *with correlation function*
$$K_3(\tau) = \sigma_\xi^2 \cdot e^{-\alpha|\tau|} \cdot \left(\cos \beta\tau + \frac{\alpha}{\beta} \sin \beta|\tau| \right)$$

Simulation algorithm: $\xi_k = a_1\xi_{k-1} + a_2\xi_{k-2} + b_1 \cdot \varepsilon_k + b_2 \cdot \varepsilon_{k-1}$

$$a_1 = 2 \cdot e^{-\alpha \Delta t} \cdot \cos(\beta \Delta t); \quad a_2 = -e^{-2\alpha \Delta t} ; \quad b_1 = \sqrt{-c_0/\vartheta_1} ; \quad b_2 = -\vartheta_1 \cdot b_1$$

$$\vartheta_1 = -\chi + sign(\chi) \cdot \sqrt{\chi^2 - 1} ; \quad \chi = c_1/(2c_0); \quad c_0 = a_{12}R_{12} - R_{11} \cdot a_{22} ;$$

$$c_1 = R_{11}(a_{22})^2 - 2R_{12}a_{12}a_{22} + R_{22}(a_{12})^2 + R_{11} ; \quad a_{11} = \rho s_1 ; \quad a_{12} = (\rho/\beta) \cdot \sin \beta_\Delta ;$$

$$a_{21} = -\rho \alpha \overline{\alpha}(1 + \overline{\beta}^2) \sin \beta_\Delta ; \quad a_{22} = \rho s_2 ;$$

$$R_{11} = 1 - \rho^2 [s_1^2 + (1 + \overline{\alpha}^2) \sin^2 \beta_\Delta];$$

$$R_{12} = R_{21} = 2\rho^2 \alpha (1 + \overline{\alpha}^2) \sin^2 \beta_\Delta ;$$

$$R_{22} = \beta^2 (1 + \overline{\alpha}^2)\{1 - \rho^2 [s_2^2 + (1 + \overline{\alpha}^2) \sin^2 \beta_\Delta\}$$

4 *The process* $\xi_4(t)$ *with correlation function* $K_4(\tau) = \sigma_\xi^2 (1 - |\tau|\alpha/2) \cdot e^{-\alpha|\tau|}$

Simulation algorithm: $\xi_k = a_1 \xi_{k-1} + a_2 \xi_{k-2} + b_1 \cdot \varepsilon_k + b_2 \cdot \varepsilon_{k-1}$

$$a_1 = 2 \cdot e^{-\alpha \Delta t} ; \quad a_2 = -e^{-2\alpha \Delta t} ; \quad b_1 = \sqrt{-c_0/\vartheta_1} ; \quad b_2 = -\vartheta_1 \cdot b_1$$

$$\vartheta_1 = -\chi + sign(\chi) \cdot \sqrt{\chi^2 - 1} ; \quad \chi = c_1/(2c_0); \quad c_0 = a_{12}R_{12} - R_{11} \cdot a_{22} ;$$

$$c_1 = R_{11}(a_{22})^2 - 2R_{12}a_{12}a_{22} + R_{22}(a_{12})^2 + R_{11} ; \quad a_{11} = \rho(1 + \alpha_\Delta); \quad a_{12} = \rho \Delta t ;$$

$$a_{21} = -\alpha^2 \rho \Delta t ; \quad a_{22} = \rho(1 - \alpha_\Delta);$$

$$R_{11} = 1 + \rho^2 [\alpha_\Delta + \alpha_\Delta^2 (\sqrt{3} - 2) - 1];$$

$$R_{12} = R_{21} = \alpha \{\rho^2 [1.5 + \alpha_\Delta(\sqrt{3} - 3) + \alpha_\Delta^2 (2 - \sqrt{3})] - 1.5\};$$

$$R_{22} = \alpha^2 \rho^2 [\sqrt{3} - 4 + \alpha_\Delta(5 - 2\sqrt{3}) + \alpha_\Delta^2 (\sqrt{3} - \Delta)] + \alpha^2 (4 - \sqrt{3})$$

The following notations are used: $\alpha_\Delta = \alpha \cdot \Delta t$; $\beta_\Delta = \beta \cdot \Delta t$; $\rho = \exp(-\alpha_\Delta)$; $\overline{\alpha} = \alpha/\beta$; $\overline{\beta} = \beta/\alpha$; $s_1 = \cos \beta_\Delta + \overline{\alpha} \sin \beta_\Delta$; $s_2 = \cos \beta_\Delta - \overline{\alpha} \sin \beta_\Delta$

Appendix 32 Equations and Frequency Responses of the Formation Filter for the Stationary Processes with Common Correlation Functions (after [12, p. 48])

1 *The process* $\xi_1(t)$ *with correlation function* $K_1(\tau) = \sigma_\xi^2 \cdot e^{-\alpha|\tau|}$

Spectral density: $G_1(\omega) = \sigma_\xi^2 \dfrac{\alpha}{\pi(a^2 + \omega^2)}$

The formation filter response: $\Phi_1(p) = \sigma_\xi \dfrac{\sqrt{2\alpha}}{p + \alpha}$

The formation filter equation: $\dot{x}_1 = a_1 x_1 + b_1 \cdot \varepsilon_h(t)$

Parameters: $a_1 = -\alpha;\quad b_1 = \sigma_\xi \sqrt{\dfrac{2\alpha}{h}}\;;\quad h = \dfrac{2\sqrt{3\delta}\,\Delta}{\alpha\sqrt{1 - \Delta^2}}$

For $\delta = \Delta = 0.1$ $h = \dfrac{0.1}{\alpha}$

2 *The process* $\xi_2(t)$ *with correlation function* $K_2(\tau) = \sigma_\xi^2 \cdot e^{-\alpha|\tau|} \cdot \cos \beta\tau$

Spectral density: $G_2(\omega) = \sigma_\xi^2 \dfrac{\alpha(\alpha^2 + \beta^2 + \omega^2)}{\pi\left[(\omega^2 - \beta^2 - \alpha^2)^2 + 4\alpha^2\omega^2\right]}$

The formation filter response: $\Phi_2(p) = \sigma_\xi \dfrac{\sqrt{2\alpha}\left(p + \sqrt{\alpha^2 + \beta^2}\right)}{p^2 + 2\alpha p + \alpha^2 + \beta^2}$

The formation filter equation: $\dot{x}_1 = x_2 + b_1 \cdot \varepsilon_h(t);\; \dot{x}_2 = a_1 x_1 + a_2 x_2 + b_2 \cdot \varepsilon_h(t)$

Parameters: $a_1 = -(\alpha^2 + \beta^2);\quad b_1 = \sigma_\xi \sqrt{\dfrac{2\alpha}{h}}\;;\quad a_2 = -2\alpha;$

$b_2 = \sigma_\xi \cdot \left(\sqrt{\alpha^2 + \beta^2} - 2\alpha\right)\sqrt{\dfrac{2\alpha}{h}}\;;\quad h = 2\Delta\sqrt{\dfrac{3\delta}{\alpha^2 + \beta^2}}$

For $\delta = \Delta = 0.1$ $h = \dfrac{0.1}{\sqrt{\alpha^2 + \beta^2}}$

3 *The process* $\xi_3(t)$ *with correlation function*

$$K_3(\tau) = \sigma_\xi^2 \cdot e^{-\alpha|\tau|} \cdot \left(\cos \beta\tau + \frac{\alpha}{\beta} \sin \beta|\tau| \right)$$

Spectral density: $G_3(\omega) = \sigma_\xi^2 \dfrac{2\alpha(\alpha^2 + \beta^2)}{\pi\left[(\omega^2 - \beta^2 - \alpha^2)^2 + 4\alpha^2\omega^2\right]}$

The formation filter response: $\Phi_3(p) = \sigma_\xi \dfrac{2\sqrt{\alpha(\alpha^2 + \beta^2)}}{p^2 + 2\alpha p + \alpha^2 + \beta^2}$

The formation filter equation: $\dot{x}_1 = x_2$; $\dot{x}_2 = a_1 x_1 + a_2 x_2 + b_2 \cdot \varepsilon_h(t)$

Parameters: $a_1 = -(\alpha^2 + \beta^2)$; $a_2 = -2\alpha$; $b_2 = 2\sigma_\xi \sqrt{\dfrac{\alpha(\alpha^2 + \beta^2)}{h}}$

$h = 2\sqrt{\dfrac{3\delta\Delta}{\alpha^2 + \beta^2}}$; for $\delta = \Delta = 0.1\,\mathrm{h} = \dfrac{0.3}{\sqrt{\alpha^2 + \beta^2}}$

4 *The process* $\xi_4(t)$ *with correlation function* $K_4(\tau) = \sigma_\xi^2 (1 - |\tau|\alpha/2) \cdot e^{-\alpha|\tau|}$

Spectral density: $G_4(\omega) = \sigma_\xi^2 \dfrac{1 + 3\omega^2/\alpha^2}{2\pi\alpha(1 + \omega^2/\alpha^2)^2}$

The formation filter response: $\Phi_4(p) = \sigma_\xi \dfrac{1 + p\sqrt{3}/\alpha}{\sqrt{\alpha}(1 + p/\alpha)^2}$

The formation filter equation: $\dot{x}_1 = x_2 + b_1 \cdot \varepsilon_h(t)$; $\dot{x}_2 = a_1 x_1 + a_2 x_2 + b_2 \cdot \varepsilon_h(t)$

Parameters: $a_1 = -\alpha^2$; $\quad b_1 = \sigma_\xi \sqrt{\dfrac{3\alpha}{h}}$; $a_2 = -2\alpha$;

$b_2 = \sigma_\xi \alpha^{3/2} \cdot h^{-1/2} \left(1 - 2\sqrt{3}\right)$; $h = 2\dfrac{\sqrt{\delta\Delta}}{\alpha}$; for $\delta = \Delta = 0.1$ $h = \dfrac{0.66}{\alpha}$

Appendix 33 **Simulation Algorithm for a Three-Dimensional Gaussian Markovian Random Process**

1 *The first component* $\xi_1(t)$

$$\xi_1(t_{k+1})=\xi_1(t_k)\cdot f(\tau)+\sqrt{D_1\big[1-f^2(\tau)\big]}\cdot\varepsilon_1(t_{k+1});\ \xi_1(t_0)=\sqrt{D_1}\cdot\varepsilon_1(t_0)$$

2 *The second component* $\xi_2(t)$

$$\xi_2(t_{k+1})=\xi_2(t_k)\cdot f(\tau)+\rho_{12}\sqrt{D_2\big[1-f^2(\tau)\big]}\cdot\varepsilon_1(t_{k+1})$$

$$+\sqrt{D_2\big(1-\rho_{12}^2\big)\big[1-f^2(\tau)\big]}\cdot\varepsilon_2(t_{k+1})$$

$$\xi_2(t_0)=\rho_{12}\sqrt{D_2}\cdot\varepsilon_1(t_0)+\sqrt{D_2\big(1-\rho_{12}^2\big)}\cdot\varepsilon_2(t_0)$$

3 *The third component* $\xi_3(t)$

$$\xi_3(t_{k+1})=\xi_3(t_k)\cdot f(\tau)+\rho_{13}\sqrt{D_3\big[1-f^2(\tau)\big]}\cdot\varepsilon_1(t_{k+1})$$

$$-(\rho_{12}\rho_{13}-\rho_{23})\sqrt{\frac{D_3\big[1-f^2(\tau)\big]}{1-\rho_{12}^2}}\cdot\varepsilon_2(t_{k+1})$$

$$+\sqrt{\frac{D_3\big[1-f^2(\tau)\big]\big(1+2\rho_{12}\rho_{13}\rho_{23}-\rho_{12}^2-\rho_{13}^2-\rho_{23}^2\big)}{1-\rho_{12}^2}}\cdot\varepsilon_3(t_{k+1})$$

$$\xi_3(t_0)=\rho_{13}\sqrt{D_3}\cdot\varepsilon_1(t_0)-(\rho_{12}\rho_{13}-\rho_{23})\sqrt{\frac{D_3}{1-\rho_{12}^2}}\cdot\varepsilon_2(t_0)$$

$$+\sqrt{\frac{D_3\big(1+2\rho_{12}\rho_{13}\rho_{23}-\rho_{12}^2-\rho_{13}^2-\rho_{23}^2\big)}{1-\rho_{12}^2}}\varepsilon_3(t_0)$$

$\varepsilon_1,\varepsilon_2,\varepsilon_3\in N(0,1)$ are independent random Gaussian variables.

Appendix 34 **Simulation Algorithms Based on the Discrete Models of Linear Systems (Formation Filters)** (after [12, pp. 45, 55])

1 *Stationary System*

Continuous equation: $\dot{x} = ax + b\eta(t)$

Discrete equation (simulation algorithm): $x_{k+1} = A_\Delta \cdot x_k + B_\Delta \cdot \varepsilon_{k+1}$

Parameters: $a = -\alpha;\ \ b = \sigma_\xi \sqrt{2\alpha}\ ;\ \ A_\Delta = e^{-\alpha \cdot \Delta t}\ ;\ \ B_\Delta = \sqrt{K_\xi} = \sigma_\xi \sqrt{1 - e^{-2\alpha \Delta t}}\ ;$

$\Delta t = t_{k+1} - t_k$

2 *Nonstationary System:*

Continuous equation: $\dot{x} = a(t)x + b(t) \cdot \eta(t)$

Discrete equation (simulation algorithm): $x_{k+1} = \Phi(k+1, k) \cdot x_k + D_{k+1} \cdot \varepsilon_{k+1}$

Parameters: $a(t) = \dfrac{\alpha}{1 + \vartheta t}\ ;\ \ b(t) = b = const\ ;\ \ \Phi(k+1, k) = \left[\dfrac{1 + \vartheta \cdot t_{k+1}}{1 + \vartheta \cdot t_k} \right]^{\alpha/\vartheta}\ ;$

$D_{k+1} = \sqrt{K_{\xi_{k+1}}}\ ;$

$$K_{\xi_{k+1}} = \dfrac{b^2 \left(1 + \vartheta \cdot t_{k+1}\right)^{2\alpha/\vartheta}}{\vartheta - 2\alpha} \times \left[\left(1 + \vartheta \cdot t_{k+1}\right)^{1 - \frac{2\alpha}{\vartheta}} - \left(1 - \vartheta \cdot t_{k+1}\right)^{1 - \frac{2\alpha}{\vartheta}} \right]$$

Appendix 35 Simulation Algorithms for the Vector Modulus and the Unit Vector (after [12, pp. 77, 79])

Table A35.1
Simulation Algorithms for the Vector Modulus

m	# of algorithm	Normalized correlation function $R_0(|x|)$	Spectral density $S_0(|u|)$	Simulation algorithm $\vartheta = |\vec{v}|$
2	1	$e^{-\alpha|x|}$	$\dfrac{1}{2\pi\alpha^2}\left(1+\dfrac{|u|^2}{\alpha^2}\right)^{-3/2}$	$\alpha\sqrt{\dfrac{1}{\gamma^2}-1}$
	2	$\left[1+\alpha^2|x|^2\right]^{-3/2}$	$\dfrac{1}{2\pi\alpha^2}\cdot e^{-|u|/\alpha}$	$-\alpha\ln(\gamma_1\cdot\gamma_2)$
	3	$e^{-\frac{a^2|x|^2}{2}}$	$\dfrac{1}{2\pi\alpha^2}\cdot e^{-\frac{|u|^2}{2\alpha^2}}$	$\alpha\sqrt{-2\ln\gamma}$
	4	$\dfrac{\sin(\alpha|x|)}{\alpha|x|}$	$\dfrac{1}{2\pi\alpha^2}\left(1-\dfrac{|u|^2}{\alpha^2}\right)^{-1/2}$	$\alpha\sqrt{1-\gamma^2}$
3	5	$e^{-\alpha|x|}$	$\dfrac{\alpha}{\pi^2\left(\alpha+|u|^2\right)^2}$	$\alpha\sqrt{\varepsilon_1^2-2\ln\gamma/\varepsilon_2}$
	6	$e^{-\frac{a^2|x|^2}{2}}$	$\alpha^{-3}\cdot(2\pi)^{-3/2}\cdot e^{-\frac{|u|^2}{2\alpha^2}}$	$\alpha\sqrt{\varepsilon_1^2-2\ln\gamma}$

$\gamma_1, \gamma_2, \gamma_3 \in \cup N(0,1); \quad \varepsilon_1, \varepsilon_2 \in N(0,1)$

Table A35.2
Simulation Algorithms for the Unit Vector

m	e_1	e_2	e_3
2	$\cos\varphi$	$\sin\varphi$	
3	$1 - 2\gamma_1$	$\sqrt{1 - e_1^2} \cdot \cos(2\pi\gamma_2)$	$\sqrt{1 - e_1^2} \cdot \sin(2\pi\gamma_2)$

$\varphi \in UN(0, 2\pi)$; $\gamma_1, \gamma_2 \in UN(0, 1)$

Appendix 36 **Algorithms to Estimate Basic Parameters of a Stationary Ergodic Random Process** (after [11, pp. 258–268])

1 *Mean:*

$$\hat{m}_\xi = \frac{1}{N} \sum_{n=0}^{N-1} x(n\Delta t)$$

2 *Variance:*

$$\hat{D}_\xi = \frac{1}{N} \sum_{n=0}^{N-1} [x(n\Delta t)]^2 - \frac{1}{N^2} \left[\sum_{n=0}^{N-1} x(n\Delta t) \right]^2$$

3 *Autocorrelation Function:*

$$\hat{K}_\xi(r \cdot \Delta t) = \frac{1}{N-r} \sum_{n=0}^{N-1-r} x(n\Delta t) \cdot x[(n+r)\Delta t] - \frac{1}{(N-r)^2} \sum_{n=0}^{N-1-r} x(n\Delta t) \cdot \sum_{n=r}^{N-1} x(n\Delta t)$$

4 *Spectral Density:*

$$\hat{S}_\xi(\omega) = \frac{\hat{K}_\xi(0) \cdot \Delta t}{2\pi} + \frac{1}{\pi} \sum_{r=1}^{N-2} K_\xi(r\Delta t) \cos(r\Delta t) \left(1 - \frac{r}{N-1} \right) \Delta t$$

Appendix 37 Algorithms to Estimate the First Four Moments of a Nonstationary Random Process (after [11, p. 232])

1 *Moment* $m_1 = m_\xi$

Algorithm: $\hat{m}_\xi(k) = \dfrac{1}{M} \sum\limits_{m=0}^{M-1} x_m(k)$;   Error: $\sigma_{\hat{m}_\xi} = \sqrt{D_\xi / M}$

2 *Moment* $M_2 = D_\xi$

Algorithm: $\hat{D}_\xi(k) = \dfrac{1}{M-1} \sum\limits_{m=0}^{M-1} \left[x_m(k) - \hat{m}_\xi(k) \right]^2$

Error:
$$\sigma_{\hat{D}_\xi} = \sqrt{\dfrac{M}{M-1}} \left[\dfrac{M_4 - M_2^2}{M} - \dfrac{2\left(M_4 - 2M_2^2\right)}{M^4} + \dfrac{M_4 - 3M_2^2}{M^3} \right]^{1/2}$$
$$\approx \sqrt{\dfrac{M_4 - M_2^2}{M}} + o\!\left(\dfrac{1}{M}\right)$$

3 *Moment* M_3

Algorithm: $\hat{M}_3(k) = \dfrac{M}{(M-1)(M-2)} \sum\limits_{m=0}^{M-1} \left[x_m(k) - \hat{m}_\xi(k) \right]^2$

Error: $\sigma_{\hat{M}_3} = \sqrt{\dfrac{M_6 - 6M_2 M_4 - M_3^2 + 9M_2^3}{M}} + o\!\left(\dfrac{1}{M}\right)$

4 *Moment* M_4

Algorithm:
$$\hat{M}_4(k) = \dfrac{M^2 - 2M + 3}{(M-1)(M-2)(M-3)} \times \sum\limits_{m=0}^{M-1} \left[x_m(k) - m_\xi(k) \right]^4$$
$$- \dfrac{3(2M-3)}{M(M-1)(M-2)(M-3)} \times \left\{ \sum\limits_{m=0}^{M-1} \left[x_m(k) - m_\xi(k) \right]^2 \right\}^2$$

Error: $\sigma_{\hat{M}_4} = \sqrt{\dfrac{M_8 - 8M_3 M_5 - M_4^2 + 16M_3^2}{M}} + o\!\left(\dfrac{1}{M}\right)$

Appendix 38 The Mathematical Expectations and Correlation Functions Based on Impulse Response Representation

1 *Nonstationary Process* $\xi(t)$

Mean: $m_\eta(t) = \int\limits_0^t m_\xi(\tau)h(t-\tau)d\tau$

Two-dimensional correlation function:

$$K_\eta(t_1, t_2) = \int\limits_0^{t_1}\int\limits_0^{t_2} K_\xi(\tau_1, \tau_2)h(t_1 - \tau_1)h(t_2 - \tau_2)d\tau_1 d\tau_2$$

n-dimensional correlation function:

$$K_{\eta n}(t_1, ..., t_n) = \int\limits_0^{t_1}...\int\limits_0^{t_n} K_{\eta n}(\tau_1, ..\tau_n) \times h(t_1 - \tau_1)..h(t_n - \tau_n)d\tau_1...d\tau_n$$

2 *Stationary Process* $\xi(t)$

Mean: $m_\eta(t) = m_\xi \int\limits_0^t h(x)dx$

Two-dimensional correlation function:

$$K_\eta(t_1, t_2) = \int\limits_0^{t_1}\int\limits_0^{t_2} K_\xi(\tau_2 - \tau_1) \, h(t_1 - \tau_1)h(t_2 - \tau_2)d\tau_1 d\tau_2$$

n-dimensional correlation function:

$$K_{\eta n}(t_1, ..., t_n) =$$

$$= \int\limits_0^{t_1}...\int\limits_0^{t_n} K_{\xi n}(t_2 - t_1, ..t_n - 1) \times h(t_1 - \tau_1)..h(t_n - \tau_n)d\tau_1...d\tau_n$$

Appendix 39 The Impulse and Frequency Responses for Common Linear Circuits (after [4, pp. 278–281])

Frequency response: $H(j\omega) = \int\limits_{0}^{\infty} h(t) e^{-j\omega t}\, dt$

Impulse response: $h(t) = \dfrac{1}{2\pi} \int\limits_{-\infty}^{\infty} K(j\omega) e^{j\omega t}\, d\omega$

1 *LR Circuit*

$H(j\omega) = \dfrac{R}{R + j\omega L}$; $h(t) = \dfrac{R}{L} e^{-\frac{R}{L}t}$

2 *RL Circuit*

$H(j\omega) = \dfrac{j\omega L}{R + j\omega L}$; $h(t) = \delta(t) - \dfrac{R}{L} e^{-\frac{R}{L}t}$

3 *RC Circuit*

$H(j\omega) = \dfrac{1}{1 + j\omega RC}$; $h(t) = \dfrac{1}{RC} e^{-\frac{1}{RC}t}$

4 *CR Circuit*

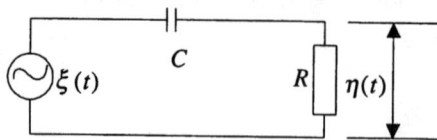

$$H(j\omega) = \frac{j\omega\, RC}{1 + j\omega\, RC}\;;\quad h(t) = \delta(t) - \frac{1}{RC}\, e^{-\frac{1}{RC}t}$$

5 RR_1C *Circuit*

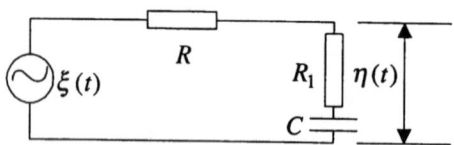

$$H(j\omega) = \frac{1 + j\omega\, T_1}{1 + j\omega\, T}\;;\quad h(t) = \frac{T_1}{T}\delta(t) - \frac{RC}{T^2}\, e^{-\frac{1}{T}t}$$

$$T = (R + R_1)C,\quad T_1 = R_1 C$$

6 RC_1C *Circuit*

$$H(j\omega) = \frac{1 + j\omega\, T_1}{1 + j\omega\, T}\;;\quad h(t) = \frac{T_1}{T}\delta(t) - \frac{RC}{T^2}\, e^{-\frac{1}{T}t}$$

$$T = R(C + C_1)\,;\ T_1 = RC_1$$

7 *CLR Circuit*

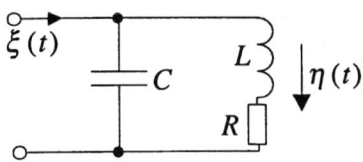

$$H(j\omega) = \frac{\omega^2}{\left(\omega_0^2 - \omega^2\right) + j2\alpha\omega}, \quad \omega_0 = 1/\sqrt{LC}, \quad \alpha = R/2L$$

$$h(t) = \begin{cases} \omega\, e^{-\alpha t} \sin\omega_0 t, & \omega_0 \gg \alpha; \\ \alpha^2 t e^{-\alpha t}, & \omega_0 = \alpha; \\ \dfrac{\omega_0^2}{\sqrt{\alpha^2 - \omega_0^2}} e^{-\alpha t}\, sh\sqrt{\alpha^2 - \omega_0^2}, & \omega_0 < \alpha \end{cases}$$

8 *CRL Circuit*

$$H(j\omega) = \frac{j2\alpha\,\omega R}{\left(\omega_0^2 - \omega^2\right) + j2\alpha\omega}, \quad \omega_0 = 1/\sqrt{LC}, \quad \alpha = 1/2RC$$

$$\eta(t) = \frac{1}{\omega_0 C} \int_0^t e^{-\alpha(t-\lambda)} \times \sin\omega_0(t-x)\xi(x)dx, \quad \omega_0 \gg \alpha$$

9 *Gaussian Filter*

$$H(j\omega) = e^{-\frac{\pi}{2}\left(\frac{\omega-\omega_0}{\Delta\omega}\right)^2 - j(\omega-\omega_0)t_0}; \quad \Delta\omega \ll \omega_0 \quad h(t) = \frac{\Delta\omega}{\pi\sqrt{2}} e^{-\frac{\Delta\omega^2}{2\pi}(t-t_0)^2 + j\omega_0 t}$$

10 *Ideal Bandpass Filter*

$$H(j\omega)=\begin{cases} e^{-j(\omega-\omega_0)t_0} & \omega_0-\dfrac{\Delta\omega}{2}\le\omega\le\omega_0+\dfrac{\Delta\omega}{2}, \\[2mm] 0, & \omega<\omega_0-\dfrac{\Delta\omega}{2},\omega>\omega_0+\dfrac{\Delta\omega}{2} \end{cases}$$

$$h(t)=\frac{\Delta\omega}{2\pi}\frac{\sin\dfrac{\Delta\omega(t-t_0)}{2}}{\dfrac{\Delta\omega(t-t_0)}{2}}e^{j\omega t_0}$$

11 *Ideal Integrator*

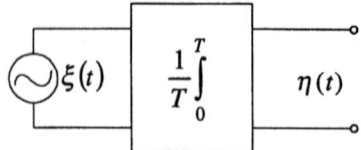

$$H(j\omega)=\frac{\sin\dfrac{\omega T}{2}}{\dfrac{\omega T}{2}}e^{-j\omega\frac{T}{2}}$$

$$h(t)=\begin{cases} \dfrac{1}{T}, & 0\le t\le T, \\[2mm] 0, & t<0,t>T \end{cases}$$

$$H(j\omega)=\frac{1}{j\omega}$$

$$h(t)=\begin{cases} 1, & 0\le t<\infty, \\ 0, & t<0 \end{cases}$$

12 *Ideal Differentiating Circuit*

$$H(j\omega)=j\omega; \qquad h(t)=\frac{d}{dt}\delta(t)=\delta'(t)$$

13 *Integrator-Differentiating Circuit*

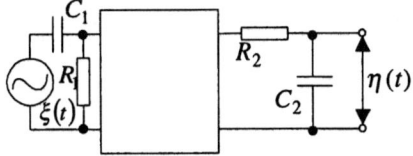

$$H(j\omega) = \frac{j\omega T_1}{(1 + j\omega T_1)(1 + j\omega T_2)};$$

$$h(t) = \begin{cases} \dfrac{1}{T_1 - T_2}\left[\dfrac{T_1}{T_2} e^{-\frac{t}{T_2}} - e^{-\frac{t}{T_1}}\right], & T_1 \neq T_2 \\[4mm] \dfrac{1}{T}\left(1 - \dfrac{t}{T}\right) e^{-\frac{t}{T}}, & T_1 = T_2 = T \end{cases}$$

$$T_1 = R_1 C_1, \quad T_2 = R_2 C_2$$

Appendix 40 The Normalized Correlation Functions and Spectral Densities at the Output of Common Linear Circuits (after [4, pp. 290, 291])

Correlation function: $R(\tau) = \dfrac{1}{2\pi} \displaystyle\int_{-\infty}^{\infty} S(\omega) e^{j\omega\tau} d\omega$

Spectral density: $S(\omega) = \displaystyle\int_{-\infty}^{\infty} R(\tau) e^{-j\omega\tau} d\tau$

1 *Low-Pass RC Filter*

$$R(\tau) = e^{-\alpha|\tau|}; \quad S(\omega) = \frac{2\alpha}{\alpha^2 + \omega^2}, \quad \alpha = \frac{1}{RC}$$

2 *Two Low-Pass RC Filters*

$$R(\tau) = \left(1 + \alpha|\tau|\right) e^{-\alpha|\tau|}; \quad S(\omega) = \frac{4\alpha^3}{\left(\alpha^2 + \omega^2\right)^2}$$

3 *Three Low-Pass RC Filters*

$$R(\tau) = \left[1 + \alpha|\tau| + \alpha^2 \tau^2/3\right] e^{-\alpha|\tau|}; \quad S(\omega) = \frac{16\alpha^5}{3\left(\alpha^2 + \omega^2\right)^3}$$

4 *Proportionally Integrating Filter*

$$R(\tau) = \frac{T_1^2}{T^2} \delta(\tau) + \frac{1}{2T} \cdot \frac{T^2 - T_1^2}{T^2} \cdot e^{-|\tau|/T}; \quad S(\omega) = \frac{1 + (\omega T_1)^2}{1 + (\omega T)^2}$$

5 *Low-Pass Gaussian Filter*

$$R(\tau) = e^{-\alpha\tau^2}; \quad S(\omega) = \sqrt{\frac{\pi}{\alpha}} \cdot e^{-\omega^2/4\alpha}$$

6 _Low-Pass Ideal Filter_

$$R(\tau)=\frac{\sin(\Delta\omega\,\tau)}{\Delta\omega\,\tau}\,;\quad S(\omega)=\begin{cases}\dfrac{\pi}{\Delta\omega}, & |\omega|\leq\Delta\omega\\[2mm]0, & \text{otherwise}\end{cases}$$

7 _High-Pass Ideal Filter_

$$R(\tau)=\frac{\pi}{\Delta\omega}\,\delta(\tau)-\frac{\sin(\Delta\omega\,\tau)}{\Delta\omega\,\tau}\,;\quad S(\omega)=\begin{cases}0, & |\omega|\leq\Delta\omega\\[2mm]\dfrac{\pi}{\Delta\omega}, & \text{otherwise}\end{cases}$$

8 _High-Pass Ideal Filter_

$$R(\tau)=\delta(\tau)-\frac{1}{2\beta}\,e^{-|\tau|/\beta}\,;\quad S(\omega)=\frac{\beta^2\omega^2}{1+\beta^2\omega^2}\,,\quad\beta=\frac{L}{R}$$

9 _Oscillatory Circuit_

$$R(\tau)=e^{-\alpha|\tau|}\left(\cos\omega_0\tau+\frac{\alpha}{\omega_0}\cdot\sin\omega_0|\tau|\right)$$

$$S(\omega)=\frac{4\alpha\left(\alpha^2+\omega_0^2\right)}{C}$$

$$C=\left[\alpha^2+(\omega-\omega_0)^2\right]\cdot\left[\alpha^2+(\omega+\omega_0)^2\right]$$

Appendix 41 **The Characteristics of Nonlinear Circuits** (after [4, pp. 344, 345])

1 *The Circuit with a Characteristic of γ th Order*

$$\eta = \varphi[\xi] = a\xi^{\gamma} \quad (\gamma \text{ is an integer})$$

$$F(ju) = \frac{a\gamma!}{(ju)^{\gamma+1}} \; ; \; L \text{ is a positive loop around zero}$$

2 *The Circuit with a Characteristic of γ th Order and a Shift*

$$\eta = \varphi[\xi] = a(\xi - B)^{\gamma} \quad (\gamma \text{ is an integer})$$

$$F(ju) = \frac{a\gamma!}{(ju)^{\gamma+1}} e^{-juB} \; ; \; L \text{ is a positive loop around zero}$$

3 *A Half-Period Linear Rectifier*

$$\eta = \varphi[\xi] = \begin{cases} a\xi, & \xi \geq 0 \\ 0, & \xi < 0 \end{cases} ; \quad F(ju) = \frac{a}{(ju)^2} \; ;$$

L: real axis u from $-\infty$, cut down from $u = 0$

4 *The Rectifier with a Characteristic of γ th Order and a Limiter*

$$\eta = \varphi[\xi] = \begin{cases} a(\xi - B)^{\gamma}, & \xi \geq B \\ 0, & \xi < B \end{cases} \quad (\gamma \text{ is an integer});$$

$$F(ju) = \frac{a\Gamma(\gamma+1)}{(ju)^{\gamma+1}} e^{-juB} \; ; \; L \text{: real axis } u \text{ from } -\infty \text{ , cut down from } u = 0$$

5 *A Linear Rectifier-Limiter*

$$\eta = \varphi[\xi] = \begin{cases} aD, & \xi \geq D \\ a\xi, & 0 \leq \xi < D \\ 0, & \xi < 0 \end{cases} ; \quad F(ju) = \frac{a\left(1 - e^{-juD}\right)}{(ju)^2} ;$$

L: real axis *u* from - ∞ , cut down from *u* = 0

6 *A Half-Period Rectifier*

$$\eta = \varphi[\xi] = \begin{cases} f[\xi] & \xi \geq 0 \\ 0, & \xi < 0 \end{cases} ; \quad F(ju) = \int_0^\infty f[\xi] e^{-ju\xi} d\xi ;$$

L: real axis *u* from - ∞ , cut down from *u* = 0

7 *An Ideal Limiter*

$$\eta = \varphi[\xi] = \begin{cases} 1, & \xi \geq 0 \\ 1, & \xi < 0 \end{cases} ; \quad F(ju) = \pm \frac{1}{u}$$

L: real axis *u* from - ∞ , cut down from *u* = 0

8 *A Smoothing Filter*

$$\eta = \varphi[\xi] = \frac{1}{\sqrt{2\pi}\,\beta} \int_0^\xi e^{-x^2/2\beta^2} dx ; \quad F(ju) = \frac{1}{ju} e^{-\frac{1}{2}\beta^2 u^2}$$

L: real axis *u* from - ∞ , cut down from *u* = 0

Appendix 42 **Stochastic Parameters of the Sum of Random Noise and Deterministic Signal**

1 *Mutual Probability Density Function of Amplitude, Phase, and Its Derivatives*

$$f_4(V,\dot{V},\psi,\dot{\psi}) = \frac{V^2}{4\pi^2 \sigma_\xi^4 \left[-\ddot{r}(0) \right]}$$

$$\times \exp\left\{ -\frac{1}{2\sigma_\xi^2 \left[-\ddot{r}(0) \right]} \left[-\ddot{r}(0)\left(V^2 + A_m^2 - 2A_m V \cos\varphi\right) + \dot{V}^2 + V^2 \dot{\psi}^2 \right] \right\}$$

2 *Mutual Probability Density Function of Amplitude and Phase*

$$f_2(V,\psi) = \frac{V}{2\pi\sigma_\xi^2} \exp\left[-\frac{1}{2\sigma_\xi^2}\left(V^2 + A_m^2 - 2A_m V \cos\psi\right) \right]$$

3 *Probability Density Function of Amplitude*

$$f_1(V) = \frac{V}{\sigma_\xi^2} \exp\left(-\frac{V^2 + A_m^2}{2\sigma_\xi^2} \right) I_0\left(\frac{A_m V}{\sigma_\xi^2} \right), \quad V \geq 0$$

$I_0(x)$ is the Bessel function of zero order

$$f_1(V) = \frac{V}{\sigma_\xi^2} \exp\left(-\frac{V^2}{2\sigma_\xi^2} \right) \quad \text{if } A_m = 0 \text{ (noise-only case)}$$

4 *Probability Density Function of Phase*

$$f_1(\psi) = \frac{1}{2\pi} e^{-a^2/2} \cdot \left[1 + \sqrt{2\pi}\,a\cos\psi\,\Phi(a\cos\psi)\,e^{a^2\cos^2\psi/2} \right] \quad -\pi \leq \psi \leq \pi$$

$a = A_m/\sigma_\xi$ is signal-to-noise ratio; $\Phi(x)$ is the error function (see Appendix 1)

$$f_1(\psi) = \frac{1}{2\pi} \text{ if } A_m = 0 \text{ (noise-only case)}$$

5 *The Mathematical Expectation of Amplitude*

$$m_V = \sigma_\xi \sqrt{\frac{\pi}{2}} \cdot {}_1F_1\left(-\frac{1}{2}; 1; -\frac{1}{2}a^2\right)$$

${}_1F_1(\alpha, \beta, x)$ is the confluent hypergeometrical function (see Appendix 1)

6 *The Variance of Amplitude*

$$\sigma_V^2 = 2\sigma_\xi^2\left(1 + \frac{1}{2}a^2\right) - m_V^2$$

7 *The Correlation Function of Amplitude*

$$K_V(\tau) \approx \begin{cases} \dfrac{\pi}{8}\sigma_\xi^2\left[r^2(\tau) + \left(\dfrac{4}{\pi}\right)^2 a^2 r(\tau)\right], & a \ll 1 \\[4mm] \sigma_\xi^2 r(\tau)\left[1 + \dfrac{1}{2a^2}r(\tau)\right], & a \gg 1 \end{cases}$$

Appendix 43 **The Coefficients for Different Types of Modulation**
(after [13, p. 84])

1 *Unmodulated Signal*

Envelope and derivatives: $A = A_0; \quad \dot{A} = \ddot{A} = 0$

Complete phase and derivatives: $\Phi = \omega_0 t - \varphi_0$; $\dot{\Phi} = \omega_0 \quad \ddot{\Phi} = 0$

$a_1\left(t, \vec{\lambda}\right) = 0; \ a_0\left(t, \vec{\lambda}\right) = \omega_0^2$

2 *Amplitude-Modulated Signal*

Envelope and derivatives: $A = A_0 \left[1 + m_{AM}\, \vec{\lambda}(t)\right]$; $\dot{A} = m_{AM}\, A_0 \dot{\lambda}(t)$;

$\ddot{A} = m_{AM}\, A_0 \ddot{\lambda}(t)$

Complete phase and derivatives: $\Phi = \omega_0 t - \varphi_0$; $\dot{\Phi} = \omega_0 \quad \ddot{\Phi} = 0$

$a_1\left(t, \vec{\lambda}\right) = -2 \dfrac{m_{AM}\, \dot{\vec{\lambda}}(t)}{1 + m_{AM}\, \vec{\lambda}(t)}$; $a_0\left(t, \vec{\lambda}\right) = \omega_0^2 + 2\left\{\dfrac{m_{AM}\, \dot{\lambda}(t)}{1 + m_{AM}\, \vec{\lambda}(t)}\right\}^2 - \dfrac{m_{AM}\, \ddot{\vec{\lambda}}(t)}{1 + m_{AM}\, \cdot \vec{\lambda}(t)}$

3 *Frequency-Modulated Signal*

Envelope and derivatives: $A = A_0; \quad \dot{A} = \ddot{A} = 0$

Complete phase and derivatives: $\Phi = \omega_0 t - m_{FM} \displaystyle\int_0^t \vec{\lambda}(t)\, dt$;

$\dot{\Phi} = \omega_0 - m_{FM} \cdot \vec{\lambda}(t)$; $\ddot{\Phi} = -m_{FM}\, \dot{\lambda}(t)$

$a_1\left(t, \vec{\lambda}\right) = \dfrac{m_{FM} \cdot \dot{\lambda}(t)}{\omega_0 - m_{FM}\, \vec{\lambda}(t)}$; $a_0\left(t, \vec{\lambda}\right) = \left[\omega_0 - m_{FM} \cdot \vec{\lambda}(t)\right]^2$

4 *Phase-Modulated Signal*

Envelope and derivatives: $A = A_0$; $\dot{A} = \ddot{A} = 0$

Complete phase and derivatives: $\Phi = \omega_0 t - m_{PM}\, \vec{\lambda}(t) - \varphi_0$;

$$\dot{\Phi} = \omega_0 - m_{PM}\, \dot{\vec{\lambda}}(t); \quad \ddot{\Phi} = -m_{FM} \cdot \ddot{\vec{\lambda}}(t)$$

$$a_1\!\left(t,\, \vec{\lambda}\right) = \frac{m_{PM} \cdot \ddot{\vec{\lambda}}(t)}{\omega_0 - m_{PM} \cdot \dot{\vec{\lambda}}(t)}; \quad a_0\!\left(t,\, \vec{\lambda}\right) = \left[\omega_0 - m_{PM}\, \dot{\lambda}(t)\right]^2$$

Appendix 44 **Mathematical Models of Radio Devices at the Block Diagram Level** (after [13, pp. 89–93])

1 *Oscillator (Local Oscillator)*

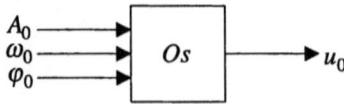

$$u_0(t) = A_0 \cos(\omega_0 t - \varphi_0)$$

2 *Voltage-Controlled Oscillator*

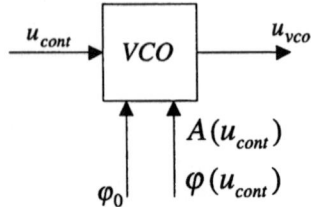

$$u_{VCO}(t) = A(u_{cont}) \cos\left[\omega_0 t - \varphi(u_{cont}) - \varphi_0\right]$$

3 *Amplitude Modulator*

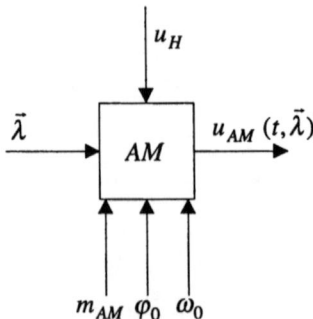

$$u_{AM}(t) = (1 + m_{AM}) \cdot u_H(t) = A_0\left(1 + m_{AM} \cdot \vec{\lambda}\right) \cos(\omega_0 t - \varphi_0)$$

4 *Phase Modulator*

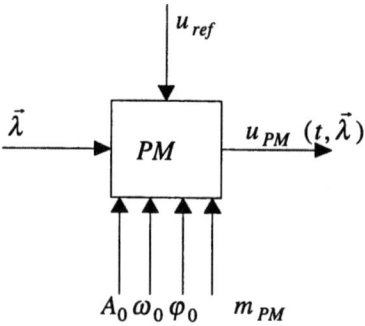

$$u_{PM}\left(t, \vec{\lambda}\right) = A_0 \cos\left[\omega_0 t - m_{PM} \cdot \vec{\lambda} - \varphi_0\right]$$

5 *Frequency Modulator*

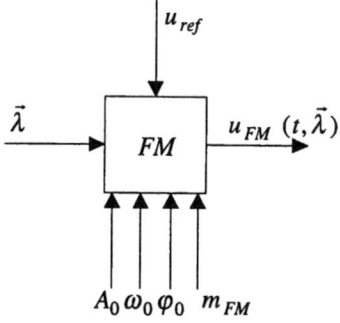

$$u_{FM}\left(t, \vec{\lambda}\right) = A_0 \cos\left[\omega_0 t - m_{FM} \int_0^t \vec{\lambda}(t)dt - \varphi_0\right]$$

6 *Adder*

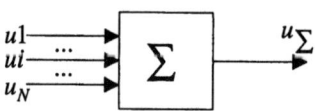

$$u_\Sigma(t) = \sum_{i=1}^{N} u_i(t)$$

7 *Multiplier*

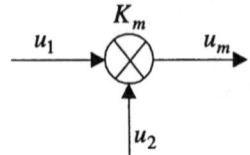

$$u_m(t) = K_m u_1(t) u_2(t)$$

8 *Integrator*

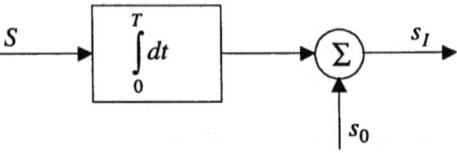

$$s_I(t,T) = s_0(t) + \int_0^T s(t)\,dt$$

9 *Correlator*

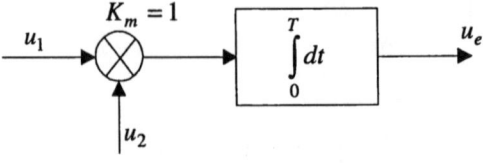

$$u_c(t,T) = \int_0^T u_1(t) u_2(t+\tau)\,dt$$

10 *Delay Line*

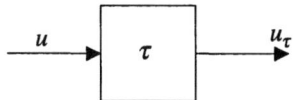

$$u_\tau = u(t - \tau)$$

11 *RF Amplifier Controlled by Automatic Gain Control*

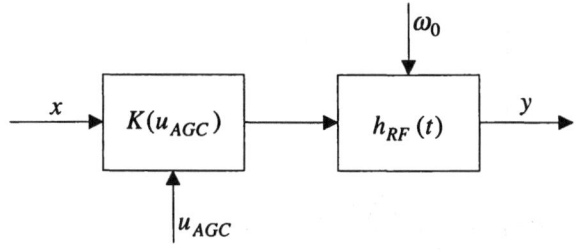

$$y(t) = K(u_{AGC}) \int_0^\infty x(t - \tau) h_{RF}(\tau) d\tau$$

12 *IF Amplifier Controlled by Automatic Gain Control*

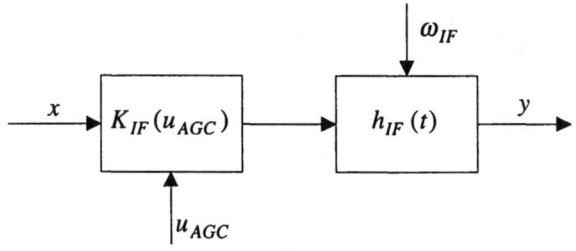

$$y(t) = K_{IF}(u_{AGC}) \times \int_0^\infty x(t - \tau) \cdot h_{IF}(\tau) d\tau$$

Appendix 45 **Simplification of the Transfer Function Model** (after [13, pp. 104–107])

1 *Resonance Amplifier with a Single Resonant Circuit*

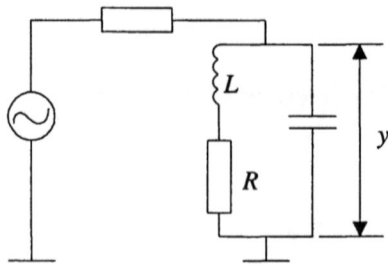

Exact model: $H(j\omega) = \dfrac{S(R + j\omega L)}{1 + j\omega CR + (j\omega)^2 LC}$

where S is a slope of amplification curve

Approximation: $H(j\omega) = \dfrac{1}{1 + j\alpha}$; $\alpha = 2Q\dfrac{\omega - \omega_0}{\omega_0}$; $Q = \dfrac{1}{\omega_0 RC} \gg 1$

2 *Resonance Amplifier with Bandpass Filter*

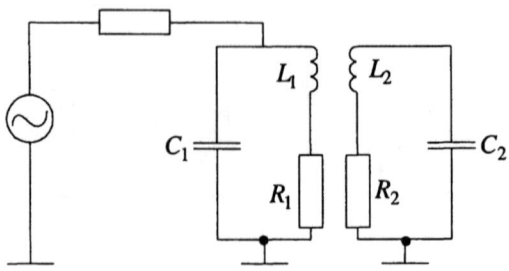

Exact model:

$$H(j\omega) = \delta^2 \left(1+\beta^2\right) \left(\frac{j\omega}{\omega_0}\right) \times \left\{ \left[\left(\frac{j\omega}{\omega_0}\right)^2 + \delta\left(\frac{j\omega}{\omega_0}\right) + 1 \right]^2 - \delta^2 \beta^2 \left(\frac{j\omega}{\omega_0}\right)^4 \right\}^{-1}$$

Approximation: $H(j\omega) = -\dfrac{1+\beta^2}{(1+j\alpha)^2 + \beta^2}$; $\beta = \dfrac{M}{\delta\sqrt{L_1 L_2}}$;

$$\delta = \omega_0 C_1 R_1 = \omega_0 C_2 R_2$$

Appendix 46 **The Coefficients for Nonlinear Circuits** (after [13, p. 111])

Nonlinear circuit	$y = G(x)$	$L_0(A)$	$L_1(A)$	$L_2(A)$
1 *Linear Amplifier*	$y = k_0 x$	0	$K_0 A$	0
2 *Square-Power Transducer*	$y = k_0 x^2$	$\dfrac{1}{2} K_0 A^2$	0	$\dfrac{1}{2} K_0 A^2$
3 *Half-Period Detector*:				
linear	$y = \begin{cases} k_0 x, & x > 0 \\ 0, & x \le 0 \end{cases}$	$\dfrac{1}{\pi} K_0 A$	$K_0 A$	$\dfrac{4}{3\pi} K_0 A$
square	$y = \begin{cases} k_0 x^2, & x > 0 \\ 0, & x \le 0 \end{cases}$	$\dfrac{1}{4} K_0 A^2$	$\dfrac{4}{3\pi} K_0 A^2$	$\dfrac{1}{4} K_0 A^2$
cubic	$y = \begin{cases} k_0 x^3, & x > 0 \\ 0, & x \le 0 \end{cases}$	$\dfrac{5}{6\pi} K_0 A^3$	$\dfrac{3}{8} K_0 A^3$	$\dfrac{5}{6\pi} K_0 A^3$
4 *Ideal Limiter*	$y = \begin{cases} k_0, & x > 0 \\ 0, & x \le 0 \end{cases}$	$\dfrac{1}{2} K_0$	$\dfrac{2}{\pi} K_0$	0

Appendix 47 **Mathematical Models for Common Devices and Circuits** (after [13, pp. 122–131])

| Block diagram of the circuit | Block diagram of the model |

---

1 *Oscillatror*

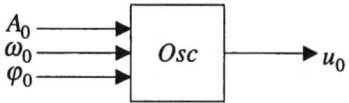

$$u_0 = \mathrm{Re}\left\{A_0 \cdot e^{j(\omega_0 t - \varphi_0)}\right\}$$

$$\overline{A_0} = A_0 \cdot e^{-j\varphi_0} = a_0 - jb_0$$

2 *Voltage-Controlled Oscillatror*

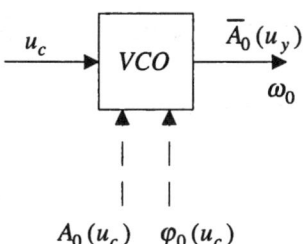

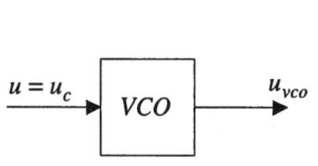

$$u_{VCO} = \mathrm{Re}\left\{A_0(u_c) e^{j[\omega_0 t - \varphi_0(u_c)]}\right\} \qquad \overline{A_0}(u_c) = A_0(u_c) \cdot e^{-j\varphi_0(u_c)} = a_0(u_c) - jb_0(u_c)$$

3 *Amplitude Modulator*

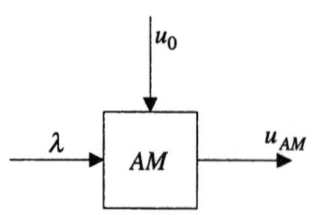

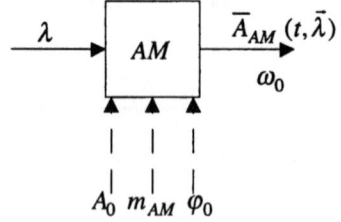

$$u_{AM}(t,\lambda) = \mathrm{Re}\left\{A_{AM}(t,\lambda) \times e^{j\omega_0 t}\right\} \qquad \overline{A}_{AM}(t,\lambda) = A_0\left[1 + m_{AM}\,\lambda\right] \cdot e^{-j\varphi_0}$$

4 *Phase Modulator*

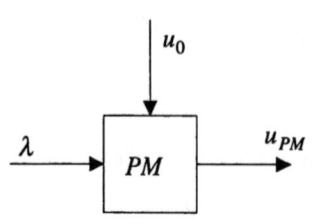

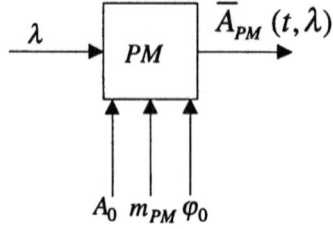

$$u_{PM}(t,\lambda) = \mathrm{Re}\left\{\overline{A}_{PM}(t,\lambda) \times e^{-j\omega_0 t}\right\} \qquad \overline{A}_{PM}(t,\lambda) = A_0\,e^{-j(m_{PM}\lambda + \varphi_0)}$$

5 *Frequency Modulator*

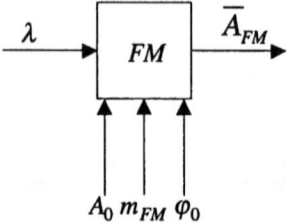

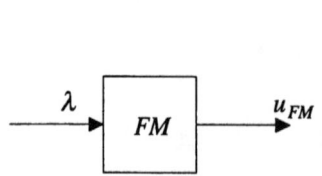

$$u_{FM}(t,\lambda) = \mathrm{Re}\left\{\overline{A}_{FM}(t,\lambda)\,e^{j\omega_0 t}\right\}; \quad \overline{A}_{FM}(t,\lambda) = A_0 \cdot \exp\left[-j\left(m_{FM}\int_0^t \lambda(t)\,dt + \varphi_0\right)\right]$$

6 *Summator*

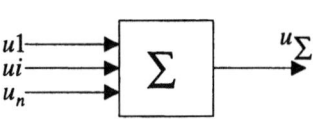

$$u_i = \text{Re}\left\{\overline{A}_i \cdot e^{j\omega_0 t}\right\}; i = 1, 2, \ldots n \qquad \overline{A}_\Sigma = A_\Sigma \cdot e^{-j\varphi_\Sigma} = \sum_{i=1}^{n} \overline{A}_i = \sum_{i=1}^{n} A_i \cdot e^{-j\varphi_i}$$

$$u_\Sigma = \text{Re}\left\{\overline{A}_\Sigma \cdot e^{j\omega_0 t}\right\}$$

7 *Multiplier*

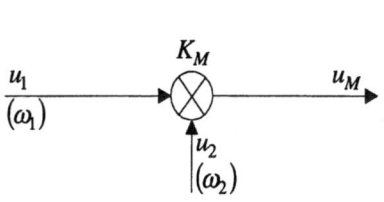

$$u_i = \text{Re}\left\{\overline{E}_i \exp(j\omega_i t)\right\} \quad ; i = 1, 2 \qquad \overline{E}_d = \frac{1}{2} K_M \overline{E}_1 \overline{E}_2^*;$$

$$\overline{E}_\Sigma = \frac{1}{2} K_M \overline{E}_1 \overline{E}_2$$

$$u_M = u_{M_d} + u_{M_\Sigma}$$

$$u_{M_d} = \text{Re}\left\{\overline{E}_d \exp[j(\omega_1 - \omega_2)t]\right\}$$

$$u_{M_\Sigma} = \text{Re}\left\{\overline{E}_\Sigma \exp[j(\omega_1 + \omega_2)t]\right\}$$

8 *Integrator*

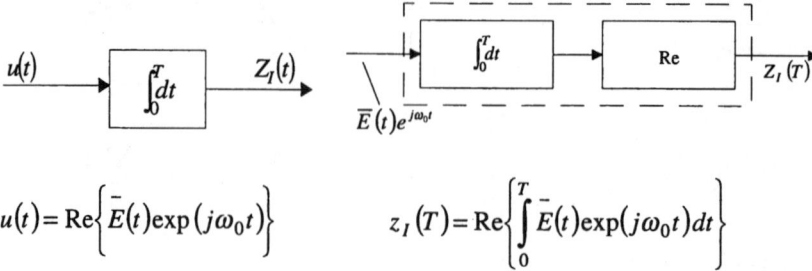

$$u(t) = \mathrm{Re}\left\{\overline{E}(t)\exp(j\omega_0 t)\right\}$$

$$z_I(T) = \mathrm{Re}\left\{\int_0^T \overline{E}(t)\exp(j\omega_0 t)\,dt\right\}$$

9 *Correlator*

$$u_i(t) = \mathrm{Re}\left\{\overline{E}_i(t)\exp(j\omega_i t)\right\}, i = 1,2 \qquad z_d(\tau,T) = \frac{1}{2}\mathrm{Re}\left\{\overline{E}_d(\tau,T)\times\exp(-j\omega_2\tau)\right\}$$

$$z_K(\tau,T) = z_d(\tau,T) + z_\Sigma(\tau,T) \qquad z_\Sigma(\tau,T) = \frac{1}{2}\mathrm{Re}\left\{\overline{E}_\Sigma(\tau,T)\times\exp(j\omega_2\tau)\right\}$$

$$\overline{E}_d(\tau,T) = \int_0^T \overline{E}_1(t)\overline{E}_2^*(t+\tau)\times\exp[j(\omega_1 - \omega_2)t]\,dt$$

$$\overline{E}_\Sigma(\tau,T) = \int_0^T \overline{E}_1(t)\overline{E}_2(t+\tau)\times\exp[j(\omega_1 + \omega_2)t]\,dt$$

10 _Delay Line_

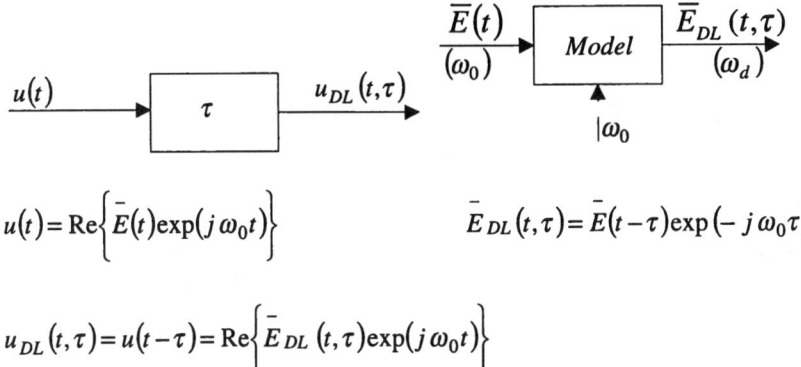

$$u(t) = \text{Re}\left\{\bar{E}(t)\exp(j\,\omega_0 t)\right\} \qquad \bar{E}_{DL}(t,\tau) = \bar{E}(t-\tau)\exp(-j\,\omega_0\tau)$$

$$u_{DL}(t,\tau) = u(t-\tau) = \text{Re}\left\{\bar{E}_{DL}(t,\tau)\exp(j\,\omega_0 t)\right\}$$

Appendix 48 **Statistical Linearization Coefficients** (after [13, p. 139])

1 *Criteria 1, 3*

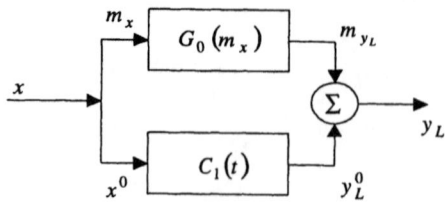

$$G_0^{(1)}[m_x(t)] = m_y(t) = \int_{-\infty}^{\infty} G(x) f_1(x,t) dx \; ; \; C_0^{(1)}(t) = \frac{m_y(t)}{m_x(t)} = \frac{1}{m_x(t)} \int_{-\infty}^{\infty} a(x) f_1(x,t) dx$$

$$C_1^{(1)}(t) = \pm \sqrt{\frac{D_y(t)}{D_x(t)}} = \frac{1}{\sigma_x(t)} \times \int_{-\infty}^{\infty} G^2(x) f_1(x,t) dx - m_y^2(t)$$

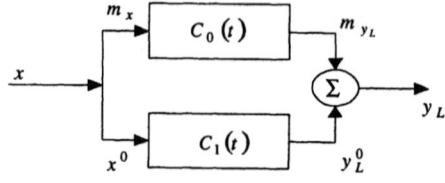

$$C_0^{(2)}(t) = C_0^{(1)}(t) = m_y(t)/m_x(t) \quad C_1^{(2)}(t) = \frac{D_{xy}(t)}{D_x(t)} = \frac{\int_{-\infty}^{\infty} x G(x) f_1(x,t) dx - m_x(t) m_y(t)}{\sigma_x^2(t)}$$

2 *Criterion 2*

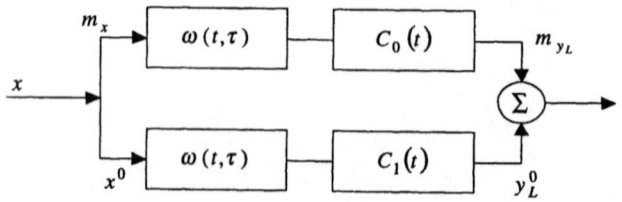

$$C_0^{(3)}(t) = \int_{-\infty}^{\infty} G(x) f_1(x,t) dx \bigg/ \int_{-\infty}^{t} m_x(t) \omega(t,\tau) d\tau$$

$$C_1^{(3)}(t) = C_1^{(1)}(t) = \pm \sqrt{\frac{D_y(t)}{D_x(t)}}$$

$$\int_{-\infty}^{t_1} \int_{-\infty}^{t_2} K_x(\tau_1, \tau_2) \omega(t_1, \tau_1) \omega(t_2, \tau_2) d\tau_1 d\tau_2 = \frac{\sigma_x(t_1) \sigma_x(t_2) K_y(t_1, t_2)}{\sigma_y(t_1) \sigma_y(t_2)}$$

3 *The Harmonic Statistical Linearization Coefficients*

$$C_0 = \frac{m_{y_{aV}}}{m_{x_{aV}}} = \frac{\langle h_{00}(t) \rangle}{m_{x_{aV}}} \; ; \; C_1 = \pm \sum_{n=1}^{\infty} \frac{\sigma_x^{2(n-1)}}{\Gamma(n+1)} \left[\langle h_{0n}^2(t) \rangle + 2 \sum_{m=1}^{\infty} \langle h_{mn}^2(t) \rangle \right]^{1/2}$$

$$C_2 = \left\{ \frac{\langle [h_{00}(t) - \langle h_{00}(t) \rangle]^2 \rangle}{[A_s(t) - \langle A_s(t) \rangle]^2} \right\}^{1/2} = \left\{ \frac{\langle h_{00}^2(t) \rangle - [\langle h_{00}(t) \rangle]^2}{\langle A_s^2(t) \rangle - [\langle A_s(t) \rangle]^2} \right\}^{1/2}$$

$$C_3 = 2 \left\{ \frac{\langle h_{10}^2(t) \rangle}{\langle A_s^2(t) \rangle} \right\}^{1/2} \; ; \; C_4 = 2 \left\{ \frac{\langle h_{m0}^2(t) \rangle}{\langle A_s^2(t) \rangle} \right\}^{1/2}$$

Appendix 49 **Mathematical Models of RCL** (after [13, p. 82])

1 *Resistor*

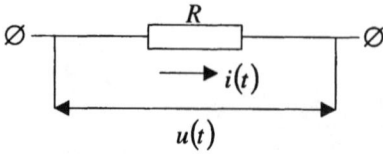

$$u(t) = R \cdot i(t)$$

2 *Inductance*

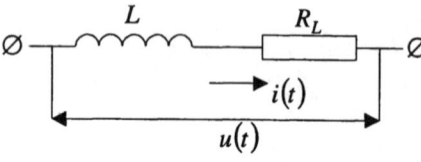

$$u(t) = R_L \cdot i(t) + L\frac{di(t)}{dt}$$

3 *Capacitor*

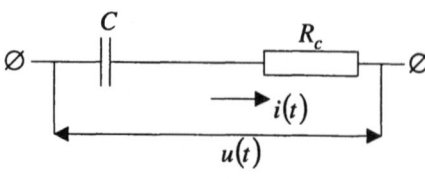

$$u(t) = R_C \cdot i(t) + \frac{1}{C}\int_0^t i(t)\,dt$$

ABOUT THE AUTHORS

Dr. Alexander I. Leonov is well known in Russia as a scientist and engineer in the field of radar. He was born in the Tambov region of Russia in 1927. For about 25 years he was a senior member of teams that designed and tested state-of-the-art radars for Soviet ABM programs, and now he is a professor at the Moscow Institute of Technology. He received his M.Sc. degrees from the Leningrad Military University of Communications (1956), and his Ph.D. and D.Sc. (Eng.) degrees from the Moscow Aerospace Institute. Dr. A. Leonov holds the "All-Russian Honorable" title in the field of science and engineering. He has authored over 100 papers and books, in both Russian and English, including *Radar in Anti-Missile Defense* (Voenizdat, 1967), *Monopulse Radar* (with K. I. Fomichev, Sovetskoe Radio, 1970, 1984; trans. Artech House, 1986), *Modeling in Radar* (with F. V. Nagulinko and others, Radio I Svayz, 1979), *Radar Test* (with S. Leonov and others, Radio I Svayz, 1990), and *Radar Technology Encyclopedia* (with D. K. Barton and others, Artech House, 1997).

Dr. Sergey A. Leonov is known both in Russia and the West as a bilingual expert in the field of radar and telecommunications. Dr. S. Leonov was born in St. Petersburg, Russia, in 1953. He received his B.Sc. degrees from the Moscow University of Radioengineering, Electronics and Automation (1976); his M.Sc. degree from Kharkov University (1980); and his Ph.D. and D.Sc. (Eng.) degrees from the Moscow Aerospace Institute. Dr. S. Leonov holds the "All-Russian Honorable" title in the field of science and engineering. He has authored over 70 papers and books, in both Russian and English, including *Air Defense Radars* (Voenizdat, 1988), *Radar Test* (with A. Leonov and others, Radio I Svayz, 1990), *Russian-English and English-Russian Dictionary on Radar and Electronics* (with

491

W. Barton, Artech House, 1993), and *Radar Technology Encyclopedia* (with D. K. Barton and others, Artech House, 1997).

INDEX

*Multitarget-Multisensor Tracking: Applications and Advances
Volume III*, Yaakov Bar-Shalom and William Dale Blair, editors

Principles of High-Resolution Radar, August W. Rihaczek

Radar Cross Section, Second Edition, Eugene F. Knott, et al.

Radar Evaluation Handbook, David K. Barton, et al.

Radar Meteorology, Henri Sauvageot

Radar Reflectivity of Land and Sea, Third Edition, Maurice W. Long

Radar Resolution and Complex-Image Analysis, August W. Rihaczek
and Stephen J. Hershkowitz

Radar Signal Processing and Adaptive Systems, Ramon Nitzberg

Radar System Performance Modeling, G. Richard Curry

Radar Technology Encyclopedia, David K. Barton and
Sergey A. Leonov, editors

Range-Doppler Radar Imaging and Motion Compensation,
Jae Sok Son, et al.

*Russian-English and English-Russian Dictionary on Radar and
Electronics,* Sergey A. Leonov and William F. Barton

Theory and Practice of Radar Target Identification,
August W. Rihaczek and Stephen J. Hershkowitz

For further information on these and other Artech House titles,
including previously considered out-of-print books now available
through our In-Print-Forever® (IPF®) program, contact:

Artech House
685 Canton Street
Norwood, MA 02062
Phone: 781-769-9750
Fax: 781-769-6334
e-mail: artech@artechhouse.com

Artech House
46 Gillingham Street
London SW1V 1AH UK
Phone: +44 (0)20 7596-8750
Fax: +44 (0)20 7630-0166
e-mail: artech-uk@artechhouse.com

Find us on the World Wide Web at:
www.artechhouse.com

DATE DUE	